EFFECTIVE MANAGEMENT OF HEALTH AND SAFETY PROGRAMS

A Practical Guide

THIRD EDITION

EFFECTIVE MANAGEMENT OF HEALTH AND SAFETY PROGRAMS

A Practical Guide

THIRD EDITION

Royce Moser, Jr., MD, MPH

Professor and Immediate Past Director, Rocky Mountain Center for Occupational and Environmental Health, Department of Family and Preventive Medicine, University of Utah School of Medicine, Salt Lake City

With 13 Contributing Authors

OEM Press
Beverly Farms, MA

Third Edition

Library of Congress Cataloging-in-Publication Data

Moser, Royce.
Effective management of health and safety programs : a practical guide / Royce Moser Jr. ; with 13 contributing authors. — 3rd ed.
p. ; cm.
Rev. ed. of: Effective management of occupational and environmental health and safety programs / Royce Moser Jr. 2nd ed. c1999.
Includes bibliographical references and index.
ISBN 978-1-883595-50-0
1. Industrial hygiene—Management. 2. Industrial safety—Management. 3. Environmental health—Management. I. Moser, Royce. Effective management of occupational and environmental health and safety programs. II. Title.
[DNLM: 1. Occupational Health. 2. Safety Management—organization & administration. 3. Environmental Health—organization & administration. 4. Occupational Health Services—organization & administration.
WA 400 M899e 2008]
RC967.M67 2008
658.3′82—dc22

2008003853

Printed in the United States of America

OEM Press® is a registered trademark of OEM Health Information
Questions or comments regarding this book should be directed to:

OEM Health Information, Inc.
8 West Street
Beverly Farms, MA 01915-2226
978-921-7300
978-921-0304 (fax)
info@oempress.com

5 4 3 2 1

To Lois, for her support, patience, encouragement, and love.
Without her contributions, the first and later editions might still be something I would do "some day."

In Memoriam

Jeffrey S. Lee, PhD, MPH, CIH

1944–1998

Co-founder of the Rocky Mountain Center
for Occupational and Environmental Health; Professor and
Director, Industrial Hygiene Program, Department of Family and
Preventive Medicine, University of Utah School of Medicine;
National and International Leader in Industrial Hygiene

Colleague, mentor, and friend.

Contents

Foreword by Bill Bunn xiii

Preface xv

Acknowledgments xvii

1
The Health and Safety Professional's Need for Management Expertise 1

2
Scope of Managerial Responsibilities 11

3
Planning and Prioritizing Health and Safety Programs and Activities: What to Do When There's Too Much to Do 19

4
Planning Health Care Responses to Natural Disasters, Terrorism, and Other Mass Casualty Situations 39
with Colleen Connelly

5
Implementing Plans: Meeting a Manager's Responsibility to "Get Things Done" 57

6
Organizing Health and Safety Activities 65
with Jeffrey S. Lee

7
Recruiting, Retaining, and Evaluating Health and Safety Personnel 85

8
Personality Aspects of Working with Others 107
by Neal Whitman

9
Team Building and Motivation 117

10
Financial Planning and Budgeting Aspects of Health and Safety Activities 129

11
Entrepreneurship: Implementing or Expanding Health and Safety Programs 143
with Karl Blake and Don Marano

12
Evaluation of Health and Safety Programs and Projects 157
with Richard E. Johns, Jr.

13
Leadership: What Works and What Doesn't 177
with Kurt Hegmann

14
Decision Making and Problem Solving 191
with Jeffrey S. Lee

15
Containing Health and Safety Costs: Health and Safety Risk Management Programs 203

16
Health and Productivity Enhancement 225
by Ronald Loeppke and Pamela Hymel

17
Health Promotion and Disease Prevention in the Workplace and Community 245

18
Health and Safety Consulting 267
by Robert J. McCunney

19
Time Management 281

20
Effective Communication 299

21
Effective Meetings and Presentations 317

22
Interacting with the News Media 337
by Ralph R. Cook

23
Legal and Regulatory Aspects of Managing OEHS Programs 351
by Carolyn S. Langer

24
Preparing for and Providing Health and Safety Legal Testimony 377
by Donald C. Sinclair II

25
Improving Health and Safety Program Quality 403
with Jeffrey S. Lee

26
Meeting Health and Safety Integrity and Ethical Challenges 417
with Jeffrey S. Lee

Appendices

A
Items for New Health and Safety Managers to Consider 429

B
Examples of Budget Categories 431

C
Examples of Audit Instruments for Occupational and Environmental Health and Safety Program Components 435

D
Summary of Parliamentary Procedures 447

Index 449

Foreword

Dr. Moser continues to be an outstanding scholar and contributor to the field of health and safety. He has assembled an exceptional group of authors for this, the third edition of the book.

The book is unique in its approach, which focuses on the practical "how to's" of managing health and safety programs. This approach contrasts with those of many other texts, which focus on the disease and/or exposure approach to occupational and environmental medicine (OEM).

This edition retains the solid base from previous editions but includes five new chapters reflecting the changes and newest trends in the field. Responses to terrorism and pandemics have become a vital part of the practice; the link between health and productivity is now a key part of practice; and legal issues continue to be a major focus.

The field of OEM and health and safety is constantly changing, and this book reflects the changes that have occurred since publication of the last edition. It provides the information needed to prepare health and safety professionals for the future. The practice of occupational and environmental medicine has evolved rapidly over the last 10 years. The focus on clinical expertise, toxicology, and epidemiology continues, but management skills and an emphasis not only on health and safety but also on productivity are vital. It is not enough to prevent illnesses and accidents in the workplace; workers must also be maximally productive. In addition, an assessment of cost effectiveness and cost avoidance must be regularly calculated to assure the stability and success of programs.

Dr. Moser and his contributing authors have compiled an excellent third edition of this classic book. I am sure you will find it a mainstay of your library.

William B. Bunn, III, MD, JD, MPH
Professor
Northwestern University School of Medicine
Vice President, Health, Safety, Security and Productivity
Navistar
Chicago, IL

Preface to the Third Edition

Like the previous two editions, this new edition is structured to provide practical guidance to help health and safety professionals effectively meet their management responsibilities. The book and its contents are based on the premise that all of us, as health and safety professionals, will eventually become managers, either by choice or by having the "opportunity thrust upon us."

The title of the third edition of this book has been changed from *Effective Management of Occupational and Environmental Health and Safety Programs* to *Effective Management of Health and Safety Programs*. This change was made because numerous students using the book in both academic and correspondence courses have commented on its usefulness even though they were not in an occupational and environmental health and safety (OEHS) field of study. Many have said that, for the most part, the content of this book had not been an integral component of their professional training programs.

For convenience, the term *health and safety* is used in the text to refer to those responsible for both areas or only one (i.e., health or safety). Some of the OEHS flavor of earlier editions has been retained where the discussion specifically addresses this particular aspect of management. A few chapters focus primarily on OEHS functions because of unique aspects relating to such activities, such as Dr. Langer's chapter on legal aspects of OEHS programs, Dr. McCunney's on consulting, Drs. Loeppke and Hymel's on safety and productivity enhancement, and the chapter on cost containment.

This edition continues to emphasize "how to" rather than "what to," although a number of the major components of health and safety programs are identified. The book stresses how to design, implement, evaluate, and obtain support for health and safety programs, examining practical management techniques and applications that have proved to be particularly helpful. The information is based on the experience of the authors as well as on that of the reviewers of the work. This edition also reflects comments and suggestions from the readers of the previous editions who kindly provided feedback.

Because of the increasingly multifaceted skills, experience, and functions required of health and safety managers, five new chapters have been added to this edition:

- Planning Health Care Responses to Natural Disasters, Terrorism, and Other Mass Casualty Situations (with Colleen Connelly)
- Entrepreneurship: Implementing or Expanding Health and Safety Programs (with Karl Blake and Don Marano)
- Leadership: What Works and What Doesn't (with Kurt Hegmann)
- Health and Productivity Enhancement (by Ronald Loeppke and Pamela Hymel)
- Preparing for and Providing Health and Safety Legal Testimony (by Donald C. Sinclair II)

As noted in the previous editions, the text does not include extensive discussions of current or prior management theories, although reference is made to a number of these. Numerous books and other publications, several of which are cited in the references and additional reading lists, provide extensive information on current and evolving management theories. A caveat is that today's "best management book ever written" will be superseded in weeks to months by another "best management book ever written," which is not to say that each may not provide some useful information, but rather, that no single book can provide "all the answers" to the multitude of management challenges.

Because my management experience has proved that an individual does, in fact, learn from mistakes as well as successes, this book includes examples of "lessons learned the hard way." Some of the scenarios in these examples may appear too extreme to have actually occurred, but readers of this book may rest assured that the examples come from personal observations or other reliable sources.

It is hoped that this third edition will provide practical assistance to health and safety colleagues as they manage programs to improve the health, safety, and well-being of those they serve.

ACKNOWLEDGMENTS

This third edition has been prepared as a result of comments and feedback from colleagues in senior health and safety management positions in corporate and other settings. In addition, the faculty of the Rocky Mountain Center for Occupational and Environmental Health, Department of Family and Preventive Medicine, University of Utah School of Medicine, stressed the need for an updated version of this work. I am grateful for this encouragement from colleagues dedicated to improving the health and safety of those they serve.

I deeply appreciate the efforts of the contributors. Their dedication to this effort, which required major expenditures of their time, has greatly enhanced the usefulness and applicability of this text. I again want to thank the publisher, Curtis Vouwie, for his belief in the first edition and his unflagging support during the preparation of this edition. Without his efforts, this text would not have been developed. Curtis also made certain that I had exceptional copyeditors for all editions. Ms. Kathy O'Brien was the copyeditor for this edition, and I have continually been impressed by her ability to clarify a confusing sentence or paragraph so that the reader readily grasps the point.

I also want to express my appreciation to Ms. Alice I. Weber, MLS, RN, Assistant Librarian, Spencer S. Eccles Health Sciences Library, and Research Librarian for the Department of Family and Preventive Medicine, University of Utah. She provided invaluable assistance in taking my vague descriptions of a health or safety subject and finding specific references so that I could provide current, valid information on the topic.

Finally, I again want to thank the graduate, continuing education, and other students who used the earlier editions of this work and provided such useful feedback. I have had the good fortune to work directly with several thousand students during my career, and it has been a great privilege to be involved with and learn from these students. This privilege remains one of the greatest rewards of serving in a teaching position.

Royce Moser, Jr.
Salt Lake City

Contributing Authors

KARL BLAKE, MBA
Principal Consultant
Strategic Solutions
Park City, Utah

COLLEEN CONNELLY, RN, BSN
Emergency Preparedness Manager
University of Utah Hospital
Salt Lake City, Utah

RALPH R. COOK, MD, MPH
RRC Consulting, LLC
Midland, Michigan

KURT HEGMANN, MD, MPH
Associate Professor and Director
Rocky Mountain Center for Occupational and Environmental Health
Department of Family and Preventive Medicine
University of Utah School of Medicine
Salt Lake City, Utah

PAMELA HYMEL, MD, MPH, FACOEM
Global Medical Director
Cisco Systems
San Jose, California

RICHARD E. JOHNS, JR., MD, MPH
Clinical Professor
Department of Family and Preventive Medicine
University of Utah School of Medicine; President and CEO
Corporate Health Systems
Bountiful, Utah

CAROLYN S. LANGER, MD, JD, MPH
Instructor
Harvard School of Public Health
Boston, Massachusetts

JEFFREY S. LEE, CIH, PhD, MPH
Deceased

RONALD LOEPPKE, MD, MPH, FACOEM, FACPM
Chief Strategic Officer and Executive Vice President
Matria Healthcare, Inc.
Marietta, Georgia

DON MARANO, PE, CIH
President
IHI Environmental
640 East Wilmington Avenue
Salt Lake City, Utah

ROBERT J. McCUNNEY, MD, MPH
Research Scientist
Department of Biological Engineering
Massachusetts Institute of Technology
Cambridge, Massachusetts
Staff Physician, Pulmonary Unit
Massachusetts General Hospital
Boston, Massachusetts

DONALD C. SINCLAIR II, JD
Sinclair PLLC
Occupational & Environmental Legal Consulting
Wheeling, West Virginia

NEAL WHITMAN, MPA, EdD
Professor
Department of Family and Preventive Medicine
University of Utah School of Medicine
Salt Lake City, Utah

CHAPTER 1

The Health and Safety Professional's Need for Management Expertise

The health and safety practitioner must have both professional capabilities and management expertise in order to meet today's evolving health and safety challenges. Whether in a corporate environment, a government agency, an independent group practice setting, or a full-time consultant position, the professional needs management skills. Those in a hospital setting also find that management capabilities help them provide more comprehensive programs than they would without this expertise. As Drucker has noted, "[T]he fundamental task of management remains the same: to make people capable of joint performance through common goals, common values, the right structure, and the training and development they need to perform and to respond to change." He emphasizes the need for management capabilities by stating "Not to know how to manage is the single largest reason for the failure of new ventures" [1]. In essence, a manager's task is to get things done, through or with others.

Without management expertise, those responsible for directing all or part of a health, safety, or health and safety program will find that their efforts are seriously compromised or possibly outright failures. These practitioners must obtain approval, funding, and support of their programs from managers who may not have health care expertise. For example, providers must convince hospital boards, typically composed of community leaders with various, nonhealth backgrounds, of the value of continuing or implementing new programs in the same manner as other health and safety managers have advocated their activities to corporate managers, legislators, and other decision makers. Similarly, those responsible for safety activities have to obtain program support from executive managers who often have limited, if any, knowledge of safety needs. In essence, senior executives expect health and safety managers to present and justify programs using the same techniques as other managers who are recommending a new process or activity directly related to the overall mission of a company or organization.

Health and safety professionals must plan programs, accomplish budgeting, and evaluate program progress and success. Problems must be recognized and resolved, and appropriate personnel selected and encouraged to accomplish needed activities. Effective communication with managers, workers, patients, the public, and others involved in or affected by the particular program's health or safety activities is essential. In today's health care environments, the director of health and safety programs regularly uses these and other management techniques to accomplish effective health and safety functions.

OCCUPATIONAL AND ENVIRONMENTAL HEALTH AND SAFETY (OEHS) ACTIVITIES AS AN EXAMPLE OF THE UNIQUE SCOPE OF A HEALTH AND SAFETY PROGRAM

OEHS professionals dedicated to providing quality health and safety programs are concerned with factors that do not challenge providers in the usual clinic or hospital health care situation—challenges such as meeting worker health and safety regulatory requirements, ascertaining worker fitness or disability, improving workplace safety, preventing adverse environmental exposures of workers and those in communities near worksites, improving worker productivity, controlling health care costs while still providing proper care of workers, and designing effective health promotion and disease prevention programs at the workplace. Other areas of concern include meeting both worker and community right-to-know requirements, recommending safe exposure levels, evaluating or conducting epidemiologic or other studies designed to define hazards, ensuring effective emergency response to incidents that may involve hundreds of people, helping design appropriate product labeling, and similar activities.

Developments such as the rapidly changing sociopolitical milieu, evolving interpretations of the Americans with Disabilities Act (ADA), an increase in activist regulatory and legal programs, expansion of managed care programs, increasing public concern about environmental effects of work activities, globalization, and changes in the work force indicate that the challenges will become even more complex in the future. In addition to coping with these workplace needs, the OEHS practitioner will have to be able to apply advances in professional knowledge in order to respond to new health and safety concerns, such as the continuing introduction of new workplace chemicals and processes, increasing numbers of workers with latex allergy, possible health effects of low frequency electric and magnetic fields, and nanotechnology [2]. These challenges call for increased expenditures, and obtaining resources for OEHS programs will continue to be of critical concern to those responsible for protecting and improving worker health and safety.

AN ORIENTATION TO MEET OEHS CHALLENGES

As Tepper [3] and others have noted, in the past OEHS professionals were frequently hired only because of their technical expertise. Both management and those being hired perceived that "experts" were included *in* the organization essentially to provide technical information when requested or as needed. Increasingly, both groups now recognize that those in OEHS positions can make much more important contributions if the incumbents are *of* the corporation, working proactively to support the goals of the organization by taking the initiative to provide health and safety services that will benefit both the worker and the company [4,5].

The OEHS professional with such a strategic view may find opportunities for improved program effectiveness that would escape a director with a narrower focus. For example, an on-site effort to detect and control nonoccupational disease, such as mental illness, which can still affect worker productivity, may be needed [6]. Efforts to reduce health risks related to lifestyle can benefit workers and their families. Innovative programs can be developed to improve both quality and accessibility of health care. As Moore [7] has noted, considering both the organization's culture and infrastructure in implementing health and safety programs can improve the probability of success. Those who are *of* a corporation will have the perspective necessary to recognize the need for just such an orientation and to put it to work.

Additionally, OEHS activities can significantly benefit local and regional "neighbors" of worksites as well as industry customers. For example, OEHS professionals helped one company redesign product containers so that customers would not have to move excessively heavy items and face other ergonomic problems presented by the original containers. These initiatives resulted in reduced risk of customer injury and benefited both the manufacturing company and its customers.

A perhaps unexpected advantage of the new strategic orientation may be a reduction in the ethical dilemmas that frequently face those who were only *in* the organization. That is, by actively working to improve both the health of the worker and the company's ability to accomplish its goals, the OEHS professional can help meet the needs of both management and employees.

The complex nature of the scientific challenges facing OEHS providers means that all of these practitioners must have extensive professional knowledge. However, they also need the unique orientation and management capabilities required to develop, implement, and evaluate comprehensive and effective OEHS programs. While OEHS members are hired for what they know, they are evaluated on what they do. Thus, significant accomplishments require a blend of professional knowledge and management skills.

This brief consideration of OEHS programs' management challenges and opportunities illustrates the types of factors that those directing any health or safety program must consider. Subsequent chapters are designed to help health and safety managers enhance their capabilities to accomplish the varied and numerous management aspects of their programs.

DEVELOPING HEALTH AND SAFETY MANAGEMENT EXPERTISE

Many health and safety graduate education programs emphasize the development of competence in clinical medicine, safety and ergonomics, environmental and industrial toxicology, or other practice areas, depending on the emphasis of the particular graduate program. Some programs, such as clinical residencies, may have little education in management techniques and applications. In other programs, management training focuses on standard management aspects such as theories, accounting, organization, and budgeting. This more traditional management education in these programs provides useful information but may not equip practitioners to develop the practical expertise needed to "sell" their programs, deal with different personality types, communicate effectively, recognize the numerous "evaluators" of their programs, and respond to the political factors that may affect their efforts. Essential techniques such as time management, organizing and conducting meetings, and making effective presentations may not be considered. Compounding the difficulty for students in academic programs is the fact that a number of recent texts on health and safety issues do not include "management" as an index topic, and there is little or no discussion of management in many of these reference works. Consequently, the management capabilities needed to direct heath and safety efforts are often not developed during students' academic work. Graduates can subsequently be at a significant disadvantage as they begin their careers.

The need for improved management skills has been recognized for some time [8–10]. For example, the American College of Occupational and Environmental Medicine, the American Association of Occupational Health Nurses, the American Industrial Hygiene Association, the American College of Physician Executives, and other health and safety organizations have offered short courses and other programs designed to increase the management expertise of professionals. New management courses that include orientation to the management needs of practitioners are being introduced into some graduate programs. Some commentators, such as Livingston, suggest that management education activities cannot develop management skills [11]. However, comments from a number of corporate, government, and other health and safety directors, as well as from health and safety professionals in other management positions, strongly support the benefits of management education. This book was designed to complement academic and continuing education courses and also to be of value to those who have management requirements but cannot attend more formal educational activities. It can also serve as a companion to other publications that discuss management requirements and techniques [12–17].

It should be noted that, in addition to health and safety management writings, the health and safety professional will benefit from an acquaintance with more standard management works. Books by Covey, Drucker, Peters, Goldratt, and others can be considered management "classics" [16–19]. Experience has demonstrated that a manager's knowledge of the core concepts of more recent texts such as *Good to Great* [20] and *Creating We* [21] will be valuable in discussions with supervisors, senior managers, and other decision makers who do not have health and safety backgrounds and perspectives. In addition, portions of a number of such books are directly applicable to health and safety programs. For example, the discussion of "bottlenecks" in *The Goal* are relevant to clinic patient scheduling and organizing a series of safety surveys [19]. Similarly, the director will find the discussion of personnel management in *First Break All the Rules—What the World's Greatest Managers Do Differently* helpful in dealing with members with different capabilities [22]. It is of note that in some organizations, senior management has made one or more of these books or other similar texts required reading for all of their junior or middle managers. The fact that the health and safety manager has recognized the significance of writings by management gurus may aid in establishing rapport and credibility with the other managers who will impact the professional's programs. In this regard, it may be of as much benefit to a health and safety program, from the standpoint of senior manager support, for the director to peruse the *Harvard Business Review, Forbes, The Wall Street Journal,* or similar publications as it is to do in-depth reading of professional journals.

Effective management techniques and applications should facilitate development of programs that will enhance the health, safety, productivity, and well-being of individuals and their families. With respect to OEHS professionals, these improvements should also translate into improved productivity, competitiveness, and profitability for the organizations that the OEHS professional supports. Similarly, the accomplishments of those in other health and safety programs should enable these organizations to expand still further, resulting in even more opportunities to provide benefits to groups, communities, or other populations they serve.

REFERENCES

1. Drucker PF. *The Essential Drucker.* New York: HarperCollins, 2003. P 8.

2. Department of Health and Human Services, National Institute for Occupational Safety and Health. Progress Toward Safe Nanotechnology in the Workplace—A Report from the NIOSH Nanotechnology Research Center. DHHS(NIOSH) Pub 2007-123, February 2007.

3. Tepper LB. The strategic dimensions of occupational medicine (1986 Gehrmann Lecture). *J Occup Med* 1987;29:325–329.

4. McCunney RJ. Occupational medical services. In: McCunney RJ, *A Practical Approach to Occupational and Environmental Medicine,* 3rd Ed. Philadelphia: Lippincott, Williams & Wilkins, 2003.

5. Tonn MD. Health care management In: McCunney RJ, *A Practical Approach to Occupational and Environmental Medicine,* 3rd Ed. Philadelphia: Lippincott, Williams & Wilkins, 2003.

6. McCunney RJ, Anstadt G, Burton WW. The competitive advantage of a healthy work force: opportunities for occupational medicine (editorial). *J Occup Environ Med* 1997;39:611–613.

7. Moore JS. Office ergonomics programs—a case study of North American corporations. *J Occup Environ Med* 1997;39:1203–1211.

8. Moser R, Meservey D, Lee JS, et al. Education in management aspects of occupational and environmental health and safety programs. *J Occup Med* 1989;31:251–256.

9. Cordes DH, Rea DF, Vuturo A, et al. Introducing management skills and issues in preventive medicine residency programs: a survey. *J Occup Med* 1991;33:977–979.

10. Snyder TB, Himmelstein J, Pransky G, et al. Business analysis in occupational health and safety consultations. *J Occup Med* 1991;33:1040–1045.

11. Livingston JS. Myth of the well-educated manager. *Harvard Bus Rev* 1971;49. Reprinted in *Harvard Business Review Business Classics* 1986;96–106.

12. McCunney RJ (ed), American College of Occupational and Environmental Medicine. *A Manager's Guide to Occupational Health Services.* Beverly, MA: OEM Press, 1996.

13. Fallon LF Jr, Dickerson OB, Brandt-Rauf PW (eds). The management perspective. *State Art Rev Occup Med* 1989;4(1).

14. Patterson WB. Establishing an occupational health program. In: McCunney RJ, *A Practical Approach to Occupational and Environmental Medicine,* 3rd Ed. Philadelphia: Lippincott, Williams & Wilkins, 2003.

15. Leone FH. The physician working with the business community. In: McCunney RJ, *A Practical Approach to Occupational and Environmental Medicine,* 3rd Ed. Philadelphia: Lippincott, Williams & Wilkins, 2003.

16. Covey SR. *The 7 Habits of Highly Successful People: Restoring the Character Ethic.* New York: Simon & Schuster, 1989.

17. Drucker PF. *Managing for the Future: The 1990's and Beyond.* New York: Truman Talley Books, 1992. See also Drucker PF. *The Effective Executive.* New York: HarperBusiness Ed, 1993 (earlier eds, 1966, 1985).

18. Peters TJ, Austin N. *A Passion for Excellence: The Leadership Difference.* New York, Warner Books, 1985. See also Peters TJ, Waterman RH Jr. *In Search of Excellence: Lessons From America's Best Run Companies.* New York: Warner Books, 1982.

19. Goldratt EM, Cox J. *The Goal—A Process of Ongoing Improvement.* 2nd Rev Ed. Great Barrington, MA: North River Press, 1992.

20. Collins J. *Good to Great—Why Some Companies Make the Leap and Others Don't.* New York: HarperBusiness, 2001.

21. Glaser JE. *Creating We—Change I-Thinking to We-Thinking: Build a Healthy, Thriving Organization.* Avon, MA: Platinum Press, 2005.

22. Buckingham M, Coffman C. *First Break All the Rules—What the World's Greatest Managers Do Differently.* New York: Simon & Schuster, 1999.

ADDITIONAL READING

Anderson WM. Managing occupational health performance: A view from the corporation. *Am Ind Hyg Assoc J* 1980;41:A53–A55.

Asfahl R. *Industrial Safety and Health Management,* 5th ed. Upper Saddle River, NJ: Prentice Hall, 2004.

Draper E. *The Company Doctor.* New York: Russell Sage Foundation, 2005.

Drucker PF. Managing for business effectiveness. *Harvard Bus Rev* 1963;41. Reprinted in *Harvard Business Review Business Classics* 1986;58–65.

Fleming LE, Herzberg JA, Bunn WB III (eds). *Issues in International Occupational and Environmental Medicine.* Beverly, MA: OEM Press, 1997.

Hagan PE, Montgomery JF, O'Reilly JT (eds). *Accident Prevention Manual for Business & Industry: Administration & Programs.* Itasca, IL: National Safety Council, 2001.

Hammer W, Price D. *Occupational Safety Management and Engineering,* 5th ed. Upper Saddle River, NJ: Prentice Hall, 2001.

Hersey P, Blanchard KH, Johnson DE. *Management of Organizational Behavior,* 8th Ed. Upper Saddle River, NJ: Prentice Hall, 2000.

Huey MA. The industrial hygiene program. In: Plog BA, Niland J, Quinlan PJ (eds), *Fundamentals of Industrial Hygiene.* Itasca, IL: National Safety Council, 1996.

Hughes RG. A safety director in the land of business. *Professional Safety* 1987; March:22–23.

Leibowitz A. Program management. In: *The Occupational Environment—Its Evaluation, Control, and Management,* 2nd Ed. Fairfax, VA: American Industrial Hygiene Association Press, 2003.

CHAPTER 2

Scope of Managerial Responsibilities

As noted in Chapter 1, a manager's primary responsibility is to get things done, usually with and through the assistance of others [1]. The effective manager helps establish goals and objectives for the organization and then facilitates the efforts of other members of the unit to achieve the goals and accomplish the objectives. In meeting this primary responsibility, managers must be able to perform a number of diverse activities. Success as a manager depends on *how well* these activities are performed. This chapter provides an *overview* of some of these responsibilities in general, and subsequent chapters discuss selected aspects in more depth.

TECHNICAL, HUMAN, AND CONCEPTUAL FUNCTIONS

Katz [2] notes that managers need a combination of capabilities, grouped into technical, human, and conceptual skills. This approach is in contrast to the concept of a "universal manager" that suggests that an individual can be an effective manager without technical (or professional) expertise in the area being managed. While some individuals may be able to function in this environment, most find it essential to have some technical skills in their areas of responsibility. One may not need to know all details of a unit's operations, but the manager who knows what questions to ask about an activity's technical aspects usually has a significant advantage over one who relies only on "management indicators" to direct a unit. The following example is illuminating. In 2007, a chief executive officer of a major clothing chain was dismissed by the board of directors because of poor performance of the chain. He was previously with a major entertainment organization involving theme parks, movies, and similar activities. He was replaced by a person who had had extensive experience in clothing apparel, including significant activities in the company he was asked to direct [3]. The subsequent upsurge in stock price indicated the "owners" recognized the value of having a manager/leader who had experience in the primary focus of the company.

For the application of technical skills, an OEHS manager's expertise is typically in one of the four OEHS disciplines (occupational and environmental medicine, industrial hygiene, occupational health nursing, and safety and ergonomics), but a manager usually has knowledge of the other three areas as well. In both graduate education programs and practice settings, there is increasing emphasis on interdisciplinary approaches to OEHS activities that utilize all four disciplines. Such interaction can significantly enhance program effectiveness, and the OEHS director who is familiar with the different disciplines enjoys a distinct advantage in leading an interdisciplinary team. The same principles apply to public health and to other medical and safety programs.

In addition to technical/professional qualifications, Katz stresses that effective managers need good "people" skills: The manager who needs to get things done through others has to be able to build teams of workers to get the work done [1,2]. Selecting the right people, motivating them, and getting them to accomplish necessary actions by linking together the "people" strategy and the operations is a daunting task [1]. Achieving desired outcomes demands effective leadership, and Chapters 7, 8, 9, and 13 address various aspects of leading an organization of productive people.

In addition to working with and through others in the unit, health and safety managers must also be able to work effectively with their supervisors as well as with middle and senior management. The broader the network of contacts with whom the manager can work, the greater the manager's effectiveness. Having the "people" skills needed to work with the other managers' different types of personalities is an important attribute for the manager. (Chapter 8, by Dr. Whitman, provides guidance in recognizing and relating to the personality types and needs of co-workers and others in the work environment.) The effective manager must have the ability to work with different types of personalities to obtain essential support for the health and safety program from other senior managers.

The third skill Katz delineates is a conceptual one. Managers need to be able to develop concepts and formulate plans to accomplish goals, achieve objectives, and enhance the activities of their organizations. Covey, in *The 7 Habits of Highly Effective People,* stresses that individuals in leadership management positions may be most effective as they move away from authoritarian, independent positions to an interdependence orientation [4]. When desired results are clearly delineated, and information, recognition, and reward systems encourage cooperation, all involved in an activity can be in a "win/win" situation. The current emphasis on participatory management reflects the fact that interactions among planners and those implementing plans can lead to synergy that produces markedly better results than would have occurred without such interchanges. Managers who foster this type of approach may accrue many benefits. For example, the process may result in flashes of insight that produce major advances in the organization's capabilities and scope of activities. Such accomplishments can be one of the most satisfying aspects of being a manager.

Managers are expected to plan their units' long-range and short-term activities. The plans for the health and safety unit must mesh with those of the organization to which the unit belongs. This requirement exists whether the unit is in a corporation, hospital, government body, consultant group, or in some other setting. Planning should not be restricted to approaches to accomplish routine objectives and cope with predictable problems. Rather, emphasis should be placed on developing and taking advantage of opportunities to make the health and safety activities more effective as well as more professionally satisfying.

A number of other management responsibilities are included in the three major aspects listed by Katz. For example, once plans have been defined, the health and safety manager must be concerned with developing methods to accomplish the plans: organizing facilities, people, and other resources; taking the steps needed to implement plans; directing and motivating those responsible for achieving goals and objectives; coordinating activities of people and functions within the health and safety unit and between the unit and other portions of the organization the unit supports; accomplishing budgeting and monitoring to ensure proper use of financial and other resources; and evaluating progress of individuals and the unit in accomplishing activities.

COMMUNICATION

Managers must communicate effectively. Efforts and programs can be severely jeopardized by poor written or oral communications. Communication concerns are not limited to interactions with supervisors, colleagues, and members of the heath and safety unit. The health and safety manager must convey information to other managers and members of the organization or entity supported as well as to those outside the organization. The latter requirement can involve publications, exchanges with the public, meetings with legislators, and interchanges with representatives of the media. Diverse skills are needed to communicate effectively in these different settings, and the manager's overall success is dependent on having the ability to communicate effectively. Of all the capabilities needed by managers, communication skills are of primary importance. The chapters discussing various communication requirements and techniques underscore the significance of communication capabilities.

PROBLEM SOLVING AND DECISION MAKING

Managers are expected to solve problems and make decisions. Problems will occur in spite of the best planned efforts and applications of valid management techniques. There are different methods to define and solve problems. These are discussed in Chapter 14. Whatever process is used, it is essential that the manager be able to recognize difficulties early and develop methods to overcome them. Effective monitoring processes and other management techniques will help in this process, and the manager has to be confident that procedures are in place to provide "real-time" information on progress or difficulties in accomplishing the organization's goals.

A manager's decision-making responsibilities are of particular significance. The discussion of decision making in Chapter 14 underscores the impact of an organization's decision-making process. A manager who repeatedly delays making a decision should not be surprised to find that the delays have a demoralizing effect on the organization. Typically, that manager's supervisor and more senior managers in the corporation will also become frustrated and disenchanted with the health and safety manager. These attitudes can seriously compromise support for the manager and even jeopardize the entire health and safety program.

These comments are not to suggest that a manager should rush into every decision without attempting to obtain and evaluate available information. Rather, the effective director evaluates the information that can be obtained in a reasonable time, carefully considers options, makes a decision, and monitors the results. All of us make mistakes, and one should not be surprised if some decisions prove to be wrong. (In struggling with a difficult decision, it may be of some solace to remember the overall guidance for managers, "If you aren't making some mistakes, you're not trying.") In almost every instance, the decision can be changed, incorporating information gained through the original course of action, with an increased probability of success with the new approach. The important consideration in decision making is to make a reasoned decision within an appropriate time. As noted in Chapter 14, "not deciding now" is actually a decision and may have a more negative impact than any other possible course of action.

ENSURING PROGRAM CURRENCY AND RESPONSIVENESS

In addition to providing the professional expertise needed to design and accomplish comprehensive health and safety programs, the manager must remain current with pertinent regulatory, legislative, and legal requirements. Ensuring this currency is a major, critically important, area of responsibility. A company, hospital, or government agency facing a multi-million dollar citation because of failure by the health and safety manager to be aware of a health or safety requirement, cannot be expected to provide strong support for other aspects of the health and safety program and activities. Conversely, helping the organization avoid such liabilities can result in beneficial recognition and support of the health and safety program. Similar requirements exist for health and safety units in other settings.

Meeting legal and regulatory requirements may also necessitate testimony during health and safety litigation or during Family Medical Leave Act, Americans with Disabilities Act, workers' compensation, discrimination, or other hearings. Occasionally, a health and safety professional may be asked to provide testimony during depositions or trials without realizing the need for extensive preparation or recognizing the rules of evidence. Results are consequently less optimal than the professional might wish. It is hoped that the material in Chapters 23 and 24 will help the professional to be better prepared to provide testimony and other legal input.

ESTABLISHING STANDARDS

The manager should participate in establishing performance and ethical standards for that activity. Often, this effort involves setting examples for others. Accepting poor quality work, such as poorly organized reports or letters with typographical errors, will suggest to other members of the unit that such work is "acceptable." Those dealing with the organization, however, may believe that the work reflects the quality of professional activities. Quality in an organization and in its "product" continues to be a management challenge that must be met if the organization is to be effective [5–7]. Total Quality Management and Six Sigma are only two of many approaches to improving quality of health and safety activities [5,7]. Chapter 25 addresses quality programs and activities in more detail.

Similarly, ethics and personal integrity can play major roles in the success or failure of health or safety programs. A director, or other member of the unit's team, who compromises personal or professional standards may jeopardize both program activities and the health and safety of those the organization serves. Compromises in integrity "to get the job done" may suggest that expedient actions that could jeopardize health or safety are acceptable. The tragic results of the apparent initial cover-up of asbestos hazards dramatically illustrate the consequences of not maintaining appropriate ethical standards. Conversely, early reporting of adverse events can benefit countless workers even if there is a short-term, adverse impact on the company.

This chapter's overview of the scope of some of the management responsibilities of health and safety directors emphasizes the diverse abilities needed by managers. Many of these management concerns obviously do not involve direct application of health and safety professional knowledge. However, programs incorporating effective management knowledge and skills can dramatically improve the health, safety, and productivity of large numbers of workers, community citizens, their families, and others who are or may be affected by the organization's activities. The remainder of this text considers these management applications and techniques more extensively in an effort to help health and safety professionals meet their management responsibilities.

REFERENCES

1. Bossidy L, Charan R. Execution—*The Discipline of Getting Things Done.* New York: Crown Business, 2002.

2. Katz RL. Skills of an effective administrator. *Harvard Bus Rev* 1974;52. Reprinted in *Harvard Business Review Business Classics* 1986;23–35.

3. Waters J. CEO Pressler is out at Gap, Inc.—struggling retailer is looking for an industry veteran to take the reins. *MarketWatch* Jan 22, 2007.

4. Covey SR. *The 7 Habits of Highly Effective People—Restoring the Character Ethic.* New York: Simon & Schuster (Fireside), 1989.

5. Walton M. *The Deming Management Method.* New York: Putman (Perigee), 1986.

6. Peters TJ, Waterman, RH Jr. *In Search of Excellence.* New York: Warner Books, 1982.

7. Welch J. Six sigma and beyond. In: Welch J, Byrne JA. *Jack—Straight from the Gut.* New York: Warner Business Books, 2001.

ADDITIONAL READING

Buckingham M, Clifton DO. *Now, Discover Your Strengths.* New York: Free Press, 2001.

Buckingham M, Coffman C. First Break All the Rules—*What the World's Greatest Managers Do Differently.* New York: Simon & Schuster, 1999.

Chowdhury S. *The Power of Six Sigma: An Inspiring Tale of How Six Sigma is Transforming the Way We Work.* Chicago: Dearborn Trade, 2001.

Drucker PF. *Managing in a Time of Great Change.* New York: Truman Talley Books/Dutton, 1995.

Drucker PF. *The Practice of Management.* New York: HarperCollins, 2006.

Fallon LF Jr, Dickerson OB, Brandt-Rauf PW (eds). The management perspective. *State Art Rev Occup Med* 1989;4(1).

Guidotti TL, Cowell JWF, Jamieson GG. *Occupational Health Services—A Practical Approach.* Chicago: American Medical Association, 1989.

Learned EP. Problems of a new executive. *Harvard Bus Rev* 1949. Reprinted in Harvard Business Review Essentials of Executive Development No. 12204:17–25.

Lorsch JW, Mathias PF. When professionals have to manage. *Harvard Bus Rev* 1987;65(4):78–83.

Managerial Excellence. *McKinsey Award Winners From The Harvard Business Review, 1980–1994.* Boston: Harvard Business Review Press, 1996.

Rose F. Now quality means service too. *Fortune* 1991;123:97–111.

Tagiuri R. Managing people—ten essential behaviors. *Harvard Bus Rev* 1995;73(1): 10–11.

Zaleznik A. HBR classic—managers and leaders: are they different? *Harvard Bus Rev* 1992;70(2):126–135.

CHAPTER 3

Planning and Prioritizing Health and Safety Programs and Activities: What to Do When There's Too Much to Do

In order to be effective, managers must be able to plan and prioritize their units' activities. They must then organize people and other resources to accomplish the planned programs. This chapter considers some planning and prioritization aspects, Chapter 4 discusses planning for emergency events, Chapter 5 addresses methods of implementing plans, and Chapter 6 focuses on organizational factors, including the position of some health and safety programs in a few typical organizations.

GENERAL CONSIDERATIONS

Planning should be the initial step in developing or modifying a health and safety program. A misstep of many managers is to simply keep doing what has always been done. In my experience, this is particularly likely when a manager moves to a new position; however, it may also occur in many long-running programs. The numerous pressing demands that the manager faces, including crises, heighten this tendency to continue in-place activities.

It also is not unusual to experience resistance to planning on the part of other health and safety members. If it is done effectively, planning takes time, and this requirement often means that other work is deferred to the point that significant extra effort is necessary to remain current with day-to-day challenges. Those who have seen extensive time and effort spent on developing plans that are then placed in thick three-ring binders that are not touched until the next planning cycle are, not surprisingly, prone to be quite vocal in their resistance to planning efforts. Another stumbling block is the difficulty of responding to variable, rapidly changing challenges and opportunities, particularly in regard to disaster planning, wherein a typical comment is "No disaster will ever follow a plan, so why plan?" The questioner is quite correct that a disaster will not follow a plan. However, in my own experience in disaster response situations or evaluating them, it has been apparent that health and safety units with a well-developed and practiced plan were able to respond in a coordinated, organized, and effective way. These units also responded readily to the rapidly changing situations that typically occur during disasters. Conversely, groups that believed they could "play it by ear," tended to take confused, ineffective actions that not only reflected poorly on the unit but, tragically, resulted in life-threatening treatment delays. The next chapter considers disaster planning aspects in more detail. The point is that if planning can produce beneficial results in a disaster situation, it should be of even more value to health and safety programs in more normal settings. As Mackay [1] and others have noted, "Failing to plan is planning to fail."

Basic concerns in planning are "Where are we going and how are we going to get there?" As Covey observes in *The 7 Habits of Highly Effective People,* it is important to "Begin With the End in Mind" and "Be Proactive" [2]. Clearly defining the outcomes and priorities for the health and safety program, and stipulating how the results are to be achieved, are crucial for managers in effectively directing their organizations.

It should be mentioned that some, although a decreasing number, of corporations and larger organizations accomplish planning in separate planning departments. However, in many other businesses, agencies, and other organizations, greater responsibility for planning is being assigned to the different directors of the functional areas of the organization. This latter approach reflects the increasing emphasis on participatory management, and in many instances the health and safety manager will have the same planning responsibilities as managers of other departments. As indicated by Covey's recommendation to be proactive, the health and safety director should take the lead in initiating the planning process, defining major areas of concern and interest, arranging for appropriate inputs from others, and overseeing development of the final plan. It is also important that the director ensure that the planning document is a dynamic one that is consulted frequently to help make the health and safety program evolve as delineated during the planning process. Regular updates will help keep the plans current and relevant.

An effective planning process also helps the manager establish priorities. This result is particularly significant in view of limited resources and the multiple demands for "everything" to be done "at once." In my experience, it has never been possible to do "everything," and appropriate prioritization, recognizing the variables that can affect assigned priorities, is essential.

The following sections consider planning and prioritization in more detail.

PLANNING STEPS

Table 3-1 depicts some planning steps that have been useful in developing effective plans.

Table 3-1. Planning Steps

1. Decide on planning format in relation to corporate planning procedures
2. Determine current status and potentials
3. Use forecasting in planning
4. Decide what the health and safety unit needs to do
5. Make plans appropriate and relevant
6. Use planning to help find opportunities
7. Use planning to help avoid crises
8. Make planning a dynamic process
9. Incorporate evaluation in the planning process
10. Have patience

1. Decide on Planning Format in Relation to Corporate or Other Organizational Planning Procedures

A number of different planning formats can be used, and planning that has a long-term, *strategic* focus may be organized differently from that concerned with near-term, *tactical* concerns. One typical approach for strategic planning is to use a "Vision, Mission, Goals" approach. Other formats may be more appropriate, depending on the organization. Most corporations, agencies, or businesses have established planning formats, and use of a similar format will facilitate incorporation of the health and safety plans into the overall efforts of the organization. Relating health and safety planning to the plans of the corporation or organization also helps emphasize the importance of the unit's contributions to achieving the overall corporate or organizational goals.

The vision portion of the "Vision, Mission, Goals" planning approach defines the desired capabilities and activities of the organization a number of years in the future. A typical period for a vision to be achieved is five years. However, depending on the situation, shorter or longer periods may be used. This statement is couched in general terms, such as "Be internationally recognized for providing quality health and safety services to major, global corporations." In essence, a vision statement is a summary of what the unit hopes to achieve and gives members an agreed-upon outcome for which to strive.

A statement of the organization's mission, or missions, is somewhat more specific. Mission statements define the major activities of the organization. At our educational center, a mission statement might be "To Protect the Worker and the Environment Through Interdisciplinary Education, Research, and Service." Mission statements can be brief, such as an earlier United States Air Force statement that its mission was to "Fly and Fight." A health and safety unit might use "Provide Comprehensive Health and Safety Services to Maintain and Improve the Health, Safety, and Productivity of the Individuals the Organization Serves." Mission statements often contain phrases that are difficult to define, such as *quality,* e.g., "Provide Highest Quality Health and Safety Services." Such statements may have a tendency to become long and cumbersome if the planning group attempts to include all possible aspects of the organization's activities. For example, the mission statement of one university consists of 12 lines of text, followed by Teaching, Research, and Public Life sections of 13 to 15 lines of text. Ideally, a mission statement should be short enough to be memorized by the organization's members. It is doubtful if even senior professors at the institution just mentioned can quote over 50 lines of text from memory. The observation that a primary criterion for a mission statement is that it can be printed on a coffee cup is not inappropriate guidance. As noted by Adams, the creator of the comic strip *Dilbert,* mission statements may be crafted to incorporate current buzz words and leave the reader of the statement unable to decipher what is actually to be accomplished. He uses the example of an actual mission statement, "We will produce the highest quality products, using empowered team dynamics in a new Total Quality paradigm until we become the industry leader" to illustrate the point [3]. Ideally, all members of the organization should be able to state readily what the group's mission is. It is said that in the early days of the National Aeronautics and Space Administration (NASA), everyone from senior scientists to housekeepers knew NASA's primary mission was "To Put a Man on the Moon." Although the current statement, "To Pioneer the Future in Space Exploration, Scientific Discovery, and Aeronautics Research" is succinct, it does not have the emotional impact of the earlier one.

Goals are the ends, that, if accomplished, will enable the organization to achieve its mission(s) and vision. Goals are stated in general terms and define the major efforts needed for each mission. For example, a goal might be stated as "To Establish an Industrial Hygiene Service That Will Ascertain and Accomplish Appropriate Monitoring of Hazardous Environmental Exposures." As suggested by this example, goals are more specific than missions but still should have a general, qualitative tone and emphasis.

Properly accomplished, the "Vision, Mission, Goals" format will enable the planners to specify the actions needed to accomplish the goals and hence the mission and vision. As suggested by Drucker, the necessary actions may be formulated as objectives [4]. In contrast to goals, objectives specify in measurable terms what is to be done, by whom, at what resource cost, and in what time frame. More detailed discussion of developing objectives are considered in Chapter 5, using the Management by Objectives process as one example of methods to accomplishing goals. In turn, other techniques, including tracking systems to monitor progress, such as Program Evaluation and Review Techniques (PERT), discussed in Chapter 12, can be applied to the efforts resulting from the planning process. Through these applications, planning can contribute directly to the overall management activities and effectiveness of the health and safety program director.

2. Determine Current Status and Potentials

A somewhat trite phrase that has applicability in planning is "It is difficult to know where you are going if you do not know where you are." This observation is particularly apt in developing long-range, strategic plans for a health and safety unit. Spending sufficient time defining current capabilities and potentials as well as current and future adverse factors facilitates the formulation of plans that are realistic and have a better chance of success. There are a number of approaches to accomplishing this part of planning. One approach is to use a Strengths, Weaknesses, Opportunities, and Threats (SWOT) analysis or any of a number of similar approaches. This process helps planners deliberately consider current strengths, capabilities, constraints, and weaknesses of the organization. It also encourages evaluation of the potentials of the unit and the chances for realizing them. A "needs assessment" (e.g., surveys, consumer interviews) to estimate demand for new or revised programs may be part of the Opportunities component of SWOT. Finally, planners try to define adverse factors, (e.g., factory closures, competition, budget reductions) that could affect the unit. By considered analysis of SWOT components, planners can ascertain which efforts have high probabilities of success and which may present significant difficulties.

3. Use Forecasting in Planning

Most planning efforts require forecasting. Although the future cannot be precisely predicted, it is possible to define some aspects in qualitative terms. Forecasting should consider anticipated changes in the health and safety manager's unit as well as in the organization the unit supports. Expansion in managed care programs is only one example of organizational and other changes that may impact the health and safety function [5–7]. Projected modifications in manufacturing processes, with potentials for exposures to new health hazards—for example, emerging infections and potential pandemics, with resultant increased demands on health and safety units at the time the units may be short of personnel due to illness or fear of exposures—obviously have to be incorporated into planning. Changes in the work force, including number and types of employees, anticipated educational levels of future employees, sex and age distribution of the workers, numbers of individuals with disabilities, projected turnover rates, and similar factors should be considered in planning. Impacts of competition, including requirements for new plants or relocation of manufacturing activities, can also affect health needs and activities. The increasing emphasis on globalization may have major impacts on future activities and thus needs to be considered in planning if the organization is multinational or becoming such [8]. The expanding uses of nanotechnology may pose special challenges for some organizations [9].

In addition to forecasting factors that directly affect company or other organizational operations, the health and safety manager should consider broader sociopolitical and economic factors that can potentially affect the organization or the population the health and safety unit supports. Global political and economic changes, such as the European Community changes in Europe, may significantly impact a manager's corporation and its health concerns [8]. Anticipating these possibilities in the planning process can help ensure that the health and safety organization is poised to respond appropriately.

As Drucker noted in 1992, "businesses will undergo more, and more radical, restructuring in the 1990s than at any time since the modern corporate organization first evolved in the 1920s" [10]. At the current time (2007) it is apparent that this statement was correct [11]. It also appears that the rate of change will not decrease in the near future. Reviewing pertinent changes that have already taken place, or are projected to take place in the near future, will help define what can reasonably be anticipated in a more distant time frame [10,11].

4. Decide What the Health and Safety Unit Needs to Do

This seemingly simple requirement has caused significant problems for many health and safety managers. Some sources of inputs include:

a. Personal interests and expertise of the health and safety director

b. Participatory inputs from colleagues in the health and safety unit

c. Regulatory requirements

d. Input from "customers," recognizing the need to define the many "customers" the unit serves

e. Recommendations and "lessons learned" from other health and safety managers

f. "Guidance" from supervisors and superiors

Although listed last, meeting with senior supervisors early in the planning process is always helpful. This permits discussion of their areas of interest so that due consideration can be given to meeting their needs and expectations. In order to provide a teaching example, I carefully demonstrated the adverse effects of not meeting early with a supervisor when I went to a new location to be in charge of a health unit. In the best participatory management fashion, I met with the professional and staff members of the unit to make plans for our activities. I contacted colleagues in other locations to ascertain their major emphases. We established a series of goals and objectives to accomplish our perceived mission. All of us worked to accomplish the plans we had made, and I was proud of our progress. After several weeks, I discussed our efforts with my supervisor, who was a senior manager with no formal training in any health field. He listened to my presentation and then observed, "That's all very well, doctor, but when are you going to get around to doing the things I want you to do?" Failure of a health and safety director to bring supervisors into a planning process early is *not* the way to instill confidence in the director's managerial capabilities. Even if they lack formal health and safety training, supervisors may have very definite ideas as to what they expect from the health and safety organization! (Appendix A provides some suggestions to help health and safety professionals who are assuming managerial positions avoid experiences similar to mine.)

Certainly, the health and safety director should try to incorporate programs and projects of particular personal interest into the unit's plans. There should be some benefit to being "in charge." It also is personally satisfying to have interchanges with colleagues in the planning effort, and such participation facilitates "buy-in" by the other health and safety members. In developing plans, it is, of course, important to meet essential regulatory requirements. The ever-changing nature of these requirements means that plans have to have sufficient flexibility to adjust to the changing needs.

In planning, it is essential to get input from the individuals or populations that will be affected by the efforts of the health and safety organization. A former vice president for occupational health and safety of a Fortune 500 company observed that success as a health and safety manager requires an appreciation of who your customers are and what they believe the unit should be providing them. Like me, he commented that he had made many mistakes by assuming that he knew what the organization wanted from his program. Other colleagues have made similar observations. They have repeatedly commented that, in order to be successful, health and safety managers need to understand the mission or function of the organization they serve. How well they then focus their efforts on the organization's strategies is often directly related to their success as managers.

Balancing the health and safety director's need to understand the goals and mission of the organization is the need to help other supervisors and managers understand and appreciate the technical aspects and requirements of an effective health and safety program. It is the responsibility of health and safety managers to express their plans in terminology free from jargon so that they are readily understood by nontechnical supervisors and upper management. In short, it is necessary to "sell" the plans to management.

Many health and safety managers have found that establishing an advisory committee of "customer" representatives can help ensure the health and safety program will meet the needs of the organization, customers, and the health and safety team members. Such a committee should be composed of senior managers, worker representatives, community members, union representatives (if applicable), and others who understand the organization's priorities and needs. Not only will such a group assist the health and safety professional in developing appropriate plans and programs, but by becoming familiar with and participating in health and safety planning, the committee members may become champions for health and safety efforts. Advisory committee members can thus be of crucial importance in selling and obtaining support for health and safety plans. However, it is important that the committee members do not feel that they are merely being used for appearance's sake and to provide support in getting plans approved. The advisory committee should be actively involved in formulating objectives and directions of the health and safety organization. Once plans are implemented, it is important to provide supervisors and advisory committee members periodic updates on progress.

None of these comments are intended to suggest that health and safety directors should abdicate their planning responsibilities. The comments are meant to emphasize the need for health and safety plans to mesh with the plans and efforts of the total organizations or other entities the program supports. Further, it must be recognized that every organization has its own corporate image, ethics, and operating procedures. The health and safety plan must be consistent with these if it is to be successful. Unless "customer" issues are included in planning, there is the obvious risk that plans, even though well constructed, will not be supported by other managers in the corporation. This lack of support may result in loss of credibility for the planner and failure to get necessary health and safety programs accomplished. Conversely, an effective long-range planning process can be a highly useful method to obtain funding and management backing for health and safety efforts. Once the health and safety plan is accepted and funding is included in the company's or other organization's plan, the frequency and number of day-to-day struggles for power and money will usually be reduced.

It is essential that the manager be able to review overall organizational plans in order to ascertain any requirements for the health and safety unit that other managers have established. Once aware of these requirements, the manager can plan to meet them or to obtain approval for appropriate modification of the requirements. Such knowledge will help the health and safety manager avoid the disquieting experience of learning on short notice of a requirement that will have a major impact on the manager's unit.

The health and safety director who is considering what the unit should do can often profit from the experiences and recommendations of others, including "lessons learned." A number of sources provide information on recommended components of health and safety programs, focusing on what to do to have an effective program [8,12–18]. Chapters 15, 16, and 17 of this text also discuss efforts that may be incorporated into comprehensive health and safety programs.

5. Make Plans Appropriate and Relevant

Plans need to be both appropriate and relevant. As Drucker has noted, it is essential that an organization do the *right* things right rather than just do *things* right [19]. A comment often heard—"Paperwork is our most important product"—suggests that this particular activity is considered doing things right but is not necessarily the right thing. Others have made similar observations, and "The Needle in the Haystack" fable underscores this requirement. The story describes the praise given to members of a mythical kingdom who produced elaborate reports of their efforts to find a needle in the haystack, even though none of the searchers had a clear understanding of what a needle was. Efforts included careful measurements of the length and thickness of pieces of straw in haystacks and similar nonproductive efforts. Rewards were directly correlated with the weight of the reports, and those who did reports "right" were honored. In contrast, one individual developed a systematic approach to the search. His final report weighed little, and he was banished from the kingdom because of his poor results. He left with his magnet and one metal sliver that had a point on one end and a small opening in the other end [20]. Effective planning will help ensure that emphasis is placed on doing the *right* things.

Well-developed plans will help establish what, when, how, and by whom items should be done. Generally, planning is oriented toward macro rather than micro aspects of a unit's future, and, as previously mentioned, the planning effort should involve the individuals who will be responsible for accomplishing parts of the plan. The plan does not have to be in the detail required for specific objectives, which can be subsequently established using the techniques described in Chapter 5. A plan, however, needs to be useful to the unit's members in accomplishing established goals. Success in the planning effort can be evaluated by how frequently the plan is actually used and how often it is updated to reflect actual steps taken to accomplish the plan.

Done effectively, planning should lead to action. The Shewhart Cycle, as used by some planners, has been summarized by Deming as "Plan, Do, Check, Act" [21]. The emphasis is on transforming plans into action. Stressing the goal of acting on plans should help encourage those involved in the planning process to develop effective, relevant plans. Of course, if the plans are then relegated to the bookcase, subsequent planning efforts will engender little enthusiasm. Planning does not equate to performance. Planning without action does not benefit anyone. This status reflects the "Noah Principle," as described by Tepper [22]: "Predicting rain doesn't help—building arks does."

6. Use Planning to Help Identify Opportunities

A proactive planning process, designed to identify new opportunities and efforts, not only enhances the effectiveness of a health and safety unit but also increases job satisfaction for its members. As in other aspects of business, an activity performed to maintain the status quo usually has a regressive effect. Conversely, proactive planning helps delineate potentials that will result in continuing progress of the health and safety program.

7. Use Planning to Help Avoid Crises

Effective planning helps avoid crisis management, wherein most of the unit's efforts are spent reacting to the latest unanticipated disaster rather than focusing on proactive efforts to help improve workers' or citizen's health, productivity, and safety. In essence, if you are always putting out fires, you cannot build many houses. Of course, there will always be unanticipated requirements, but effective planning helps reduce these to acceptable levels. Effective planning also facilitates the formulation of optimal responses to unexpected events, such as a corporate-wide budget reduction. As in the previously mentioned disaster planning efforts, appropriate planning helps the manager establish and implement organized, effective responses to the latest unprogrammed health and safety requirement. However, if the health and safety function is constantly being "overtaken by events" (OBE'd), evaluation and revision of planning procedures is warranted.

8. Make Planning a Dynamic Process

As previously mentioned, carefully formulated, detailed plans that are put on the shelf until the next planning session are useless. Unless plans are used to provide overall guidance to activities, much planning time and effort has been wasted. While it may be appropriate to completely redo plans only once a year, they should be reviewed and revised much more frequently. Scheduling updates on specific dates (e.g., quarterly on the first Monday of the month), may help keep the plans active. It may be appropriate to take key health and safety members away from the organization during some of these reviews or at other times in the planning process. Staying at an off-site or "retreat" location usually not only helps members focus on planning without constant interruptions but also helps enhance team spirit while simultaneously emphasizing the importance of the effort.

9. Incorporate Evaluation in the Planning Process

As managers, we want to turn our plans into results. Often, we concentrate on planning and implementing plans but fail to remember the importance of evaluation of the outcomes. Specifying evaluation processes and time lines in the unit's plans will help prevent omission of this important aspect.

The health and safety manager's programs will be evaluated by others besides the program members and the director's immediate supervisors. Evaluation considerations are of such importance and sufficiently complex that Chapter 12 is devoted to this topic.

10. Have Patience

A good health and safety manager has to keep long-range goals in focus when circumstances preclude doing everything that the manager wants. During adverse economic times, programs will often have to be cut back. Even some of the most important activities may have to be deferred or stretched out over longer periods of time than desired. When faced with obstacles, the health and safety manager has to accept the fact that it may be necessary to start over again in obtaining approval and funding for sound, important projects. Repeated efforts may be necessary before the project or effort is finally approved and completed. For a health and safety manager to be effective, "patience is more than a virtue, it is an absolute necessity," as one corporate health and safety director observed. In some circumstances, delays may be such that a brilliant idea reaches fruition only after the originator has moved, with the result that someone else gets the credit. However, the manager should have a sense of satisfaction from knowing that the overall health and safety of the affected group has been improved. The situation is somewhat analogous to planting trees by a home even though you know you will have to move long before they are grown. Someone else may enjoy your efforts, but the overall benefits balance your own loss.

PRIORITIZATION

The planning steps to this point should help the manager define a comprehensive set of goals for the health and safety unit, and ideally, the unit should work to accomplish every desired action. Unfortunately, there are usually never sufficient people, funding, or other resources to do everything that one would wish. In essence, the manager will *always* have limited resources whether it is money, personnel, equipment, or other program resources. In these circumstances, it is imperative that the manager establish priorities for goals and actions. In some situations, a particular goal may be a priority due to the need to meet regulatory requirements. In other instances, a priority may be dictated by near-term projected needs. As planning evolves, it may become apparent that some priority projects, if fully implemented, would require essentially all resources and thus other needed efforts could not be accomplished. Deciding what must be done with the limited resources available represents a continuing challenge for many health and safety managers.

In considering action options and priorities, application of the Pareto principle may be of assistance. The current version of this principle is a modification of the one originally developed by Vilfredo Pareto. It evolved from sales activities and experiences and can be summarized as "20 percent of your effort will produce 80 percent of your results." Some would argue that the appropriate figures are 10 and 90 percent, respectively. The principle is also stated as "a way to sort out the 'vital few' from the 'trivial many' " [21]. Application can be illustrated by decisions regarding efforts to reduce lost time and attendant costs due to occupational injuries. An OEHS director might want to implement a program that would address all such injuries but could not do so due to resource limitations. If analyses of injury data demonstrated that the majority of lost duty days were due to back injuries, it would be appropriate to address prevention of these injuries as a first priority. Once accomplished, resources could be directed to other high-cost injuries, establishing action priorities in the same manner. (See further discussion of the Pareto approach in Chapters 15 and 25.)

It also is important to remember that supervisors may require that they help set priorities. This factor is another reason to meet with one's supervisor early rather than delaying the interchange as I did in the previously described instance. If a supervisor or senior manager indicates an area is of particular concern or priority, the wise manager will either incorporate the area as a priority in the planning process or get agreement with the supervisor to lower the priority.

Effective prioritization planning not only establishes what is to be done but also identifies what is not being done, why it is not being done, and what actions the director is taking to address the deferred items. In evaluating programs, it is usually easy to support a manager who can describe the rationale for doing certain things and not doing others. This is particularly the situation if the manager can also delineate actions taken to obtain the resources or other items needed to accomplish the deferred, lower priority activities. Conversely, a program director who is performing numerous activities of varying priorities but is not aware of projects that are not being accomplished even though they were designated high priority by senior executives, should not be surprised by a poor-performance evaluation from his or her supervisor.

NEGOTIATION

The health and safety manager will typically find that negotiation is an essential aspect of obtaining approval and support for the unit's plans and activities. Senior managers may have different priorities than the health and safety director, and it will be essential to agree on both the content of unit's plans and the priorities of the various components of the plan. Implementing plans may also require negotiations. For example, colleagues may consider some essential activities to be of minor significance, with the result that the required actions are continually deferred, necessitating negotiation on work activities. In hiring new health and safety members and establishing expected responsibilities, negotiation will usually be required. Similarly, agreeing on evaluation criteria for the health and safety unit, as discussed in Chapter 12, may require extensive negotiations by the health and safety manager.

Too often, negotiations are approached from a "win-lose" standpoint, wherein someone has to be the loser if the other person "wins." Covey emphasizes the importance of a "win-win" strategy, wherein the outcome is perceived by both parties as a fair one that benefits both members [2]. If negotiations begin with this end in mind, there is greater opportunity for a beneficial outcome than if the win-lose approach is taken.

If negotiation is to provide the best possible results, it is essential that the negotiator prepare for the effort. Preparation includes defining clearly what the manager's desired outcome is before negotiations begin. Are certain segments of the health and safety plans so essential that they have to be approved? Are other aspects of less importance so that they could be deferred? In essence, if you know what you want, you are much more likely to get what you want. Effective planning will help the manager delineate what is wanted (the goals) and the priority of each component.

Preparation also includes attempts to agree on the scope of negotiation before interactions begin. For example, in one negotiating session discussing the scope of our planned health and safety activities, my counterpart suggested that we also discuss the salaries and proposed raises for the health and safety members. I considered this a separate topic for negotiation, and, fortunately, was able to reach an agreement to focus only on the main thrust of the session. If the diversion of discussion had occurred, it could have been much more difficult to reach decisions concerning the scope of the health and safety program.

It is essential that negotiators maintain objectivity during a negotiation session. Emotional interplays that get out of hand can rapidly wreck any effort to get desired agreements. Avoiding such situations may take considerable restraint on the part of the health and safety director. Recognition of the personality type of the other party, as discussed in Chapter 8, can be especially useful in avoiding conflicts.

It is also possible to fail to recognize that one is negotiating until the process has advanced significantly, and such a failure can set the stage for significant difficulties. (I have experienced this situation during discussions of projected activities with associates. It was only after some discussion time had passed that I realized that I had made a number of concessions with little return from the other person.) Recognizing the potential for this to happen can help prevent undesirable outcomes.

Various authors have provided guidance on negotiation. Covey's "win-win" approach has been mentioned [2]. Drucker emphasizes the need to avoid "I am right and he is wrong" but, instead, focus on defining and understanding the real reasons for disagreement [19]. Steinberg stresses the importance of working in a climate of cooperation rather than conflict, learning to listen, being comfortable with silence, developing long-lasting relationships rather than conquests, and setting your limits before beginning negotiations [23]. Dawson provides extensive guidance on successful negotiating [24]. Perusing these writers' works and others' will assist health and safety managers in developing negotiating skills of great benefit in gaining both approval and support of their programs.

REFERENCES

1. Mackay H. Failures need to plan, then begin again. *Salt Lake Tribune,* January 18, 1998. E6.

2. Covey SR. *The 7 Habits of Highly Effective People.* New York: Simon & Schuster (Fireside), 1989.

3. Adams S. *The Dilbert Principle: A Cubicle's Eye View of Bosses, Meetings, Management Fads & Other Workplace Afflictions.* New York: HarperCollins, 1996. P 36.

4. Drucker PF. *The Practice of Management.* New York: Harper and Row, 1964.

5. Sparks PJ, Feldstein A. The success of the Washington Department of Labor and Industries managed care pilot project: the occupational medicine-based delivery model. *J Occup Environ Med* 1997;39:1068–1073.

6. Green-McKenzie J, Parkerson J, Bernacki E. Comparison of workers' compensation costs for two cohorts of injured workers before and after introduction of managed care. *J Occup Environ Med* 1988;40:568–572.

7. Bernacki EJ, Guidera JA. The effect of managed care on surgical rates among individuals filing for workers' compensation. *J Occup Environ Med* 1998;40:623–631.

8. Fleming LE, Herzstein JA, Bunn WB III (eds). *Issues in International Occupational and Environmental Medicine.* Beverly, MA: OEM Press, 1997.

9. Department of Health and Human Services, National Institute for Occupational Safety and Health. Progress Toward Safe Nanotechnology in the Workplace—A Report from the NIOSH Nanotechnology Research Center. DHHS(NIOSH) Pub 2007-123, February 2007.

10. Drucker PF. *Managing for the Future: The 1990s and Beyond.* New York: Truman Talley Books/Plume, 1992.

11. Drucker PF. *Managing in a Time of Great Change.* New York: Truman Talley Books/Dutton, 1995.

12. McCunney RJ (ed), American College of Occupational and Environmental Medicine. *A Manager's Guide to Occupational Health Services.* Beverly, MA: OEM Press, 1996.

13. Guidotti TL, Cowell JWF, Jamieson GG. *Occupational Health Services—A Practical Approach.* Chicago: American Medical Association, 1989.

14. Fallon LF Jr, Dickerson OB, Brandt-Rauf PW (eds). The management perspective. *State of Art Rev Occup Med* 1989;4(1).

15. Committee Report. Scope of occupational and environmental health programs and practice. *J Occup Med* 1992;34:436–440.

16. Patterson WB. Establishing an occupational health program. In: McCunney RJ, *A Practical Approach to Occupational and Environmental Medicine,* 3rd Ed. Philadelphia: Lippincott, Williams & Wilkins, 2003.

17. LaDou J. *Current Occupational and Environmental Medicine,* 4th Ed, Section I: Occupational Health. New York: McGraw-Hill, 2007.

18. Levy BS, Wegman DH. *Occupational Health: Recognizing and Preventing Work-Related Disease,* 2nd Ed: Section VI: Approaches to Occupational Health. Boston: Little, Brown and Company, 1988.

19. Drucker PF. *The Effective Executive.* New York: HarperBusiness, 1965 (Harper Business edition, 1993).

20. Bearden JC. The needle in the haystack (a fable). *Perspect Biol Med* 1977; spring:355–359.

21. Walton M. *The Deming Management Method.* New York: Putman (Perigee), 1986.

22. Tepper LB. The strategic dimensions of occupational medicine (1986 Gehrmann Lecture). *J Occup Med* 1987;29:325–329.

23. Steinberg L, with D'Orso M. *Winning With Integrity—Getting What You're Worth Without Selling Your Soul.* New York: Villard, 1998.

24. Dawson R. *Secrets of Power Negotiating: Inside Secrets from a Master Negotiator.* Hawthorne, NJ, Career Press: 2006.

ADDITIONAL READING

Conner DR. *Managing at the Speed of Change: How Resilient Managers Succeed and Prosper Where Others Fail.* New York: Willard Books, 1992.

Emmett EA. What is the strategic value of occupational and environmental medicine? Observations from the United States and Australia. *J Occup Environ Med* 1996;38:1124–1134.

Kanter RM. *On the Frontiers of Management.* Boston: Harvard Business Review Books, 1997.

Matthias SJ, May R, Guidotti TL. Occupational health and safety: a future unlike the present. In: Fallon LF Jr, Dickerson OB, Brandt-Rauf PW (eds), *State of Art Rev Occup Med* 1989;4(1).

Naisbitt J, Aburdene P. *Megatrends 2000: Ten New Directions for the 1990's.* New York: William Morrow, 1990.

Negotiation (Harvard Business Essential Series). Boston, MA: Harvard Business School Publishing, 2003.

Peddecord KM. Planning for health improvement: models for communities and institutions. In: Wallace RB (ed), *Maxcy-Rosenau-Last Public Health and Preventive Medicine,* 15th Ed. New York: McGraw-Hill, 2008.

CHAPTER

4

Planning Health Care Responses to Natural Disasters, Terrorism, and Other Mass Casualty Situations

with Colleen Connelly

Of all the events that can create doubts about the management and capabilities of a health and safety unit, inappropriate or disorganized response to a disaster situation ranks near, if not at, the top of the list. Many of us have consequently made a practice of reviewing the organization's disaster response plans within the first day or two of arrival as a new director. A colleague who followed this practice was able to respond effectively to a major disaster event two days after his arrival because he was familiar with the unit's response plans and capabilities. The tornados that have struck numerous towns this year (2007) have demonstrated that health care organizations might have little or no warning of an event that may produce mass casualties. The rapidity with which crisis situations can arise underscores the need for health care and safety organizations to plan to respond before such an event occurs.

The Joint Commission on Accreditation of Healthcare Organizations (JCAHO) emergency management standards have long required accredited facilities to plan for emergency situations, but the potential for terrorism has made planning even more complex [1]. Planning has typically been oriented toward training members of treatment and response teams, stockpiling equipment, and conducting "walk through" exercises that involve few members of response teams with a small number of simulated patients rather than the large numbers of people who would seek treatment after a major emergency incident [2]. Other health organizations, such as public health departments, have requirements separate from care of the ill or injured but, nonetheless, are called upon to provide immediate responses to health threats created by disasters.

As noted in prior publications, the potential for mass casualties may result in the belief that there would be so many injured that "nothing can be done" [2,3]. Chapter 3 mentions the comment heard during health facility planning sessions: "No disaster will ever follow a plan, so why plan?" Obviously, each disaster situation poses unique challenges, but health personnel who have planned will be able to respond effectively and in a coordinated way to the changing situations. Conversely, members of those facilities who have not developed, and practiced, plans often make disorganized, ineffective responses that may produce additional casualties among the responders [2,3].

Experience by the authors and others in actual and simulated mass casualty situations in peacetime and combat settings has repeatedly demonstrated that planning must focus not only on providing care for the ill or injured [3–5]. Prior publications by the authors have stressed the importance of the 4 Cs: Command and Control, Communications, and Coordination in response planning [2,6]. Experience during the devastation of New Orleans by Hurricane Katrina and in other situations has demonstrated the need to add a fifth C: Contingencies. In order for response plans to be effective, attention must also be directed to these "4 Cs plus 1." This chapter discussion assumes that plans have been made to provide care and to conduct the other usual activities of a health organization during a disaster and thus primarily considers the significance of the 4 Cs plus 1. Planners must also ensure that the disaster response plans of health and safety organizations are compatible with, and integrated into, the incident command plans of parent organizations.

THE THREATS

An initial step in disaster response planning is to define, to the extent possible, the nature of the threats, since different situations require somewhat different responses by medical and safety organizations.

Natural Disasters

Recent hurricanes, particularly Katrina, and tornados have brought the major impacts of natural disasters into sharp focus. Although the nation has not experienced a major earthquake for some time, such events will undoubtedly occur in the future, possibly in areas where earthquakes are rare. Flooding and other severe weather can also produce challenges for medical organizations. During the past few winters, communities across the nation have been without power for prolonged periods due to ice, broken tree limbs, or other adverse effects of weather. Considering the prolonged drought conditions in many states, wildfires are of particular concern. Some natural threats can be forecast, although a sudden shift in a severe storm's path can result in a disaster with little or no warning. Similarly, tornado warnings may aid in preparing for the possibility, but an actual event, and its path, are not predictable. Other events occur without any warning; earthquakes are the classic example. In addition, health care facilities are often viewed by the community as a place of refuge during disasters, and careful planning and preparation for responses to natural disaster events could be negated if members of the surrounding community attempt to seek shelter primarily at one particular hospital, as was witnessed in at least one hospital in New Orleans during Hurricane Katrina.

Terrorism

Recent events have made it very clear that terrorism will continue to be a threat for the foreseeable future. Chemical, biologic, radiologic, nuclear, and explosive (CBRNE) agents are usually designed to induce terror to achieve a political end. Chemical agents may produce their effects within minutes, or the symptoms may be delayed for hours. Symptoms from biologic agents can be expected to occur days to weeks after exposure, making their recognition difficult. Dr. Gerberding, Director of the Centers for Disease Control and Prevention (CDC), has stressed that the astute physician or other health provider is a crucial resource in recognizing a biologic event [7]. The six CDC Category A agents (anthrax, botulism, smallpox, tularemia, pneumonic plague, and the viral hemorrhagic fevers) are considered as the major bioterrorism threats, but multiple other agents may be used. In addition, "agriterrorism," with contamination or destruction of major food supplies, could produce effects that would be more widespread than those of a more typical biologic attack. The recent contamination of spinach with *E. coli* illustrates how extensive an attack could be.

Radiologic concerns have primarily been directed toward "dirty bombs" that scatter radioactive material over a limited area by an explosive device. However, the polonium 210 event in the United Kingdom demonstrated that the hazard could be much more widespread. In that event, multiple locations and aircraft were apparently contaminated by those using or contaminated by the agent. Effects from a nuclear weapon would be widespread, with immediate effects from the blast and then areas measured in miles affected by fallout. One report suggested that if a "small" 10-kiloton bomb were detonated in a city with a population like that of Washington, DC, over 200,000 people would be killed, over 90,000 individuals would be injured or ill immediately, and fallout would spread over 3,000 square miles [8,9]. Such an event would require support across the nation, with an estimated need for approximately 300,000 military and civilian responders, and medical facilities need to consider this aspect in planning [8,9]. To date, the majority of terrorism incidents have been explosions. Although these produce casualties with no warning, the area affected is limited. Of course, the potential exists for use of multiple agents by terrorists, which would make response planning even more complex.

Major Accidents

In many cities, major transportation arteries run through heavily populated areas. In Salt Lake City a railroad car leaking toxic chemicals resulted in closure of an important intercity road for a number of hours. In another recent event, a truck carrying explosives detonated, producing a crater that blocked a canyon road for days. If either of these events had occurred in heavily populated areas, large numbers of casualties would have resulted. In some cities, airports are located adjacent to population centers. The most hazardous times in routine flight operations are during take-offs and landings, and an aircraft experiencing problems will attempt to reach an airport, which could also result in a sudden influx of injured people in nearby medical facilities. Of course, aeromedical evacuation helicopters using landing pads on or near hospitals also pose a concern. As previously noted, if the health and safety unit supports an organization whose activities may result in an explosion, release of toxic agents, or similar disastrous event, the unit's director must ensure that the health and safety response plans mesh with the disaster or incident command plans of the parent organization.

Emerging Infections and Pandemics

Infectious disease experts have appropriately noted that the question is not "if" but "when" we will have another influenza pandemic [10,11]. Such an event could be due to mutation of the bird flu (H5N1) virus so that it could be transmitted from person to person. With the rapid transportation capabilities present today, widespread outbreaks will occur. However, the SARS (severe acute respiratory syndrome) episode demonstrated that rapid public health and other medical responses can mitigate or control spread of contagious agents and diseases.

The diversity of threats facing medical and safety units underscores the need for planning that will not only include training of health and safety professionals in treating and responding to resultant casualties but also the need to address the 4Cs plus 1.

COMMAND AND CONTROL

In a disaster situation, it is absolutely essential that *all* individuals in a health organization clearly understand the chain of authority, including replacements for those directing the responses. Both the person with overall responsibility for the organization's response and those responsible for directing subordinate units, and their replacements, must be clearly defined and included in practice scenarios. This helps ensure that order will be maintained if any of those in charge are unable or unavailable to perform their duties [2]. Failure to identify the person(s) responsible for making decisions during an emergency situation can produce considerable confusion and a disorganized response [12,13]. During a simulated bioterrorism incident that involved "top" government officials (referred to as *TOPOFF I*), up to 100 individuals participated in conference calls and attempted to make response decisions during the simulated plague attack [5]. Not surprisingly, decisions were made that were reversed shortly thereafter, and new decisions were made that were subsequently overturned. The Mayor of New Orleans, in an interview after Katrina struck the city, summarized confusion resulting when command and control is not properly structured by noting "[T]here were too many . . . cooks in the kitchen" [14]. The New Orleans chief of homeland security commented that four days after the event "FEMA has been here three days, but there is no command and control" [15]. Conversely, when a gunman began shooting people in a Salt Lake City shopping center, command and control of the police and emergency medical technicians responding from multiple communities in the area was quickly established [16]. As a result, the situation was rapidly controlled. These examples are directly applicable to the need for effective command and control in medical and other facilities that are responding to a mass casualty situation.

"Table top" exercises have identified potential conflicts between local, state, and federal agencies, and, in particular, confusion between provision of care and the need to protect criminal evidence [3]. It is essential that the person(s) in a health and safety organization responsible for making decisions have the authority clearly defined when such conflicts arise. Other required decisions include but are not limited to the following:

1. Scope of recall, including use of modules, recognizing the potential need for 24-hour, 7-day a week (24-7) operations (as discussed in the following Communications section)

2. Shift assignments for staff, again recognizing the potential for 24-7 activities

3. Any changes to previously designated patient triage, treatment, "overflow" areas

4. Changes in standards of care, especially during an influenza pandemic or other mass casualty situation

5. Assurance of documentation of all patients seen and the treatment provided, meeting forensic requirements in known or suspected terrorist events

6. Any necessary revision of previously arranged methods of crowd or vehicular control outside the facility

7. Any necessary changes in methods of control of patient, visitor, and staff flow within the facility

8. Decisions regarding any requests for assistance from state and local public safety members, other emergency responders, the National Disaster Medial System, the Strategic National Stockpile, or other national agencies

9. Decision to evacuate the health facility due to damage; threat of mass exposures to a chemical, biologic, or nuclear agent; or other reasons

Having a designated person in charge is particularly important in containing the hazard. The natural tendency of responders is to rush to the scene and help rescue those affected; however, this response may produce additional hazards that seriously compromise response efforts. For example, in the World Trade Center attacks, many of the most experienced disaster mangers placed themselves in jeopardy by rushing to the scene of the tragedy [12]. Response capabilities could have been seriously compromised if these individuals had been caught in the collapse of the towers. In the same way, providing multiple ambulances with numerous health care providers at an evolving scene could produce serious loss of capability if secondary explosions occur. A decision to rapidly move potentially contaminated patients onto wards could place other patients and staff at risk. During the Tokyo sarin gas exposures, some of the people being rescued were placed on sidewalks outside the subway, and when some of them began experiencing increased symptoms, it was realized that they had been placed over areas that ventilated the subway tunnels containing sarin. The person responsible for allocation of health care resources during a disaster must maintain control of the decision-making process to avoid additional risks and to ensure the most effective use of limited resources.

Unfortunately, as noted by others, the question of authority in many medical organizations is unresolved; thus, who is coordinating activities at the hospital may be unclear [17]. Confusion regarding who is responsible for making decisions may also exist at other health organizations, including public health agencies. Fortunately, the planning for the 2002 Winter Olympics was of important assistance in helping many Utah hospitals and public health organizations avoid this problem. As discussed below, regular, realistic practice of disaster plans should help ensure that all members of the organization know who the persons are that have responsibility for directing the responses of the organization.

In larger hospitals, it may be appropriate to designate a hospital administrative authority as well as a medical authority [6]. These two individuals (and their replacements) should have familiarity with each other's roles and responsibilities so that when an event occurs, they know how to collaborate and what the scope of decision responsibilities should be. The hospital administrative authority should have responsibility for determining the overall facility's management plan and response to an event, whereas the medical authority must have control over the triage, emergency treatment, and ultimate clinical care activities. Having readily available job descriptions for each member of the incident command team and notebooks with printed roles and responsibilities available in the command center, will help to reduce and avoid problems due to a "lack of clarity" or a "lack of knowing" who has ultimate authority and accountability.

Familiarity and adequate training for appropriate staff on the National Incident Management System (NIMS) is a *must.* This approach is the first attempt on a federal level to mandate that everyone involved in any response use the same language. In any large-scale event, multiple agencies will be involved, and ensuring that staff members know who is in charge of what, the role of their organization, and their own personal role is imperative.

COMMUNICATIONS

Experience has repeatedly confirmed that difficulties in maintaining effective communications have presented major problems in providing effective responses. For example, during a tornado in Salt Lake City, incompatibility of communications equipment hampered responders. Reportedly, some hospitals and other health care organizations had to obtain information from watching local television news broadcasts. Communications problems at the time of the World Trade Center attacks may have contributed to the loss of responders. Police personnel in helicopters recognized the potential for structural collapse of the second tower [18]. The information was relayed to police and many of them escaped, but because the firefighters were on a different communication net and their radios did not work well in the tower, approximately 200 of the 343 firefighters who died perished because they did not receive the warning of the tower's pending collapse [19]. Similar communication problems have been noted in other emergency response situations. Conversely, during the previously mentioned shooting at a shopping mall in Salt Lake City, police, fire, command and control members, and other emergency responders were able to communicate rapidly and effectively with each other [16]. This capability was instrumental in preventing further casualties.

Effective communications must be considered in developing recall processes for medical, public health organizations, and support staff. For example, some organizations have relied on a "one person calls all" system. If that person is not available, the recall system may not function. Even if the person is available, the time involved may delay notification so that response times are unacceptable. A "pyramid" or a "phone tree" system is more efficient, with clear designation of the person responsible for notifying others, who then notify still others in each "branch" or segment. Alternate individuals are clearly identified so that if one person is missing, the notification process will continue without interruption. Such a recall system can be readily adapted to a "modular" recall system so that only a portion of the staff is recalled in case of emergency. This capability is particularly important if 24-7 operations are necessary. In addition, having pre-designated modular units can be helpful if assistance is dispatched to the disaster site [2,6].

Some organizations have relied on cellular phones to provide communications during disasters. However, as Katrina demonstrated, cellular phones require power to recharge batteries [20]. In the same manner, many "land-line" phones require power to operate. Past disasters have shown that during disasters both systems can be nonfunctional due to circuit overload, even if power is available [21]. Web-based communications; 140, 400, and 800 MHz radios; HAM (amateur) radios; and satellite phones may be helpful, but, again, these require power. If auxiliary power is not available, the equipment may be useless. In a 2005 peer-reviewed journal that included discussion of the development of a state medical surge capacity plan, the authors recommended that medical organizations provide for HAM capabilities in or adjacent to command posts to facilitate communication during an event [22]. During Katrina, an additional advantage of cellular phone capabilities was the capability to transmit photos back to the parent medical facility. When medical responders encountered a number of dermatologic disorders, they were able to obtain treatment recommendations by transmitting photos of the skin lesions. Appropriate treatment could then be provided even though no dermatologist accompanied the responding team [2].

Those involved in making radio transmissions should be trained in basic radio discipline so that information to and from medical units or other responding agencies can be transmitted concisely and accurately. Due to the emotional impact of a mass casualty situation, some organizations have developed "fill in the blank" cards for medical responders to facilitate transmittal of information regarding location of the event, number and triage category of patients, and similar information. Hospitals may also consider incorporating the use of handheld radios, cell phones, or satellite phones for all inpatient units, enabling the Command Center to more easily assess needs, available resources, and better direct the hospital response.

Members of the health organizations must have contact numbers and Web addresses readily available [1,2,6]. As noted during the TOPOFF I exercise, exchanging telephone and facsimile transmission numbers *during* an event is far too late [5]. Contact numbers must include local and state public health agencies so that any suspicion of a terrorism event can be reported on a priority basis. In some states, the local, regional, and state health departments within the state are interconnected through the Health Alert Network. This system makes it possible to contact all departments instantaneously if a disaster, terrorist attack, or other emergency occurs [6].

Another necessary aspect of communication is to know the "jargon" of the other, nonhealth agencies that will be working with the medical groups in responding [6]. The concept of an Incident Command System underlies disaster responses and delineates responsibilities for involved organizations, Investing in NIMS training for appropriate staff will ensure your institution being able to communicate effectively with others. Additionally, as noted during one major bioterrorism exercise, "No one knew what a JIC (Joint Information Center), a JOC (Joint Operations Center), or DMORT (Disaster Mortuary Assistance Team) were" [5]. EOCs (Emergency Operation Centers) may be established at federal, state, and/or local levels. These centers are structured to coordinate responses, but communication problems between EOCs and medical and safety organizations usually occur if appropriate procedures have not been developed before the event.

It is also important to recognize the need to have contact points for the media [1,2,6]. Although a public information officer or media spokesperson may provide some information, media representatives typically want to interview health professionals when a major event with casualties or fatalities occurs. Designating one health professional as a media contact can help assure authoritative, consistent information is provided and can aid in reducing interruptions of other providers. Basic training of the designated representative in media interviewing is advisable, and the person *must* be versed in the provisions of the Health Insurance Portability and Accountability Act (HIPAA) of 1996.

In any disaster or mass casualty situation, additional consideration should be given to high-profile media events regardless of the number of casualties. The previously mentioned highly publicized multiple shooting that occurred in 2007 at a mall in Salt Lake City resulted in four patients requiring transport to hospitals. Two hospitals received two patients each, and in one of the hospitals, over 50 family, visitors, and supporters arrived at the hospital within two hours of the incident. Having a plan to deal with an influx of persons not requiring care but wishing to be close to their loved ones and each other is essential for minimization of interruption of hospital operations.

COORDINATION OF DISASTER RESPONSES

Health organizations require assistance from multiple agencies when a disaster happens. It is essential that planners coordinate with these other agencies before an event occurs and then maintain contact with the organizations during simulation exercises to assure that appropriate responses will occur during an actual disaster [3]. Calling the police department to assist in crowd control when several hundred people arrive at medical facilities wanting decontamination will likely be ineffective. A terrorism event adds additional concerns in that the Federal Bureau of Investigation (FBI) and police departments will be involved. Specific requirements, including chain of custody and obtaining the names and addresses of victims, will occur as part of the criminal investigation. Laboratory results of patients who have been exposed to a chemical or biologic agent may be valuable to the investigation or used as evidence. Hospitals must work with their local FBI and other law enforcement constituents to develop a procedure for obtaining and relaying this information expediently so the investigation is not delayed, but they must continue to exercise their ethical and legal responsibilities for protecting patients' rights. Establishing contacts with these groups before an event, and agreeing on procedures that will be followed, will help prevent problems in the midst of responding to the emergency. Failing to coordinate effectively can lead to major difficulties, as illustrated by the reported antagonism between the New York City fire and police departments that is currently being resolved [12]. Depending on the scope of activities of the organization, coordination with federal programs such as the Strategic National Stockpile or the National Disaster Medical System may be required. Since public heath agencies play major roles in any disaster or terrorism situation, prior coordination between treatment facilities and public health organizations is essential. Signed memoranda of agreements documenting expected support are advised.

Coordination is also necessary to meet the heavy logistical demands that an event will generate [2,3,6]. Re-supply of fuel pharmaceuticals, general supplies (e.g., beds, linens, gowns, masks, gloves, foot coverings, decontamination materials, and similar items), equipment (e.g., respirators and personal protective equipment), food and laundry services, and housekeeping supplies will be required. Ascertaining before the event what re-supplies are available locally is particularly important, since there may be delays in obtaining shipments from distant locations. Katrina clearly demonstrated that health care facilities may be "on their own" for many more days than previously thought. JCAHO now requires hospitals to assess how long they can function without fuel or other supplies and is encouraging them to increase their "stand alone" capabilities to 96 hours.

In addition to logistics, it is necessary to coordinate sources of assistance for additional staff and support personnel, especially if prolonged 24-7 activities are required. A number of cities have coordinated plans among all hospitals to provide support if one facility is overwhelmed. However, the experience in New York during the terrorist attack on the World Trade Center confirmed earlier observations that, in a disaster, patients tend to go to the hospitals they know best [23]. This response can result in severe overloading of a facility, necessitating transport of a considerable number of individuals to other facilities. In addition to coordinating with hospitals to accept patients, it may be necessary to coordinate with stand-alone clinics, nursing homes, and similar facilities to accept stable "overflow" patients. Naturally, plans for command and control, communications, staffing, and resupply of overflow facilities have to be coordinated in advance of sending patients to off-site locations. Similarly, planning should consider alternate sites if it becomes necessary to evacuate the primary medical facility. Again, command and control, communications, and coordinated plans must be assured if evacuation is required.

Terrorism events may produce large numbers of "worried well" individuals who come to treatment facilities requesting evaluation, decontamination, prophylactic treatment, or treatment for symptoms that are attributed to an exposure even though no exposure actually occurred [9,23]. For example, during the sarin gas exposure in Tokyo, 74 percent of the more than 5,000 people seeking care from emergency departments did not have any clinical findings of sarin exposure [4]. Many of these individuals may benefit from basic psychologic support, and provision should be made to provide such care.

CONTINGENCIES

In developing and practicing disaster plans, it is imperative that those involved consider and plan for contingencies. In essence, this consideration and planning require "what if" thinking. How would patients be managed if an earthquake left roads to the hospital 30 feet below the hospital entrance? How would the medical facility operate if gas, water, and electrical utilities were disrupted for a prolonged period or at night? What would happen if the disaster occurred during severe winter weather or in the midst of a major series of windstorms? How would re-supply of equipment, pharmaceuticals, and household supplies be accomplished if major transportation arteries were cut by explosions, earthquakes, floods, or other events? How would traffic and crowds be controlled if someone called a radio station to report all attending a major sporting event had been exposed to a bioterrorism agent? Considering these and other contingencies will enable planners to consider options, even if no "perfect" plan can be developed, before such situations arise. They will thus be better prepared to respond to the Murphy's law events that are likely to occur during any mass casualty event.

PRACTICING DISASTER PLANS

An apparently well-constructed disaster plan that delineates effective response activities may be useless unless it is practiced. Practice must be realistic, involve all of those who would be responsible for the multiple actions required, and be done frequently enough so that staff and support personnel know their responsibilities and how to accomplish them. "Table top" and "walk through" exercises, although useful, cannot replace field exercises with a reasonable number of simulated patients as a means of acquainting multiple members of a health organization with their roles and responsibilities. Rare "down times" in an organization provide the opportunity for face-to-face meetings with representatives of the agencies that have responsibilities for support during an emergency. Developing personal relationships among key players is of great benefit in responding to the rapidly changing situations during a mass casualty event and in dealing with the inevitable gaps and omissions that inevitably exist in the most well-conceived plans.

An important piece of practicing the plan is the use of an After Action Review (AAR) [2,6]. On-site observers can be particularly helpful in accomplishing an AAR review, since the required activities of the participants may preclude their effective evaluation of the exercise. Nevertheless, the AAR should involve all participants so that there can be a critical review of command and control, communications, and coordination as well as overall exercise management. Discussions that cover what went well and what did not go well provide opportunities for improvements to increase future functionality. In addition to bettering the plan, each participant becomes invested in the process and overall outcomes improve.

SUMMARY

Health care units and safety groups face many difficult challenges and we are being asked to do more and be prepared for more than ever before. However, disasters and other emergency events, including terrorism, may occur without warning. The potential for large numbers of patients and concerned citizens descending on health care facilities requires planning *before* an event occurs. In addition to providing the usual medical and public health support, it is essential that plans address the 4 Cs plus 1. Attention to command and control, communications, coordination, and contingencies in planning and practice will enable health and safety organizations to make effective responses to a disaster and to respond efficiently to rapidly changing situations as the event evolves.

REFERENCES

1. Centers for Disease Control and Prevention (CDC). Biological and chemical terrorism: strategic plan for preparedness and response: recommendations of the CDC Strategic Planning Workgroup (2000). *MMWR Morb Mortal Wkly Rep* 2000;49 RR4:1–14.

2. Moser R Jr, Connelly C, Catinella AP. Management aspects of medical preparation for mass casualty incidents. *J Emerg Manag* 2006;4:27–32.

3. Moser R Jr, White GL, Lewis-Younger CR, Garrett LC. Preparing for expected bioterrorism attacks. *Mil Med* 2001;166:369–374.

4. Macintyre AG, Christopher GW, Eitzen E Jr, et al. Weapons of mass destruction events with contaminated casualties: effective planning for health care facilities. *JAMA* 2000;283:242–249.

5. Inglesby T, Grossman R, O'Toole T. A plague on your city: observations from TOPOFF. *Biodefense Quarterly* 2000;2(2):1–10.

6. Moser R Jr, White GL Jr, Connelly C, Amos LK. Planning for health care responses to natural disasters, terrorism, and other mass casualty situations. *Utah's Health: An Annual Review* 2003;9:20–23.

7. Gerberding JL, Director, Centers for Disease Control and Prevention. CDC Efforts in Implementing a Smallpox Vaccination Program. (Testimony before the Senate Subcommittee on Labor, HHS, Education and Related Agencies, Committee on Appropriations.) January 29, 2003. Available at: www.hhs.gov/asl/testify/t030129.html; accessed August 24, 2007. (See also: Hughes JM, Gerberding JL. Anthrax bioterrorism: lessons learned and future directions. *Emerg Infect Dis* 2002;8:1013–1014.)

8. Gordon GC. Nuclear attack: U.S. unprepared. McClatchy News Service. As printed in *Salt Lake Tribune* March 3, 2007. A-6.

9. Homeland Security Council. National Planning Scenarios—Created for use by National, Federal, State, and Local Homeland Security Preparedness Activities. 2005. Available at: http:media.washingtonpost.com/wp-srv/nation/nationalsecurity/earlywarning/NationalPlanningScenariosApril2005.pdf; accessed April 3, 2007.

10. Freyman J. Preparing for influenza pandemic: it's not if, but when. *Occupational Hazards* 2006(July 13);33–36. Available at: www.occupationalhazards.com/

11. Homeland Security Council. National Strategy for Pandemic Influenza. Implementation Plan, May 2006. Available at: www.whitehouse.gov/homeland/nspi_implementation.pdf; accessed August 24, 2007.

12. Baker A. In Crisis—New York's emergency office was thwarted by its past. *New York Times* September 9, 2002.

13. Flynn K, Dwyer J. Fire department lapses on 9.11 are cited. *New York Times* August 3, 2002.

14. Nagin R, during CNN interview with O'Brien S. August 31, 2005, commenting on similar comments made during interview with WPAT-TV, Jackson, MS, August 30, 2005.

15. Ebbert T, as quoted in McKinnon JD, Dreazen Y, Shmukler E. Amid chaos, Louisiana calls for help. *Wall Street Journal* September 2, 2005. A1, A12.

16. Jensen DP. Police praised for timely response. "They did a terrific job." *Salt Lake Tribune* February 17, 2007. A-1,1–13.

17. Bardi J. Aftermath of a hypothetical smallpox disaster. *Emerg Infect Dis* 1999;5:547–551.

18. Dwyer J, Flynn K, Fessenden F. Fatal confusion: A troubled emergency response; 9/11 exposed deadly flaws in rescue plan. *New York Times* July 7, 2002. 1, 10–11.

19. Dwyer J, O'Donnell M. 9/11 Firefighters told of isolation amid disaster. *New York Times* September 9, 2005. Available at: www.nytimes.com (Archives link); accessed August 24, 2007.

20. Fordahi M, Meyerson B. Dead radio slowed New Orleans recovery efforts. Associated Press, as reported in *Salt Lake Tribune* September 14, 2005. A-3.

21. Garrett LC, Magruder C, Molgaard CA. Taking the terror out of terrorism: planning for a bioterrorist event from a local perspective. *J Public Health Manag Pract* 2000;6(4):1–7.

22. Moser R Jr, Connelly C, Baker L, et al. Development of a state medical surge plan, Part II: Components of a medical surge plan. *Disaster Manag Response* 2006;4:19–24.

23. Miller CW (CDC). Potential threats and the public health response: radiation. Presentation at the American College of Preventive Medicine Pre-Conference Institute on Emergency Response and Preparedness. San Diego, CA, February 19, 2003.

ADDITIONAL READING

Bigley GA, Roberts KH. The Incident Command System's high reliability organizing for complex and volatile task environments. *Academy of Management Journal* 2001;44: 1281–1300.

Health Systems Research. *Altered Standards of Care in Mass Casualty Events.* Rockville, MD: Agency for Healthcare Research and Quality, April 2005. AHRG Pub 05-0043.

Henderson DA, Inglesby TV, O'Toole T. *Bioterrorism—Guidelines for Medical and Public Health Management.* Chicago: AMA Press, 2002. (Collection of consensus articles by recognized experts for the Category A agents and for other aspects of managing bioterrorism events.)

Hick JL, O'Laughlin DT. Concept of operations for triage of mechanical ventilation in an epidemic. *Acad Emerg Med* 2006;13(2):223–229.

Hogan DE, Bustein DE. *Disaster Medicine.* Philadelphia: Lippincott, Williams & Wilkins, 2002.

Institute of Medicine, National Research Council. *Chemical and Biological Terrorism—Research and Development to Improve Civilian Medical Response.* Washington, DC: National Academy Press. 1999.

Keyes DC (editor-in-chief). *Medical Responses to Terrorism—Preparedness and Clinical Practice.* Philadelphia: Lippincott, Williams & Wilkins. 2005.

Occupational Safety and Health Administration. How to plan for workplace emergencies and evacuations. Available at: http://osha.gov/Publications/osha3088.html.

CHAPTER 5

Implementing Plans: Meeting a Manager's Responsibility to "Get Things Done"

One outcome of planning is the delineation of several organizational goals. Although goals may steer an activity in a particular direction, they are typically stated in very general terms. For example, some health and safety goals have included "improve the quality of health care provided to workers (underserved groups, or some other population)," "reduce the numbers of injuries and illnesses in the work force," or "improve the rates of use of passive restraints in vehicles in the community." Such vagueness can make it difficult to determine whether progress in accomplishing a goal is actually being made. How should *quality* be defined? Reducing injuries and illnesses appears straightforward, but does the effort translate into the actual number of injuries, severity of injuries, lost days, or some other aspect? "Increase the rates of use of passive restraints" may apply to drivers, to passengers, or to both. Does the goal apply to use of car seats for children? What percentage increase and in what age group is it desired?

Another problem in attempting to use only goals to direct organizational efforts is that it is difficult to establish priorities among the competing goals. If planning is properly done, all the goals are relevant to the mission of the health and safety unit and, hopefully, to that of the organization or other population the unit supports. However, it is unusual to have all the personnel, financial, equipment, and other resources needed to accomplish all goals simultaneously. As a result, the manager typically has to emphasize some goals and relegate others to a lower priority. It is difficult to establish priorities for the competing goals unless the manager has some information on approximate resource costs, including time, money, equipment, and personnel, to meet the different goals.

Managers can use a number of techniques to assign priorities to goals and also to measure progress in achieving goals. One approach that many have found useful is the Management by Objectives process. Peter Drucker created the concept of Management by Objectives (MBO) to help overcome the problems of working from goal statements [1]. (In today's management settings, MBO may also stand for *Management Buy Out* [2,3], so it is important to ensure that the proper meaning be used in discussions or plans.) Other, similar processes such as Project Management [4] can be just as effective, depending on the preferences of the manager.

The MBO process requires that specific objectives necessary to meet a particular goal or accomplish a planned activity be defined. Each objective must be stated so that progress in achieving the objective can be measured. Thus, a goal might be to reduce corporate health care costs, whereas the objective might be to require a 10 percent reduction in reimbursed direct health care costs within 12 months after a program is implemented.

THE MBO PROCESS

In developing objectives to accomplish a goal, the following requirements must be met:

1. *All objectives necessary to meet a goal or accomplish an activity must be defined.* This will help avoid starting a major project only to subsequently discover that the effort cannot be completed due to unexpected costs, the need for information that cannot be obtained in available time, or other constraints. Thus, a goal to improve worker health, for example, would require the completion of a series of objectives, and each of the objectives would have to be delineated during the initial stages of the MBO process.

2. *Objectives must be measurable.* MBO requires that each objective be stated in terms that allow success in accomplishing the ob-

jective to be measured. For example, an objective to determine the unmet health needs of a particular population group might include the target population, the minimum number of people to be surveyed, and the time frame for completion of the effort. As noted in numerous management publications, "If it can't be measured, it can't be managed." As is typical of most broad statements, this one does not apply universally; however, it does emphasize the underpinnings of MBO.

3. *Responsibility for an objective should be identified.* Each objective should specify an individual, group or organization that will be responsible for accomplishing the objective. Stating an otherwise well-constructed objective without specifying responsibility can result in misunderstandings, conflicts, and failure to meet the objective.

In defining responsibilities, it is imperative that the person or groups accountable be involved in setting the objective. Assigning an objective without input from those expected to do the work obviously sets the stage for failure; in fact, it is a frequent cause of problems in the MBO process. As Jeffrey Lee, a founder of our center, repeatedly noted, "lack of involvement results in lack of commitment," and the end result can be total failure of the effort.

4. *The time to accomplish each of the objectives must be stated.* To estimate the time required, the manager will use personal experience as well as input from those who will accomplish an objective. The estimate can be refined as steps to accomplish the objective are established, but it is essential that a reasonable time estimate be established as an initial step in the MBO process. This effort will be of particular value in delineating specific schedules to work on each objective and also in developing priority rankings for the overall effort.

5. *Costs to achieve the objective must be stipulated.* All management actions involve a resource cost, and resources are typically limited. Estimating personnel, equipment, and other costs is a particularly useful aspect of the MBO process. Projecting costs not only helps define an effort's overall requirements but also enables

the manager to establish the effort's priority. As costs are specified, it may become apparent that requirements are so weighty that accomplishing the objective would keep higher priority work from being done.

6. *Decide whether to include evaluation activities in the MBO process.* Thus, it might be specified that accomplishing all facets of the objective would be considered outstanding performance, while accomplishing 70 percent would be considered satisfactory. Participation by those who are to be evaluated is obviously essential if evaluation is included in the MBO process.

Applying MBO techniques to one objective designed to help meet a goal of improved health and safety marketing might look like this:

> Dr. Smith will design and conduct a survey of health and safety requirements in at least 100 businesses located within 30 miles of our home office, within four months of approval of the survey instrument by Mr. Brown, at a cost not to exceed 30 hours of Dr. Smith's time, 50 hours of administrative support time, and costs for supplies, mailing, telephone, and other requirements not to exceed $1,100.00.

MBO ADVANTAGES

A major advantage of MBO, or a similar technique, is that delineating all the objectives necessary to accomplish a goal not only helps define resource costs and develop priorities but also aids in tracking progress in meeting objectives and reaching the goal. The definition of necessary objectives and the sequence in which they must be accomplished ties in quite well with such tracking and evaluation procedures as Program Evaluation and Review Techniques (PERT), which will be discussed further in Chapter 12.

Defining time and resource costs through MBO is particularly valuable to managers in setting schedules for an effort. For example, members of our faculty were asked to develop analytical techniques and accomplish analyses for a major National Cancer Institute study. The requestor estimated that the effort would require three to four months. When the task was broken down into the necessary objectives, with reasonable times established for each step, it became apparent that the actual time required was over twice as long. Furthermore, specifying the objectives identified the need for information beyond that which the requestor could initially provide, necessitating a delay in beginning our faculty members' portion of the work. The MBO process also clearly defined the times when other investigators would have to provide data if the new schedule were to be met. Specifying time needs, other resource costs, and responsibilities precluded blame being placed on our professionals if the initial deadlines were not met. The process also delineated responsibilities for each step in the process; thus, the source of any delays could be readily identified.

As a result of using the MBO process, the project was completed according to the new schedule. In addition, due in large part to the extensive and ongoing reviews built into the revised schedule, the results were of such high quality that our faculty members were highly commended for their efforts.

MISUSE OF MBO

Some argue that MBO can cause serious problems by encouraging short-range performance at the expense of long-term progress. For example, Deming includes MBO as his third "deadly disease" [5]. He suggests a better term would be *Management by Fear* and notes that the MBO process can result in employees' working for themselves rather than advancing the objectives of the company. He also believes that risk taking is discouraged and that people even work against each other to accomplish personal objectives when the MBO system is implemented.

Like any other management technique, MBO can be misused and abused. Corrupting the MBO process so that everyone works to accomplish easily attainable objectives that have little relevance or significance is worse than worthless [6]. In some organizations, this approach has enabled everyone to "look good" to senior management, even though the organization was stagnating and regressing. The health and safety manager must ensure that objectives are relevant to appropriate goals and, if accomplished, will achieve the goals. Remember the "Needle in the Haystack" and doing the *right* thing right rather than simply doing things right, discussed in Chapter 3.

As previously discussed, assigning responsibility for an effort without the responsible party being involved in defining the objective can lead to major problems. The recent emphasis on participatory management underscores the benefits that can accrue as a result of effective interchange among managers and between managers and workers. Conversely, arbitrarily assigning objectives increase the chances for failure, with adverse impacts on both the manager and those assigned the task of meeting the objectives. Similarly, the MBO process can be misused in evaluating performance. Establishing unreasonable time lines, resource limitations, or other aspects of an objective, coupled with blaming the tasked individual for failure, is an obvious misapplication of the process.

McConkey has listed a number of other ways to sabotage the MBO process [7]. For example, assuming that MBO will solve all management problems and that no other management actions are necessary, sets the stage for failure of both the process and the manager. Failing to provide feedback on progress, financial expenditures, and other necessary factors to those responsible for objectives can make it impossible for the objectives to be accomplished. Requiring that every objective be precisely quantified so that progress can be measured with 100 percent accuracy can result in detailed but insignificant objectives. The result might be, for example, an objective to reduce copying costs by $125.00 but no mention of ways to enhance overall communication, since, for the latter, it would not be possible to provide precise quantification for this much more useful and beneficial objective.

Requiring that the MBO process be implemented overnight, often with only a few introductory comments to those responsible for accomplishing the objectives, can be expected to result in failure. Not tying the process to an effective planning effort can result in a number of comprehensive, well-stated objectives that, when accomplished, do not provide any real benefits to the organization.

Other ways of compromising the MBO process are discussed in the McConkey article.

There is little doubt that MBO, like other management techniques, can be misused, to the detriment of organizations and people. Appropriately applied, however, it, or a similar process, can be very beneficial to the manager responsible for achieving organizational goals.

Admittedly, MBO takes considerable management time. It is unreasonable to expect that the process will be used for each management task. However, in meeting broader management responsibilities, the management professional will find MBO or a similar process especially helpful. As discussed in Chapter 12, clearly defined objectives greatly facilitate the manager's evaluation of the progress being made in the health and safety program.

REFERENCES

1. Drucker PF. *The Practice of Management.* New York: HarperCollins, 2006. (Initially published in 1954; introduced the concept of management by objectives.)

2. Nash A. *The MBO Guide for Management Teams: Real Life Lessons from the Front Line.* Petersfield, UK: Harriman House, 2005. (Discussion of aspects of management buy-out and management buy-in [MBO/MBI].)

3. Hill J, Gambaccini J. *Winning Legal Strategies: Implementing a Successful Management Buy Out (MBO).* Boston: Aspatore Books, 2005.

4. Lewis JP. *Fundamentals of Project Management.* Saranac Lake, NY: AMACOM, 2006.

5. Walton M. *The Deming Management Method.* New York: Putman (Perigee), 1986.

6. Hersey P, Blanchard K, Johnson DE, et al. *Management of Organizational Behavior—Leading Human Resources,* 8th Ed. Upper Saddle River, NJ: Prentice Hall, 2000.

7. McConkey DD. 20 ways to kill management. *Management Review* 1972; Oct: 5–13.

ADDITIONAL READING

Lewis JP. *Project Planning, Scheduling, and Control—A Hands-On Guide to Bringing Projects in On Time and On Budget,* 4th Ed. New York: McGraw-Hill, 2005.

CHAPTER

6

Organizing Health and Safety Activities

with Jeffrey S. Lee

Those in charge of all or portions of health and safety programs must organize people and other resources to produce results. Although some health and safety professionals might prefer to avoid dealing with organizational factors, especially organizational charts, skills in this area can greatly facilitate the director's efforts.

THE ORGANIZATIONAL POSITION OF HEALTH AND SAFETY PROGRAMS

There have been a number of changes in organizational thought and structure over the past few years. Expanding information capabilities and emphasis on both decentralization and *teams* and *teamwork* indicate changes will continue into the near future [1–3]. The place of health and safety activities in an organization is also undergoing change [4,5]. Some of these programs are still in more traditional settings, while others are placed in newer or still evolving organizational positions [6–9]. After several years of declining in-house care of workers, an increasing number of companies are now reinstating clinics to provide acute and preventive care through on-site facilities. A recent study found that 22% of organizations surveyed provided such clinics in 2005 or 2006 and an additional 5% planned to do so in 2007 [10]. If the health and safety unit is hospital based or affiliated, part of a specialty clinic, or in a free-standing clinic, other organizational factors apply [11].

Managed care approaches can be expected to expand still further during the coming years [12–14]. The following section focuses on a few examples of organizations and considers health and safety's place in those organizations.

Figure 6-1 illustrates a typical organizational chart. Such charts depict lines of authority, supervisory responsibility, and formal channels of communication. In organizations, individuals usually report to one or more supervisors who, in turn, report to a chief executive officer (CEO), organizational president, mayor, or other senior official, and that individual may ultimately report to a board of directors, council, or similar group. In a local manufacturing plant, the senior executive may be designated as plant manager or have a similar title.

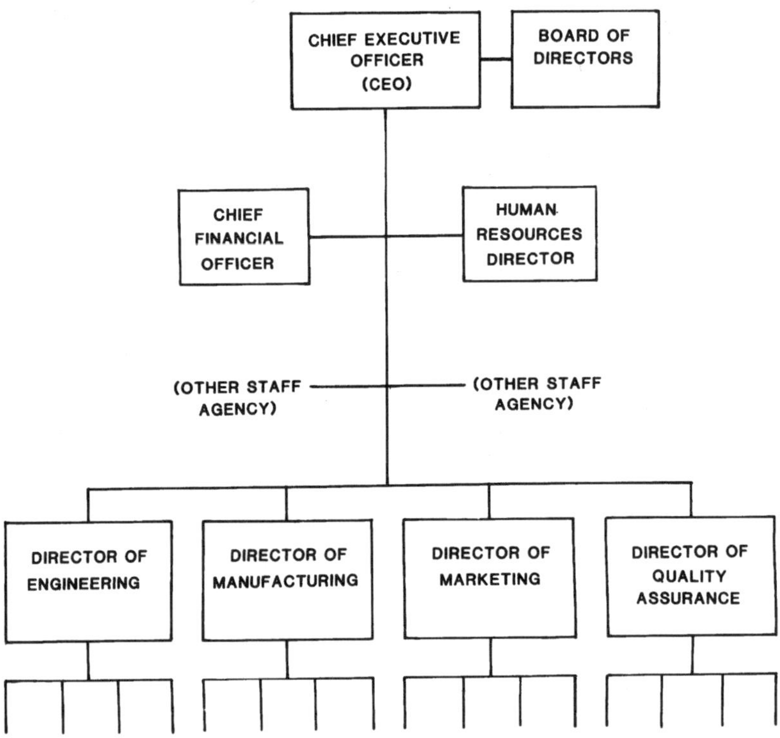

Figure 6-1. Example of organizational chart. (Only selected portions shown.)

Whether a full-time member of an organization, a representative of a group providing contract support to the organization, or an independent consultant, the director of a health and safety program must have a clear picture of the place of the program within the overall organization. This knowledge assists the manager in determining the formal lines of communication and authority. The format of the organization depicts who reports to whom and also illustrates the layers of management involved in obtaining approval for a new project. Each person's supervisor is depicted, and this enables a manager to know the individual to contact to appeal a decision by a lower level manager. The following sections focus on organizational aspects of occupational and environmental health and safety (OEHS) programs as examples of organizational impacts and considerations. However, many of the factors discussed apply to hospitals and other health and safety programs. For example, the director of a state, county, or city health department will find that the position of the health department in the overall organization, the lines of communication, and relationship to government councils or boards directly affect the activities of the health department.

The position of health and safety functions within organizations can vary widely. In some instances, OEHS functions may be grouped into a single OEHS department such as Occupational Health and Safety or Health, Safety, or Environment; in other organizations, all OEHS activities may be grouped within another department, for example Human Resources (which may also be called *Personnel* or *Employee Relations*). In still other instances, OEHS activities may be split, with units in separate parts of the organization. For example, employee health may be the purview of Human Resources, Safety in Engineering, *and* Industrial Hygiene in Environmental Affairs. Similar grouping and splitting may occur in health departments. In Utah, the state's department of health and the department of environmental quality are separate entities reporting directly to the governor. At local levels, both entities (and their activities) are usually in the local health department. If consulting or providing contract services, OEHS members are often supervised by directors of human resources or other departments. In some instances, however, consultants may report directly to the plant manager, CEO, or other senior executive.

If the OEHS activities are divided between several departments, there is significant opportunity for power and "turf" struggles, not only between the members of the OEHS factions but also among the departments and supervisors involved. Negotiating the "minefields" in such circumstances can challenge even the most experienced manager. Unfortunately, even if all health and safety components are in one department, problems of power and position can still arise, regrettably, among the members of the department. For example, many organizations emphasize that the "best qualified person" will be in charge of an activity. Such approaches have led to severe problems when a nurse or safety member was selected to head a health and safety unit that also contained physicians and industrial hygienists. Similarly, selecting a physician or industrial hygienist can produce conflicts and resentment by members of other disciplines. These circumstances have led to a "swimming with sharks" situation and an amazing number of inappropriate activities by the professionals involved. Fortunately, more health and safety functions are increasingly being performed by members of a team, which appears to be fostering more understanding of the need for interaction and cooperation. Hopefully, this "win-win" approach, which has been seen ever more frequently over the past few years, will become even more typical in the future. (Win-win is discussed further in Chapter 26.)

The organizational charts in Figures 6-2, 6-3, and 6-4 are based on actual ones and depict examples of positions of OEHS functions in organizations. If working in a plant or activity separate from corporate headquarters, the OEHS director usually has the authority to communicate directly with corporate health and safety members, even though the director reports to, and is supervised by, someone at the local plant. This capability is essential when the local OEHS manager needs to consult with colleagues on health and safety concerns, obtain professional information, or accomplish similar interactions. The ability to interact directly with corporate representatives is illustrated in Figure 6-4 by a dotted-line channel. Until recently, such direct interactions were particularly useful in discussing evaluation or treatment of patients with complex health problems. This "telemedicine" approach is facing in-

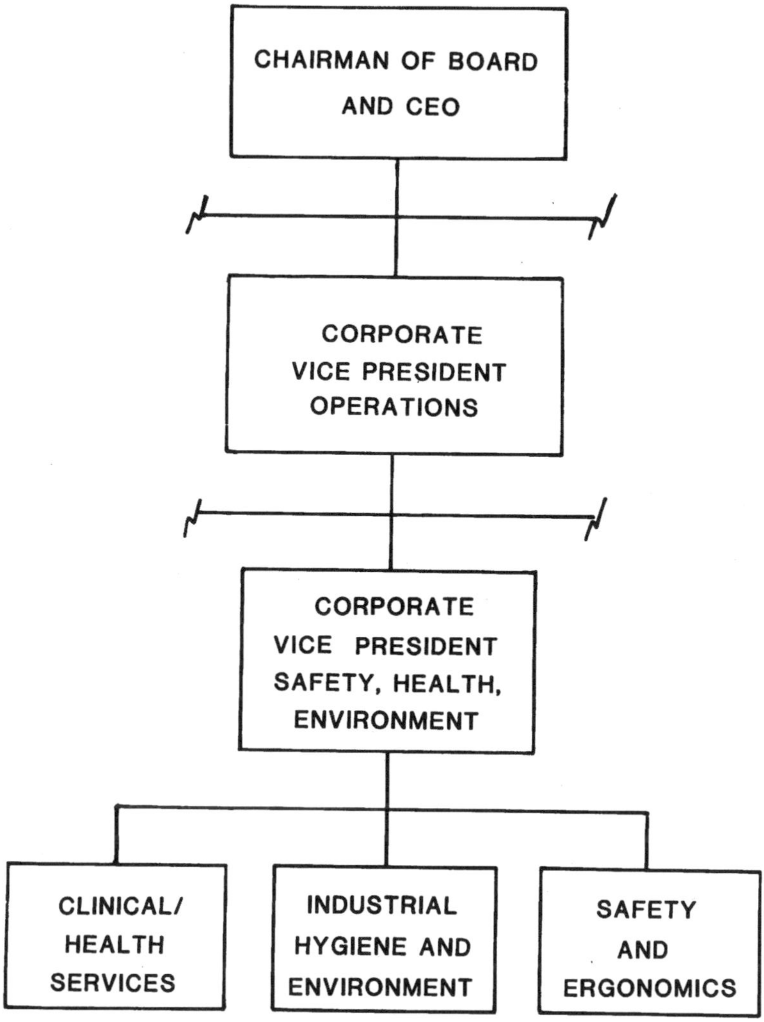

Figure 6-2. Organizational chart with OEHS director in senior management position.

creasing restrictions as many state licensing groups believe that any provider discussing a patient or providing treatment recommendations must be licensed by the state in which the evaluation or treatment is being accomplished. This restriction could mean that a senior corporate medical director of an organization with plants in

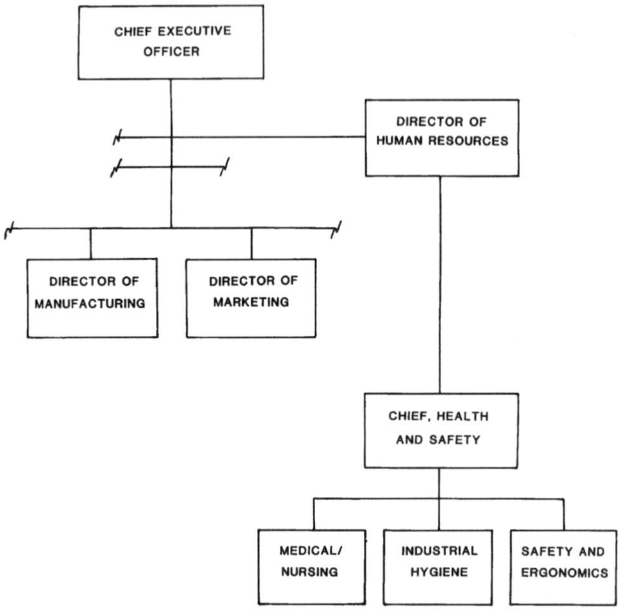

Figure 6-3. More typical OEHS positions.

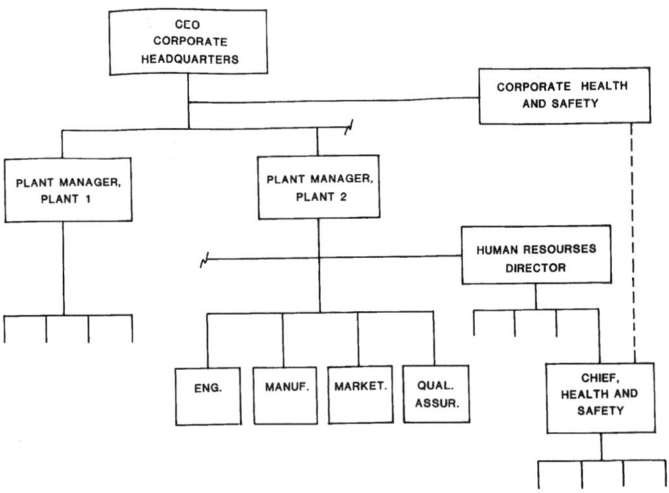

Figure 6-4. Example of relation of local plant OEHS unit to plant human resources and corporate OEHS function.

all 50 states would have to be licensed in 50 states in order to discuss treatment of patients with more junior physicians at the outlying sites.

THE SIGNIFICANCE OF LINE AND STAFF CONSIDERATIONS, AS ILLUSTRATED BY OEHS ACTIVITIES

A number of different organizational structures have evolved over the years, but some aspects are common to all of them. (This is discussed in more detail in a subsequent section of this chapter.) Two particularly important terms and concepts for the health and safety professional are *line* and *staff*. Although military in origin, these terms are now widely used in industry and in other settings. Individuals and groups that are responsible for executing the major activities of the organization, such as producing automobiles, are considered line. Those who assist line personnel are staff. Staff members perform important support functions for the organization, but the organization would not have to shut down if staff activities were disrupted or unavailable.

Note that the position of OEHS functions in Figures 6-3 and 6-4 are to the side of the straight vertical lines from the CEO to the production and marketing activities. OEHS is a staff function and does not have direct authority over the production or other line activities. Other staff activities are similarly situated.

Staff members typically provide advice, accomplish control functions, or otherwise assist line managers in getting a job accomplished. In essence, the primary purpose of staff members is to support the efforts of line personnel. Examples of staff groups include human resources (personnel), health, budget (or financial), and legal. Obviously, OEHS activities are typically a staff function within a company.

OEHS staff professionals should remember that line managers are the individuals who have authority over the work force. The line is also responsible for getting the "product" produced. The OEHS manager can make recommendations to the line manager, but acceptance or rejection of the advice is the purview of the line executive. Few actions will create a more rapid and adverse response to the OEHS manager's directives than bypassing the line organization to introduce a program or effort that affects the work force. For example, one newly arrived OEHS director was walking through an assembly area when he noted the workers were not using ear protection. The noise levels seemed, on the basis of speech interference, to be well above the 85 dB levels requiring hearing protection. The director promptly contacted the assembly line supervisor and told him to shut down the line until ear plugs or muffs were available. The shutdown actions were accomplished, and the director went back to his office, pleased that he had been able to take such direct action to prevent adverse health effects. He had barely returned to his office when the senior production manager called to advise him to meet immediately with the production manager and the director's supervisor. During the meeting, it was noted that the assembly line supervisor had requested the OEHS unit to do a noise survey several weeks earlier when the new line was installed, but the survey had not yet been done. Further, the OEHS unit was still reviewing a draft hearing-conservation policy that was supposed to have been delivered at least two months previously. Finally, even if protection had actually been necessary, the actions taken by the OEHS professional would have corrected only one line during one shift. If contacted by OEHS members, the production manager could have ensured that all assembly workers were advised and any necessary policy enforced for all shifts. When a subsequent survey did not disclose levels above 85 dB, and when the cost of the shutdown was calculated, the OEHS director received a written reprimand from a senior executive in the company. Although the failure of the OEHS unit to accomplish necessary evaluations compounded the problem, it was the OEHS director's bypassing line management that overwhelmingly concerned the other managers. Remembering the difference between line and staff, and accomplishing proper coordination with line members before the fact, could have prevented both embarrassment and loss of credibility for the OEHS manager. A similar situation can exist when an OEHS "expert" from corporate headquarters visits a plant and attempts to impose health and safety requirements without coordinating with the plant manager or other appropriate line manager. OEHS consultants or contractors have also experienced serious difficulties because they forgot the significance of line and staff distinctions.

Usurping a line manager's authority not only creates problems at the time but also affects future efforts and performance of OEHS personnel. Regardless of past accomplishments, a single incident of bypassing a line manager may have a major negative impact and be long remembered. It may undermine support for OEHS programs and even create antagonism toward future efforts. Conversely, if the OEHS director works to assist the line executives in accomplishing their activities and in solving their problems, good line–staff relationships will result, thus facilitating ongoing OEHS efforts and proposals.

Although a staff member with no direct authority over workers, the OEHS specialist must be able to deal directly with them. For example, mechanisms should be in place to enable workers to raise questions about health and safety directly to the OEHS staff. Plant industrial hygiene investigations will typically involve air monitoring of exposed workers. Occupational medicine and nursing personnel often accomplish routine medical surveillance of workers exposed to regulated substances. Ergonomics and safety professionals must be able to evaluate the activities of individuals while they are performing actual work actions, underscoring the necessity of workers' cooperation even when they may initially object to an OEHS requirement, such as the use of hearing protection or compliance with safety procedures. In such situations, it is important that the OEHS member work through the workers' supervisors. For example, if a member of the OEHS staff is in a corporate office and planning an industrial hygiene survey at one of the company's plants, it is essential that the survey be arranged through the plant manager. Then, it is usually appropriate and wise to advise the subordinate managers of the need for the effort and scope of the survey before starting the investigation. Remember that line managers are usually very possessive of their "turf," and good communication before the fact is essential in gaining their support. Once their support is obtained, the line managers and supervisors can help the OEHS member gain the cooperation of the workers, not only through the direct supervisor of the work force but also by facilitating communication with the workers to increase their understanding of the need for the effort.

ORGANIZATIONAL CONCEPTS

Programs, personnel, and activities are organized to accomplish tasks more efficiently and effectively. If efficiency and effectiveness are not the outcome of organizational efforts, something in the process is faulty.

The health and safety manager has to direct attention to at least two major organizational concerns. The first is the organization of the health and safety program itself; the other is the place of the program within the overall organization. For example, the organization of an OEHS unit is often significantly affected by the organization of the parent company. Some typical corporate organizational structures, including vertical, horizontal, and functional structures, are depicted in Figures 6-5, 6-6, and 6-7. The or-

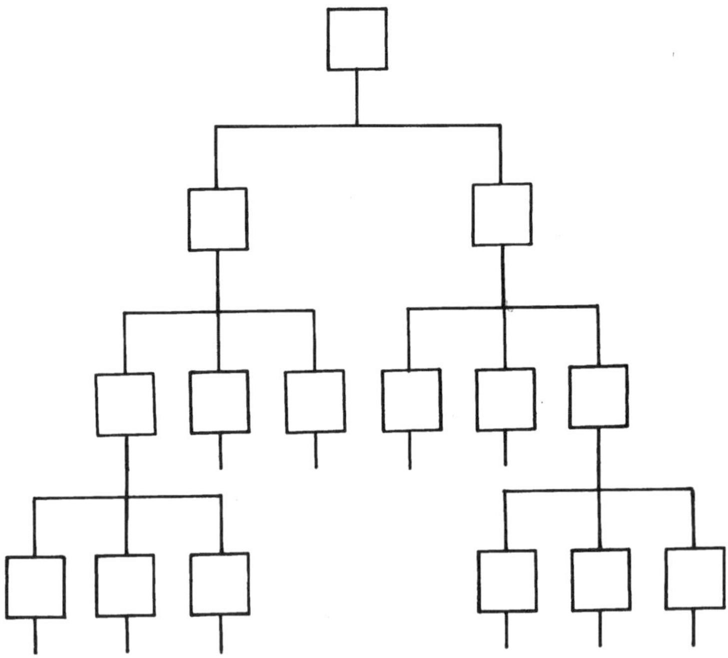

Figure 6-5. Vertical organization.

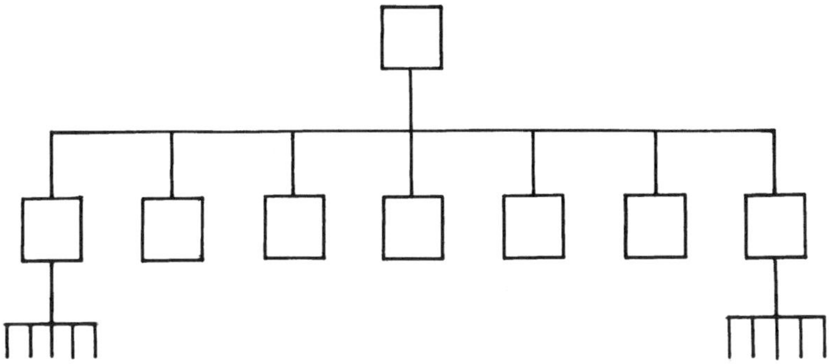

Figure 6-6. Horizontal organization.

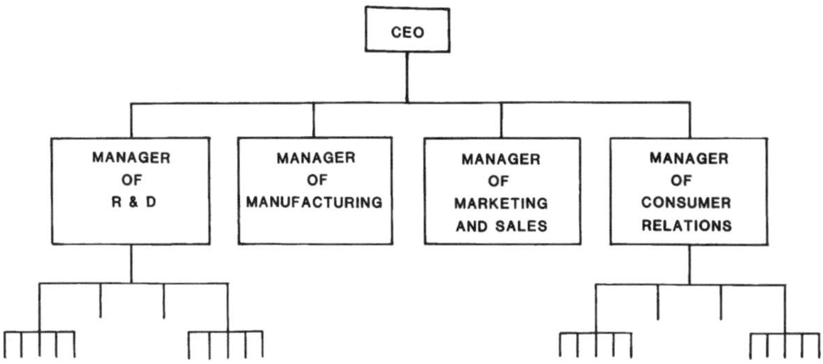

Figure 6-7. Functional organization.

ganizational structure may be quite formal, as in many larger, established companies. It may also be fairly informal, as in many new and entrepreneurial companies.

Henri Fayol observed that no individual in an organization should report to more than one supervisor, reflecting the earlier Biblical observation about the difficulties of working for more than one master [15,16]. Although reporting to only one supervisor does help avoid confusion and conflicts, rigid adherence to the principle can restrict initiative and impede other vital corporation activities. These concepts are discussed more extensively in works such as *In Search of Excellence,* which emphasizes the advantages of allowing more flexibility in supervision, using companies like Minnesota Mining and Manufacturing as examples [17].

In contrast to reporting to only one supervisor, the matrix organization marks the other extreme of the organizational structure continuum (Figure 6-8). In this type of organization, individuals are permanently assigned to a department such as engineering, research, or manufacturing. However, they also work on a particular product or effort under the direction of the person, who, though responsible for the product or effort, works in a different department. In this structure, if problems are to be avoided, it is important to specify how workers will be evaluated and how time will be allocated. As might be expected, effective communication among all involved is critical in matrix organizational activities.

In organizations that attempt to have an individual report to only one person, several people may report to a supervisor (see Figure 6-9). It has been repeatedly demonstrated that having too many subordinates reporting to a supervisor can cause a number of problems, however. The "span of control" of the supervisor becomes extended, and, as a result, there is insufficient time to furnish guidance, inability to provide timely responses, poor tracking

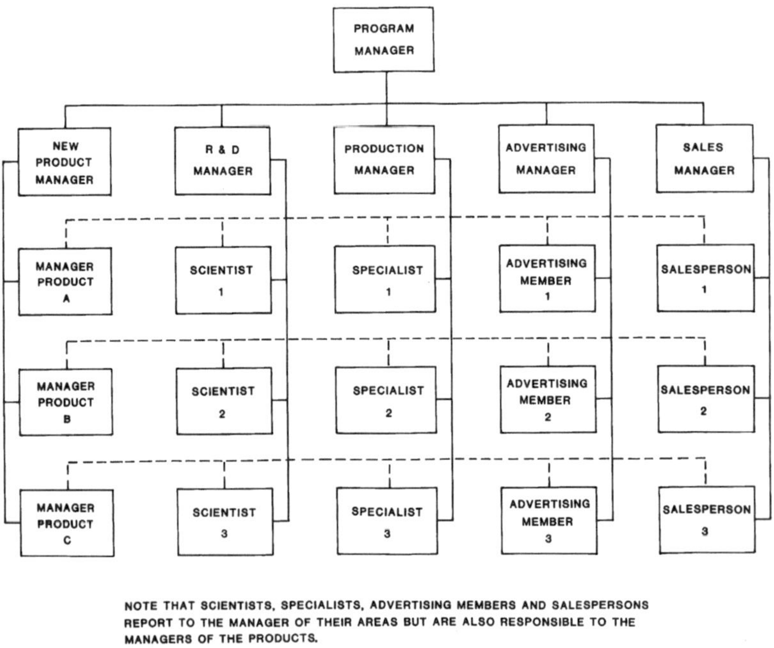

Figure 6-8. Example of matrix management organization.

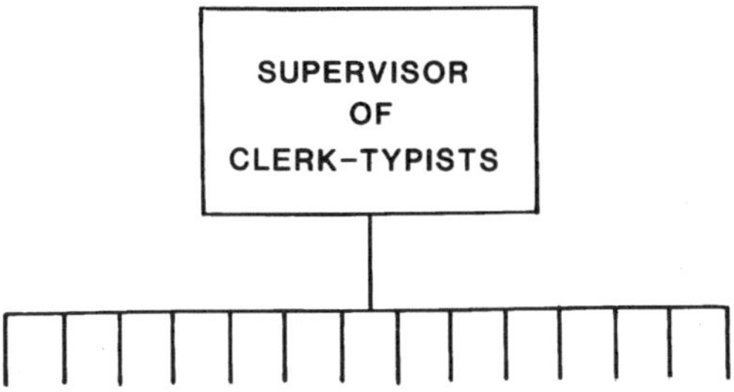

Figure 6-9. The number of people reporting to a supervisor varies with the complexity and diversity of the tasks.

of progress, and similar difficulties. Limiting the supervisor's span of control helps prevent these problems. The difficulty is in defining how many is "too many" to supervise. Management studies have demonstrated that the appropriate number depends on the type of person being supervised. It might be possible to supervise 20 clerical people without difficulty if most tasks are routine. Conversely, the number drops rapidly as the professional qualifications of those supervised and the task complexities increase. One survey found that the number of senior executives reporting to the president ranged from one to more than 20, with seven to 10 being the most frequent range. Due to the complexity of activities, the health and safety manager whose span of control exceeds approximately seven individuals may experience problems in properly supervising and managing the unit.

Knowledge of organizational structure can assist the health and safety director in getting actions accomplished. For instance, there may be situations in which the appropriate line supervisor does not accept an OEHS manager's recommendation. If the manager believes it is essential to adopt the recommendation, the decision can usually be appealed to the line manager's supervisor. In theory, this process can be continued up the chain of command to the top executive. While most managers expect some decisions to be challenged, repeatedly "going over the head" of a line manager to a superior can create resentment and opposition. Consequently, it is usually advisable to work with the line manager to see if an acceptable compromise can be reached.

SIGNIFICANCE OF ORGANIZATIONAL POSITION

Because many health and safety functions are staff activities, it is beneficial for directors to have ready access to senior managers. This capability facilitates prompt decision processing and promotes approval and support of health and safety proposals. The question "To whom do I report?" is particularly significant. If the health and safety director reports directly to the senior decision makers, which rarely occurs, the chances to present and obtain support for new programs are greater than they are if the director is in a lower position in the organization. However, if the manager has to cross several intervening layers to reach a department head, obtaining approval of recommendations is usually more difficult.

Regardless of the position of the health and safety unit and the type of organization, effective communication (see Chapters 20 and 21) is essential in advancing the program through the formal organization structure. In addition, the health and safety manager should recognize the informal organizational and communication lines and structure that exist in every organization, including the most formally and rigidly structured. Even if a health and safety department is situated on a relatively low rung on the company "ladder," or organizational structure, the health and safety manager may have opportunities to contact senior executives through informal channels. These range from joint membership in committees or groups outside the organization to interchanges related to a specific problem or project. For example, if the health and safety director is a physician or nurse, this individual might be expected to arrange for and help provide health care for senior managers. Frequent contacts in informal situations such as these can provide opportunities for significant discussions of health and safety's activities and plans.

Knowledge of informal organization structures and communication channels can benefit a manager in other ways. For example, the secretary or administrative assistant of a CEO or other senior executive often plays a crucial role in scheduling meeting time with the executive. Provided there is no breach of confidence involved, the secretary may be able to apprise the health and safety chief of particular interests or concerns of the CEO. As emphasized in Chapter 20, being aware of informal communication channels and maintaining good relationships with significant assistants and secretaries can significantly benefit both the health and safety program and the manager.

ORGANIZATION OF HEALTH AND SAFETY FUNCTIONS

The standard organizational phrase "Form follows function" emphasizes the need to structure the organization to best accomplish health and safety functions. However, some organizations specify the organizational structure of these functions, while others give the director latitude, permitting a less rigid, more functional organizational structure. As noted by a number of management experts, the simpler the organization the better. However, it must also be acknowledged that many health and safety efforts span an increasingly diverse set of functions. Figure 6-10, again based on actual organizations, depicts some of the functions that may be involved in a comprehensive OEHS unit. Other types of health and safety programs may have different functions, but the concept of organizing these functions to facilitate the overall direction and symbiosis of health and safety activities is as valid for these programs as they are for comprehensive OEHS units. In smaller organizations, several functions may be grouped together, and some, such as research, may be omitted. As previously noted, some activities may be assigned to other departments, requiring coordination of the functions of these separate units. Regardless of the size of the health and safety organization, separating the functions in a manner similar to that depicted in Figure 6-10 usually enables the different sections to function more effectively and efficiently.

Structuring functions so that the span of control of the director does not become an impossible situation can be a major challenge for the director. In these circumstances, appropriate delegation of functions, which is discussed in Chapter 9, may be particularly important.

A final note: Figure 6-11 is an actual organization chart depicting the position of a health and safety unit in a parent organization. It can serve as an example of how *not* to organize activities. Even though the occupational medicine and industrial hygiene components are functional parts of the occupational and environmental health and safety center, the chart indicates that the directors of both components report directly to the CEO of the health institute, bypassing the center's director. Fortunately, the directors of the two components within the center recognize the depicted organization is dysfunctional, especially because the director of the center is responsible for allocating funding, prioritizing interdisciplinary efforts involving both components, and exercising other

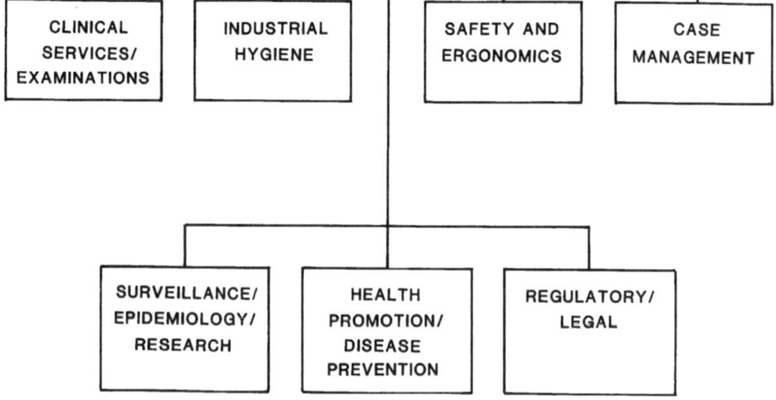

Figure 6-10. Example of organization of health and safety functions.

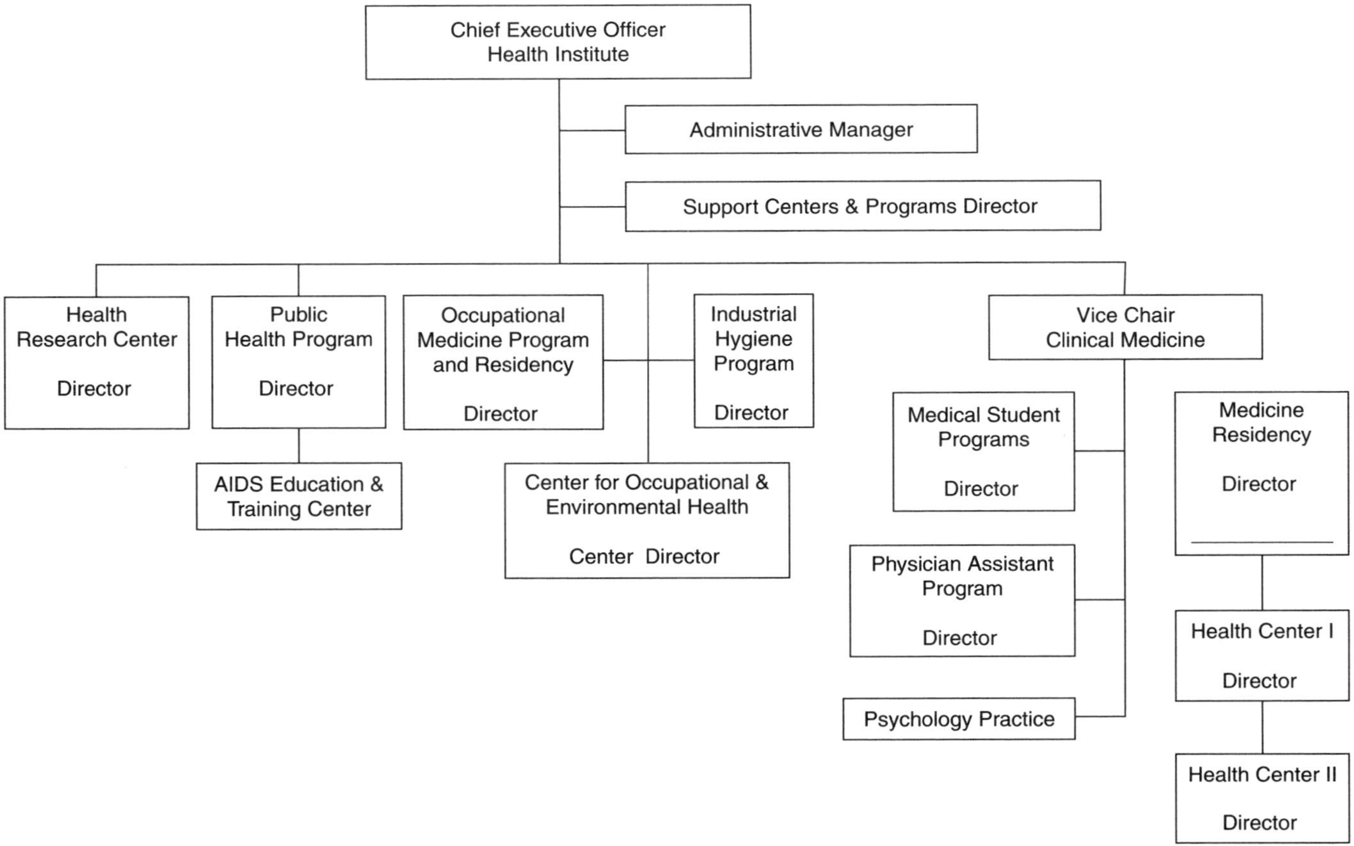

Figure 6-11. Organizational Chart, February 6, 2007.

managerial duties. If the component directors were not so astute and they had attempted to operate independently in accordance with the organizational chart, major problems could have occurred.

In summary, effectively meeting the challenges of organizing functions and activities greatly facilitates the health and safety manager's development and implementation of comprehensive programs that benefit citizens, workers, patients, and other "customers"; senior executives and senior managers, the parent organization or community; and members of the health and safety team.

REFERENCES

1. Drucker PF. *Managing in a Time of Great Change.* New York: Truman Talley Books/Dutton, 1995.

2. Kanter RM. *On the Frontiers of Management.* Boston: Harvard Business Review Books, 1997.

3. Marshall EM. *Transforming the Way We Work—The Power of the Collaborative Workplace.* New York: American Management Association, 1995.

4. Ducatman AM, Forman S, Teichman R, et al. Occupational physician staffing in large US corporations. *J Occup Med* 1991;33:613–618.

5. Guidotti TL, Cowell JWF. The occupational health care system: an overview. In: Fallon LF Jr, Dickerson OB, Brandt-Rauf PW (eds), The management perspective. *State Art Rev Occup Med* 1989;4:(1).

6. McCunney RJ. The academic occupational physician as consultant—a 10-year perspective. *J Occup Med* 1994;36:438–442.

7. Huey M. The industrial hygiene program. In: Plog BA, Niland J, Quinlan PJ (eds), *Fundamentals of Industrial Hygiene,* 4th Ed. Itasca, IL: National Safety Council, 1996.

8. Stelluto JF. Management style/system. In: Garrett JT, Cralley LJ, Cralley LV (eds). *Industrial Hygiene Management.* New York: John Wiley & Sons, 1988.

9. Moore CE. The organization of occupational health in international settings. In: Fleming LE, Herzstein JA, Bunn WB III (eds), *Issues in International Occupational and Environmental Medicine.* Beverly, MA: OEM Press, 1997.

10. Katz PS. Big employers bring health care in-house. *American College of Physicians Observer* 2007;27:1,28–29.

11. Leone FH. The physician working with the business community. In: McCunney RJ (ed), *A Practical Approach to Occupational and Environmental Medicine,* 3rd Ed. Boston: Little Brown and Company, 2003.

12. Sparks PJ, Feldstein A. The success of the Washington Department of Labor and Industries Managed Care Pilot Project: the occupational medicine-based delivery model. *J Occup Environ Med* 1997;39:1068–1073.

13. Green-McKenzie J, Parkerson J, Bernacki J. Comparison of workers' compensation costs for two cohorts of injured workers before and after the introduction of managed care. *J Occup Environ Med* 1998;40:568–572.

14. Bernacki EJ, Guidera JA. The effect of managed care on surgical rates among individuals filing for workers' compensation. *J Occup Environ Med* 1998;40:623–631.

15. Fayol HJ. *General and Industrial Administration.* London: Sir Issac Pitman and Sons, 1949.

16. Matthew 6:24. *The Bible, Containing the Old and New Testaments.* Revised Standard Version. New York, American Bible Society, 1990.

17. Peters TJ, Waterman RH Jr. *In Search of Excellence—Lessons from America's Best Run Companies.* New York: Warner Books, 1992.

ADDITIONAL READING

Blake RR, Mouton JS, Barnes LB, et al. Breakthrough in organizational development. *Harvard Bus Rev* 1964;42(6). Reprinted in Harvard Business Review Business Classics 1986;159–181.

Guidotti TL, Cowell JWF, Jamieson GG. *Occupational Health Services—A Practical Approach.* Chicago: American Medical Association, 1989.

McCunney RJ (ed), American College of Occupational and Environmental Medicine. *A Manager's Guide to Occupational Health Services.* Beverly, MA: OEM Press, 1996.

CHAPTER 7

Recruiting, Retaining, and Evaluating Health and Safety Personnel

The health and safety manager will have significant personnel responsibilities, including hiring, supervising, directing, and terminating employees. Depending on the organization, the manager may be assisted by members of a personnel, or human resources, function or department. However, primary responsibility for selection and management of health and safety professional and support members will remain with the program director.

The following sections consider some typical personnel responsibilities.

ESTABLISHING PERFORMANCE STANDARDS

Job Descriptions

Job descriptions are often available for health and safety professionals but are frequently so general that they provide little specific guidance to the individual in the described position. Job descriptions typically include identification information, including location(s) where the job is performed; a summary of the basic functions of the job; the principal duties or work to be performed; job specifications, including specific skills, efforts, or responsibilities; and any supervisory responsibilities [1]. Job descriptions should be based on appropriate and current analysis of the job itself. They must also comply with pertinent federal and state laws, especially with respect to absence of any suggestion of discriminatory practices. Poorly written job descriptions have been considered to be prima-facie evidence of noncompliance with the

Americans with Disabilities Act (ADA) or other legal requirements. Further discussion of this aspect is considered in Chapter 23. The health and safety manager should carefully coordinate any preparation of job descriptions with the organization's human resources or personnel department, especially if the descriptions are to be used in a hiring process.

Performance Standards

Performance standards provide much more specific information about actual responsibilities, activities, and criteria for acceptable and superior performance. Clearly defining what is expected of an individual in a particular position greatly facilitates hiring the right person for the job. For example, stating that an individual will be required to develop, implement, and monitor a hearing conservation program for employees in all noise hazard areas; accomplish annual air sampling of manufacturing sites to ensure compliance with OSHA and other regulatory requirements; and similar statements helps ensure that an individual who can do the work is hired. By contrast, a job description stating the "incumbent will accomplish all industrial hygiene duties, as assigned" does not identify the specific expertise that is desirable in candidates for the position.

In addition, standards are valuable in evaluating performance, supporting recommendations for awards or promotions, providing a basis for counseling when performance is at unacceptable levels, and documenting reasons for termination. Standards also help employees know what good performance "looks like." For example, a performance standard stated as "Submitting two professional publications a year is satisfactory performance; three or four is above average; and five is outstanding" helps an individual weigh priorities in programming work activities. Such standards enable individuals to evaluate their performance prior to annual reviews and to avoid surprises during reviews.

Some organizations may specify the format and content of employee performance standards; others may expect each function manager to develop performance standards appropriate for the activities of the function. The latter situation can obviously require more initial effort on the part of the manager, but tailoring standards to a specific department or activity can be particularly beneficial to the supervisor. During hiring or recruitment activities, it may be appropriate to negotiate the content of some of the individual standards with the potential hire, and the specificity of the standards can facilitate this process as well.

Some have suggested that it is inappropriate to establish standards for professionals such as those involved in health and safety activities. Advocates of this position note that these professionals have broad and diverse responsibilities that cannot be specified in detail, and consequently, specific performance standards cannot be developed for them. It is apparent that the detail in standards for professionals and other senior people in an organization should not be as rigid as that for a clerk-typist or other junior staff person; however, delineating areas of responsibility and expected performance will help avoid confusion, or even confrontation.

RECRUITING AND SELECTING HEALTH AND SAFETY PROFESSIONALS

A number of steps are involved in selecting a new member of a health and safety team. In some instances, a personnel recruiting firm may perform initial search efforts and recommend a limited number of individuals for final consideration. In other situations, the manager will be primarily responsible for finding applicants and making the final selection. Typically, written applications and interviews are important aspects of the selection process. Reports from references and testing may also be used.

1. *Written Applications.* An organization may require use of a standard application format, or the manager may have considerable leeway in designing one. Although not included in many applications, some have found it useful to delineate expected work activities along with education, experience, and other pertinent information in the application. Early definition of expectations helps prevent applicants' misunderstanding of the nature and extent of projected activities. If job performance standards have been developed for the position, they will assist in summarizing the scope of work and responsibilities.

All application procedures must meet current equal opportunity, affirmative action, Americans with Disabilities Act (ADA), and similar legal requirements. An organization's human resources/personnel department can usually provide guidance on such aspects as specific questions that can or cannot be included in an application. Meeting these requirements is essential to avoid future legal or administrative difficulties.

Responses in applications can be particularly illuminating. Fairly or not, many of us make decisions on further processing on the basis of the content of the application. At our center, I frequently have to review a significant number of applications, and when an applicant has made a number of spelling, grammar, or similar errors, I, admittedly, feel appreciative, since I can decrease the number of applications requiring my further review by that one. "Fill-in-the-blank" submissions on obviously standard forms that the applicant has prepared, along with usual trite statements like "the desire to find challenging opportunities to apply extensive expertise gained through education and experience," are not particularly appealing. An extreme example is that of a physician who was applying for a residency. In his application, under "becoming certified in;" he wrote in a specialty and closed with the remark that becoming such a specialist had been "a life-long ambition." He submitted separate applications to our institution for a preventive medicine residency, an occupational medicine residency, and a family practice residency, not realizing that the same people would be reviewing each application. Not surprisingly, he was not accepted in any of the programs.

Increasing numbers of companies now require "on-line" fill-in-the-blank applications, and these may have to be considered differently [2]. Some advantages of such applications are that each applicant provides the same type of information and that the interview is likely to be more efficient and streamlined. Disadvantages of this type of application are that it is likely to fail in capturing information regarding a specific achievement that would be of great benefit to the company and to result in "depersonalized" hiring decisions and other similar problems.

It is essential to check certifications, licensure, degrees, and similar evidence of qualifications. A company in Utah recently experienced major difficulties because an engineer who certified waste disposal processes was found not to have the state license indicated on the engineer's application. The company's excuse—that the individual somehow "slipped through" the credential-checking process—has not placed the company in a favorable light with the Utah Department of Environmental Quality nor with the public.

2. *Interviews.* The interview is often particularly important in selecting a professional for a senior position. In many instances, the individual will be part of a team. It is essential that the interview process demonstrate that the candidate is a good "fit" with other members of the team in terms of personality, orientation to health and safety, and other areas influencing an operation's success.

As with the written application process, the interview must meet certain legal requirements. Asking questions about age, number of children, marital status, or similar data can result in responses ranging from administrative complaints to litigation by the interviewee. The director of a state department recently cost the taxpayers $75,000 by missteps during interviews. In one instance, he reportedly told an applicant that he had a problem with the Family Medical Leave Act (FMLA) when she noted that she may need leave under FMLA because her son had an inoperable brain tumor. When she was not selected for the position and subsequently filed a discrimination suit, the state settled for $50,000. During another interview, the director reportedly stated, "I can't ask you this question, but would you offer to tell me how old you are?" When that person was not selected for the position and subsequently sued, the settlement was for $25,000 [3]. Current restrictions on interview content should be verified with the organization's human resources department before any interview process begins.

Regardless of the components used in the hiring process, the interview plays a major role in hiring decisions. It is as important for the manager selecting the health and safety professional to prepare for this interchange as it is for the one being interviewed [4].

Preparing well will help the manager avoid pitfalls, such as suggesting the response desired. For example, "Tell me how you would deal with this situation" is preferred over "You wouldn't have any trouble dealing with the following situation, would you?"

Buckingham and Coffman provide significant guidance on interviewing in their book *First Break All the Rules,* particularly in the chapter "The Art of Interviewing for Talent" [5]. They emphasize the need to select for talent, not simply experience, determination, or even intelligence. When interviewing, the interviewer should try to elicit specific responses, not general statements, to such questions as "How did you overcome resistance to a proposal?" or "How do you feel when someone doubts you?" and remain silent after asking such open-ended questions [5].

The interviewer should also prepare appropriate responses to queries about salary, malpractice insurance, weeks of vacation, payment of professional organization dues, parking, and similar issues. Appropriate preparation by the interviewer will also help ensure that every candidate is asked essentially the same questions, thus helping to avoid failing to obtain desired information from some candidates. Such preparation will also help avoid complaints of bias by candidates who were not selected. Properly structured and conducted, the interview will provide much of the information needed to make a reasoned selection decision.

3. *References.* Legal concerns may inhibit responses by individuals listed as references. In some instances, a verbal conversation, which is structured to avoid legally prohibited questions, may provide a better picture of a candidate than one given in a written recommendation. Written comments by references may be so general that they are of little value in candidate selection.

4. *Testing.* Some companies rely heavily on personality testing, while others do not use this technique at all. The fact that standard psychiatry texts list over 40 different personality tests suggests that

there is no one test that will clearly provide all desired personality information. However, despite the fact that skills and aptitudes determine what a person *can do,* personality and social factors often determine motivation and *what will be accomplished.* (See further discussion of psychological testing in Chapter 8.)

In considering selection processes, it must be recognized that each of us will use different criteria for selection, based on our own experiences and personalities. For example, during a meeting of directors of different medical residency specialties in which I participated, criteria used for selection of residents were discussed. Some colleagues observed that their criteria included the Dean's letter, medical school rankings, performance in various clinical rotations, and letters from preceptors. I stipulated that the applicant had to graduate satisfactorily from medical school and have no significant negative comments in his file; however, my primary criteria were initiative, good communication skills, and the ability to work well with others because I considered these capabilities absolutely essential in becoming an effective occupational medicine specialist. In my experience, the chances of an individual's successful completion of our residency were dramatically reduced if that person lacked one or more of these qualities. These criteria had worked well for me, whereas the other directors enjoyed the same level of success with the criteria they had used, illustrating how individualized criteria are for each director or manager.

Regardless of the processes and techniques used, it is essential that the heath and safety manager recognize that a hiring decision is one of the most significant decisions he or she will make. Typically, managers "spend more time on managing people and making people decisions than on anything else, and they should" [6]. If the organization is to be successful, the manager must "get the right people in the right seat on the bus" [6]. Too often, the pressure to fill a position results in mistakes that can seriously damage an organization. As Collins notes, "When in doubt—keep looking." I have found that any time I had reservations about a person but did not "keep looking," I soon realized I had made a hiring mistake [7].

ORIENTATION OF NEW EMPLOYEES

Too frequently, employees are expected to perform their duties with little formal orientation. This is particularly likely to occur in smaller units and especially when the new member is a health and safety professional. In addition, what orientation is provided may be informal or "on-the-job" and may be conducted by whoever takes the initiative to furnish information. This approach sets the stage for misunderstandings and difficulties that can severely affect the unit's work.

At the very least, there should be a structured orientation to the organization, including its mission and goals; personnel policies, including conflict of interest, sexual harassment, discrimination, and similar aspects; company policies; organizational standards and expectations; and health and safety programs. A unit or departmental manual—with information on the organization's mission; organizational structure; relationship to other departments, program elements, and functions; and recent achievements—may be of particular value in orienting new members of the heath and safety unit.

EMPLOYEE HANDBOOKS

Employees may be provided employee handbooks or similar material to assist in orientation. The director of health and safety programs should be cognizant of the contents of the organization's employee handbook. Handbooks have presented significant problems when comments in the publication have been considered to be implied contracts between the company and the employee. For example, one employee interpreted a handbook statement as "guaranteeing" continued employment as long as the company was in existence. He brought suit when he was terminated for documented poor performance, citing a statement in the handbook that was vague and open to misinterpretation. A discussion of the structure and content of an employee handbook is beyond the scope of this summary of personnel concerns; however, the manager should ensure that both legal and personnel representatives have approved the language in an employee handbook before it is given to prospective, new, or in-place heath and safety employees.

EVALUATING PERFORMANCE AND PROVIDING FEEDBACK

Employees should be regularly advised of the progress they have made in their work. Performance standards can help provide this information, but if these have not been established, the supervisor should still ensure that employees know how their efforts are perceived. Such interactions provide the opportunities to clarify any misunderstandings and to develop new approaches that might increase the organization's effectiveness and efficiency. Another benefit of appropriate feedback is that it demonstrates the manager's interest in the health and safety unit members and in helping them achieve their goals as well as the goals of the unit.

In accomplishing evaluations, it is important to have the person being evaluated participate in the process. The following example illuminates this point. Recently, senior members of a department were surprised to learn that the department manager had appointed a committee of junior members to evaluate the performance of the senior members. The report of the committee's evaluation was submitted to the CEO of the organization without the senior members being aware of the committee's appointment or having the opportunity to provide input into the performance-evaluation process. Accomplishing evaluations of professionals and reporting results to the senior executive without the opportunity for those being evaluated to participate is *not* a recommended evaluation process.

It is particularly important to recognize outstanding efforts. Neglecting to praise an individual or group for outstanding work is an omission that can have a negative impact on morale, productivity, and retention. The standard guidance to "praise in public and criticize in private" is especially pertinent in recognizing outstanding efforts. Individuals appreciate thoughtful, sincere recognition of superior work. However, improperly done, praise may detract from, rather than improve, performance of individuals or teams. Many of us have received the "standard" letter of praise, "I want to express my sincere appreciation to you for your outstanding efforts during the past year that have contributed so much to the success of this organization." It is often disturbing to get such items, especially when they are obviously computer addressed with a copied signature.

The advice of Blanchard and Johnson is particularly applicable with respect to praise [8]. Their book, *The One Minute Manager,* describes a manager who recognizes that it would be easier for an employee to do well if the employee got crystal-clear feedback on how work was progressing. The manager decides to let the employee know, in no uncertain terms, when he was doing well and when he was doing poorly, with an emphasis on the former, in keeping with the motto of the manager's organization: "To Help People Reach Their Full Potential, Catch Them Doing Something Right." In contrast to organizations in which a manager focuses on finding out what people are doing wrong, the one-minute manager "catches" people doing something right and immediately tells them precisely what they did right and how good the manager feels about what they did. Comments are not saved until the annual performance review but are provided as near as possible to the time of the event.

Blanchard's emphasis on specificity illustrates the difference between compliments and positive feedback. The previously mentioned annual "well-done" form letters, or telling someone how glad you are that they are in the health and safety unit, are compliments. Conversely, telling someone how glad you are that they are part of the organization because they did "such an exceptional job in accomplishing the air sampling program" is informative positive feedback that usually has a much stronger favorable impact than simple compliments.

Regularly recognizing specific outstanding performance gives the manager credibility if it becomes necessary to discuss unsatisfactory performance. It is essential that the health and safety manager tell employees and co-workers when their work is not satisfactory. The "criticize in private" is particularly appropriate with respect to a private discussion of unsatisfactory performance. However, the manager will be well served by recognizing that it is more appropriate to provide feedback than to criticize. Feedback focuses objectively on performance; while criticism focuses on the individual. ("Your report had a number of errors. Let's discuss how you can correct them and prevent them in the future" versus "Your report is terrible! I don't want you to ever send me such poor work again.") Concentrating on the specific difficulties, and attempting to determine how the manager and employee can work together to resolve them, will result in more desirable outcomes than if the manager attributes the poor work to lack of interest or poor effort on the part of the employee.

In discussing poor performance by health and safety members, it is particularly appropriate to begin by asking employees how they perceive their work. In some instances, this may uncover the fact that an individual is well aware of work problems but may have legitimate reasons for the poor performance. Such information helps preclude a supervisor's chastising an individual only to find that valid reasons for the inadequacy exist. For example, there may have been factors beyond the worker's control or that had even been brought to senior managers' attention, with requests for help that went unheeded. Until I learned to take this approach, I repeatedly found myself in embarrassing situations. Prior to this insight, I found myself beginning a counseling session by telling individuals how disappointed I was in their performance and how I expected them to do better in the future. It was only after I completed my discussion of "their" poor efforts that I asked them for their comments. When one reminded me that I had not provided promised secretarial support, and a few months later, another pointed out that a personal friend of mine had refused to give promised information for a needed project, I realized that asking employees how they perceive their work to be progressing, as a first question, produces dramatically better results in performance evaluations. Thus, if an employee does not appear to be doing well, it is important to "begin with the learner" to determine the underlying cause(s) of the problem. ("Begin with the learner" is an adage I have used as an educator to emphasize the importance of beginning at the level of knowledge and understanding of the learner, an educational theory popularized in the 1960s by Malcolm Knowles.)

Failing to advise an individual of unsatisfactory performance and to give the person an opportunity to correct it is a management failure. Unfortunately, this error frequently occurs. We enjoy praising the work of others, but it is often personally stressful to discuss their performance when it is poor. Our natural tendency is to avoid such discussions, but this works to the detriment of both the employee and the organization. In addition to holding a discussion of unsatisfactory performance in private, in general, it should take place as close in time to the event as possible. In some instances, it may be appropriate to delay for a brief period if the manager and the worker would not have sufficient time for the meeting or if either party needs time for emotions to "cool off." In general, delays should not exceed 48 hours, however. It is important that the discussion be objective and clear in providing feedback and agreeing on what is expected and who is responsible. Time should then be provided for the individual to improve his or her performance. Terminating an individual who has not been appropriately counseled about the need for improvement is not only unfair but may also have adverse legal consequences. Because of potential grievance or legal actions, the director should obtain guidance from the human resources/personnel department about the need for, and processes involved in, documentation of counseling sessions and other efforts to assist the employee. Chapters 8 and 9 provide additional information on working with people. The discussions in these chapters include pertinent aspects of Maslow's and Herzberg's theories on motivation as well as a more detailed discussion of other aspects of dealing with health and safety team members and with other individuals inside and outside of an organization.

DEALING WITH THE PROBLEM HEALTH AND SAFETY PROFESSIONAL

One of the most challenging personnel areas facing the health and safety manager is that of dealing with problem professionals. These individuals may be friends as well as colleagues, which further compounds the problem. Unfortunately, many managers will have to respond to this situation during their management careers.

One essential aspect of managing a problem professional is to identify the problem early [9,10]. Most of us have multiple responsibilities, and we trust our team colleagues to maintain appropriate standards. However, if the manager becomes aware that the quality of a team member's work is deteriorating, that the individual is "expected in soon" even though the work day is well underway, or that others are doing work that was assigned to that person, it is time for the manager to take prompt, deliberate action to learn more about the situation. Waiting "a few days" to see if the situation improves frequently extends into several weeks, as other priority items consume the manager's time. This extension of time makes it even more difficult to correct the situation, since it appears that the manager has accepted the below-average performance for some time.

In dealing with colleagues, it is especially important to begin by asking the professional how he or she perceives work is progressing. As previously noted, "begin with the learner" in the same manner as when dealing with a staff member. The individual may identify an unusual set of circumstances, such as the serious illness of a family member, which may be the cause of the problem. However, if no acceptable cause is put forth, or, worse, the person believes all is going well, the manager has to identify the problem objectively and explain why it is of concern: For example, "Your repeated late arrivals are affecting your work and also the morale of people in your unit. I want to work with you to see how you can best resolve this problem."

Typically, the manager will find it difficult to be this objective and clear in providing negative feedback to a professional. It is particularly stressful to have to advise colleagues that their work is not of expected quality. My natural tendency has been to discuss how "we" had a problem, and "we" needed to improve some aspect of "our" activities. Conversations with others indicate that I am not alone. On more than one occasion, my obtuse comments have resulted in the other individual assuming that I would work to improve his material after he submitted it to me—definitely not the result I wanted! Awareness of this tendency to avoid specificity may help the manager bypass similar difficulties.

An objective discussion of unsatisfactory employee performance with the individual may identify one or more factors causing the problem, including [8–10]:

1. Lack of skill
2. Lack of knowledge
3. Factors beyond the individual's control
4. Poor attitude or motivation
5. Lack of basic capabilities

The first three items can be directly addressed, with the aim of resolving the difficulties. Skill and knowledge may be improved with short courses or additional training. The manager may be able to take action to control factors beyond the individual's control, such as lack of computer support or supplies. However, poor attitude or motivation is much more difficult to address. In some instances, changes in the work environment can result in improved attitudes or motivation; in other situations, changes in assignments may be beneficial. The discussion of motivation in Chapter 9 may be of assistance in dealing with motivational problems. The fifth item, lack of basic capabilities, is particularly difficult to confront and remedy. The presence of this factor may reflect a major error in hiring or promotion.

As specific factors are identified, it is appropriate to ask for ideas from the employee about how best to deal with the situation. The individual can then possibly develop a specific action plan to resolve the problem. In this regard, it is important for the manager to offer reasonable assistance. Once the plan of action to resolve the problem is agreed on, summarizing what is expected and who is responsible will help avoid future misunderstandings. When planned solutions have been developed, the manager should schedule follow-up sessions to determine how successful the approach has been.

Throughout the process, it is especially important for the manager to maintain calmness, objectivity, and self-esteem. As previously considered, it is essential to focus on the problem and provide feedback rather than focus on the person and give personal criticism. Although all managers would like to avoid an undesirable outcome, it is just as important to document the meetings, actions taken, and efforts to resolve the situation as it is with other members. Lack of appropriate documentation could make a bad situation far worse if it becomes necessary to terminate the person.

It should be noted that if the manager experiences difficulties with multiple members of the unit, it is time for that manager to review his or her personal management techniques and skills.

FAIRNESS AND DISCRIMINATION

Most managers rapidly develop excellent knowledge of equal opportunity, affirmative action, sexual harassment, and other antidiscrimination requirements. However, even if all legal and administrative specifications are met, more subtle problems may occur if health and safety members perceive one person is being treated more favorably than others. For example, a secretary may request permission to leave two hours early each Friday to take a graduate course toward achieving a business degree, offering to work through three one-hour lunch periods to "more than make up" for the absence while reminding the manager of departmental policy to support employee efforts to take university courses and noting that two other secretaries would be available to provide coverage during her absences. Although the secretary's request seems reasonable, granting it can be expected to pose major "challenges" for her manager. Typically, the other two secretaries will also rapidly find legitimate reasons to request time off, and Friday afternoon may prove a particularly requested time. Denying these subsequent requests can immediately create reactions ranging from obvious disgruntlement to formal complaints that may require numerous hours of the manager's time to resolve, if resolution is possible. When this scenario actually occurred in an academic health unit, the director and his supervisor spent the equivalent of two full weeks attempting to resolve the situation.

While perceived favoritism in treating of staff may produce major difficulties, its impact may be markedly amplified if health and safety team professionals are involved. If the health and safety program director is perceived as "playing favorites" with colleagues in his or her discipline, serious problems, which can jeopardize the entire program, are likely to arise. Because of the potentially serious adverse effects of this situation, it is discussed more extensively in Chapter 9.

Recognizing concerns about the appearance of favoritism and the desire to provide equitable opportunities and working conditions, some argue that these concerns do not mean that everyone is treated the same, although all should be treated fairly [5]. A survey of 80,000 successful managers found that they spent the most time with their best people, believing that they should treat people as they deserve on the basis of merit. In contrast, less successful managers focused on less productive people, trying to help them improve. The take-away lesson of these survey results is that people do not change that much, so managers should not waste time trying to "put in what was left out" but, instead, try to "draw out what was left in" [5].

The ADA has created a number of difficulties for managers. Accommodating individuals with progressive health problems can be of particular concern. Retaining an employee with gradually deteriorating capabilities is usually of marked benefit to the organization as well as the individual; however, it is important to balance this action with the need to avoid imposing on other employees in the unit. Achieving an appropriate balance can present personally trying situations for the manager who wants to be fair to both the individual with the illness and that person's co-workers [11]. Dealing with such situations is considered more extensively in Chapters 17 and 23.

LABOR RELATIONS

The health and safety manager may be in an organization whose employees have been organized into one or more labor unions. In other organizations, there may not be union representation of employees, but there still may be specific personnel policies that affect all members of the company, including health and safety members. It is essential that the chief of the health and safety unit be well versed in personnel policies and contract requirements that apply to employees.

The manager is specifically responsible for ascertaining the personnel factors that directly impact, or are affected by, health and safety activities. If the program director works in a unionized organization, it is particularly important to become acquainted with union stewards and other individuals representing those who might be affected by health or safety activities or programs. Obviously, this requirement is especially significant if one of the health and safety groups is represented by a union or another group. Establishing effective relationships with worker representatives enables the manager to identify requirements or potential problems before implementing a new effort or program. For example, obtaining input from these representatives during development of a new health-promotion/disease-prevention program serves not only to enhance program content but also to increase worker support and participation. Similar benefits from effective coordination can accrue in many other situations. Conversely, implementing a new activity or establishing new requirements without appropriate coordination can create major problems for the manager and even compromise essential health and safety programs.

In addition to providing assistance in selecting new health and safety team members, an organization's human resources/personnel department can provide important assistance in labor relations. Consulting frequently with personnel representatives may prevent difficulties arising from failure to recognize labor or employee relations requirements.

STAFFING LEVELS

Various authorities have recommended the number and types of health and safety professionals needed to support various kinds of organizations. Obviously, these recommendations can only be considered guides, since the amount of expertise needed varies, depending on the extent of hazards present at a worksite, number of shifts, geographic isolation, local or nearby health and safety capabilities, country involved, administrative requirements, and other similar factors. However, in attempting to justify the number and types of health and safety professionals needed, it has often been found effective to reference an "expert" when presenting the requirements to senior management. For example, Guidotti, Cowell, and Jamieson [12] suggest that an occupational health nurse is needed for 300 employees and each additional 750 employees require another nurse. A work force of 500 employees requires a full-time industrial hygienist, and an occupational medicine physician should be hired, at least on a part-time basis, when the number of employees reaches approximately 1,000. If the worksite has more than 2,000 employees, a full-time occupational medicine physician is needed. Similar figures are available for other health and safety activities and may serve as an excellent starting point to determine the actual number of professionals, staff, and support members needed for a particular organization or activity.

While the numbers and capabilities of professionals in the health and safety program are, of course, important, the significance of administrators, secretaries, other staff, and support members should not be overlooked. These members of the team can play critical roles in enhancing both the effectiveness and efficiency of the entire unit. Often, one of them is the first member of the unit to meet potential clients, senior visitors, and others with health and safety interests. The courtesy and professionalism that the first-contact staff member exhibits usually sets the tone for the rest of the interactions between the visitor and the members of the unit. Administrative or secretarial individuals with superior capabilities are invaluable, and the wise manager will be as concerned about their selection as about choosing other members of the organization.

Finally, it should be noted that in recruiting and retaining employees, the emphasis has changed over the years: It has increasingly been placed on creating a succession plan and methods to retain top talent [13]. For many health and safety professionals, primary emphases are placed on family lifestyle and opportunities, compensation methodology, children/community (housing, schools, recreational), professional challenges and working conditions, opportunities for professional growth, and academic opportunities [13–15]. Those recruiting and attempting to retain quality health and safety personnel must recognize that extrinsic factors (remuneration) may play a secondary role to intrinsic factors (family satisfaction, cultural opportunities, opportunities for professional growth, and similar factors) in recruiting the right person for the right position at the right time. Returning to the survey of successful managers, four key elements seem to emerge as essential in getting and retaining quality employees [5]:

1. Select for talent, not experience, determination, or intelligence (as mentioned earlier).
2. Set expectations for the right outcome, not the right steps.
3. Motivate someone by focusing on strengths, not on weaknesses.
4. Help people find the right fit, not simply the next rung on the ladder.

Finally, returning to the concept of getting the right person in the "right seat on the bus" [7], it is worth nothing that Thomas Jefferson observed, while a member of President Washington's cabinet, "No duty the Executive had to perform was so trying as to put the right man in the right place."

THE "BOTTOM LINE" IN RECRUITMENT AND RETENTION

The Gallup Organization surveyed over 1 million workers and over 80,000 managers and found that there are 12 key questions that measure the core elements needed to attract, focus, and retain the most talented employees [5]. The questions are [5]:

1. Do I know what my supervisors expect of me?
2. Do I have the tools, materials, and equipment I need to do my work?
3. Am I able to do what I do best every day at work?
4. Have I received recognition for doing good work during the past seven days?
5. Does anyone at work care about me as a person?
6. Is there someone at work who helps me develop professionally? (See discussion of mentorship in Chapter 13.)
7. Do my opinions at work seem to count?
8. Does the mission of my company make me feel my work is important?
9. Are my co-workers committed to doing quality work?

10. Do I have a best friend at work?

11. Has someone talked to me about my progress during the past six months?

12. Have I had the opportunity to learn and grow at work?

It is obvious that positive answers to these questions will play a major role in how well motivated the worker is. Thus, these questions can serve as another way to help people stay motivated as work, as discussed in Chapters 8 and 9.

REFERENCES

1. Diorio JA Jr, Fallon LF Jr. Compensation and benefits. In: Fallon LF Jr, Dickerson OB, Brandt-Rauf PW (eds). *State Art Rev Occup Med* 1989;4:(1).

2. Stafford D. "Don't update your resume—it's already a dinosaur."—Hiring: Online processes replace face-to-face contact. *Kansas City Star* March 13, 2007. A-1, A-5.

3. Rolly P. "Big mouth costs the taxpayers." *Salt Lake Tribune* March 16, 2007. B-1.

4. Jenks JM, Zevnik BLP. ABCs of job interviewing—preparation, via a scenario, is the key. *Harvard Bus Rev* 1989;67(4):38–42.

5. Buckingham M, Coffman C. *First, Break All the Rules: What the World's Greatest Managers Do Differently.* New York: Simon & Schuster, 1999.

6. Drucker PF. Picking People—The Basic Rules, In: *The Essential Drucker.* New York: HarperCollins, 2003. Chap 9.

7. Collins J. *Good to Great—Why Some Companies Make the Leap and Others Don't.* New York: HarperBusiness, 2001.

8. Blanchard K, Johnson S. *The One Minute Manager—The Quickest Way to Increase Your Own Prosperity.* New York: Berkley Books, 1982.

9. Hersey P, Blanchard KH, Johnson DE. *Management of Organizational Behavior—Leading Human Resources,* 8th Ed. Upper Saddle River, NJ: Prentice Hall, 2000.

10. Welch J. *Jack: Staight from the Gut.* New York: Warner Books, 2001.

11. Banas GE. Nothing prepared me to manage AIDS. *Harvard Bus Rev* 1992; 70(4):26–33.

12. Guidotti TL, Cowell JWF, Jamiesom GG. *Occupational Health Services—A Practical Approach.* Chicago: American Medical Association, 1989.

13. Blouin AS, McDonagh DJ, Neistadt AM, et al. Leading tomorrow's healthcare organizations: strategies and tactics for effective succession planning. *J Nurs Adm* 2006;36:325–330.

14. King K, Camfield P, Breau L. Factors influencing successful physician recruitment in pediatrics and internal medicine. *Health Manag Forum* 2005;18(3):29–34.

15. Weber DP. Hiring, engaging, and retaining the A-team. *Physician Exec* 2005; 31(5):6–12.

ADDITIONAL READING

Covey SR. *The 7 Habits of Highly Successful People.* New York: Simon & Schuster (Fireside), 1989.

Diorio JA Jr, Fallon LF Jr. Compensation and benefits. In: Fallon LF Jr, Dickerson OB, Brandt-Rauf PW (eds). *State Art Rev Occup Med* 1989;4(1).

Herzberg F. One more time: how do you motivate employees? *Harvard Bus Rev* 1968;46(1). Reprinted in *Harvard Business Review Classics* 1986:13–22.

Lieblich J. Managing a manic-depressive. *Harvard Bus Rev* 1994;72(3):20–32.

Paul WJ Jr, Robertson KB, Herzberg F. Job enrichment pays off. *Harvard Bus Rev* 1969;47(2). Reprinted in *Harvard Business Review Classics* 1986:44–50.

Peters TJ, Austin N. *A Passion for Excellence: The Leadership Difference.* New York: Warner Books, 1985.

Peters T, Waterman RH. *In Search of Excellence.* New York: Warner Books, 1982.

CHAPTER

8

Personality Aspects of Working with Others

Neal Whitman

People are different. Although this seemingly simple statement reflects a basic fact of nature, it is surprising how often we, as managers, experience serious problems because we forget this truism. Managerial approaches that work exceedingly well with one individual may generate resistance, or even hostility, with another person. To further complicate matters, a manager's actions that generated outstanding work by an individual last week may produce a dramatically different response this week. Such seeming paradoxes reflect the fact that each one of us has a distinct personality, composed of different components, and our personality is not static over time.

An effective manager recognizes that personality orientations of colleagues are different and constantly changing and understands that his or her own personality will affect personal responses to others as well as to the diverse managerial challenges that occur each day.

Personality development and structure are obviously complex. The fact that numerous personality tests are in use today highlights the difficulties in clearly defining the parameters of an individual's personality. Yet managers have to interact with people and obtain their support on a daily basis. They need to recognize that different members may require different approaches in order to obtain desired outcomes. Except in rare circumstances, managers are neither psychiatrists nor psychologists. However, recognizing and adapting to different personality types does not require a degree in one of these disciplines. While a manager without extensive psychometric training may not be able to delineate all the nuances of colleagues' personalities, this limitation does not preclude identifying major aspects of personality that, properly considered, will facilitate getting desired results.

Various approaches have been developed to assist managers in defining personality characteristics of members of an organization. For example, in spite of strongly negative psychometric critiques, a number of companies still use Myers-Briggs evaluations. Respondents to Myers-Briggs answer a series of questions, and their responses are assigned to one of four categories indicating (1) introversion or extroversion, (2) intuitive or sensing, (3) feeling or thinking, or (4) perceiving or judging. The four categories are then combined to define 16 different types of individuals: For example, one type is the ISTJ (Introspective, Sensing, Thinking, Judging) person, who is purportedly quiet, serious, thorough, well organized, and responsibility-taking and works steadily and tenaciously toward accomplishing a project. Companies using this psychometric test for employee selection may look for a manager who demonstrates extroversion, intuitive, thinking, and perceiving traits. Aside from the negative critiques of Myers-Briggs evaluations, the number of categories makes it difficult for many managers to try to ascertain the orientation(s) of a colleague or another manager with whom they interact. Further complicating the situation is the fact that most managers reflect a combination of personality types rather than just one type. Eight or more types might be used to define a person's personality according to Myers-Briggs criteria.

Unfortunately, it is easy for a person who has limited familiarity with the Myers-Briggs process and its constraints to inappropriately categorize another individual as one specific type and then act incorrectly on the basis of the false assessment [1]. A comparison of seven personality surveys, including the Myers-Briggs instrument, concluded that one style or type of person is not better than another and that each style or type has its own strengths and weaknesses. In fact, the findings using these tools demonstrate the heterogeneity of organizations and the value of learning to work with individuals with different personalities and styles [2]. Although the publisher of the results of the comparison of the various surveys cautions against using the instruments to judge applicants or candidates for promotion, a report in *Fortune* magazine found that it is an open secret that surveys are used in just this way [3]. Another example of misuse is to require managers to wear color-coded name badges to reveal how they scored on a personality test. This practice was actually supposed to enhance communication among those involved [3]! Today's health and safety managers should resist such inappropriate applications of these tools and should use personality evaluation techniques to improve counseling, team building, and personal development, while recognizing the limitations of any personality evaluation process.

As discussed in the earlier editions of this book, the work of Stuart Atkins, a psychologist who has served as a consultant to large and small companies, may provide some practical assistance to the health and safety manager in recognizing and working with different personality types [4–6]. Dr. Atkins has developed a Life Orientation model with only four major categories, although it is emphasized that an individual may manifest two or more of these classifications at any one time. There is no one "right" type for any particular position in a health and safety unit. Further, the predominant style or styles of a person may change over time or in different circumstances. The four categories are summarized as follows:

1. People who are *supporting-giving* like to feel appreciated, understood, accepted, and trusted. They tend to be thoughtful, idealistic people who value cooperation and loyalty.

2. People who are *controlling-taking* like to be seen as capable and competent. They tend to be forceful, self-confident people who seek competition and like to take risks.

3. People who are *conserving-holding* like a reputation for being practical and methodical. They tend to be steadfast, reserved people who seek data and like to analyze problems.

4. People who are *adapting-dealing* like to be liked! They tend to be animated, socially skillful people who like to get along with others and to socialize.

The professional goals of these four Life Orientations are different. If you, as a manager, are supporting-giving, your greatest goals may be to prove your worth and to be helpful. If you are controlling-taking, your greatest goals may be to take action and get things done. If you are conserving-holding, your greatest goals may be to use your head and be sure of yourself. Finally, if you are adapting-dealing, your greatest goals may be to get to know people and to get along.

If you hate disappointments, take people at their word, expect others to live up to their commitments, and are a stickler for quality, you probably are supporting-giving. If you do not like beating a dead horse, like juggling a lot of activities, and seek new opportunities, you probably are controlling-taking. If you do not like highly charged emotions, tend to look before you leap, and take life one step at a time, you probably are conserving-holding. If you do not like "all work and no play," are sensitive to the needs of others, and prefer harmony, you probably are adapting-dealing.

It is important that you identify your Life Orientation(s). As previously noted, no orientation is better than another, and many of us are a combination. Whatever style or combination of styles best describe you, there are two keys to improving your interaction with others. The first key is to build on your strengths and to minimize your weaknesses. The second key is to diagnose the Life Orientation of others so that you can communicate on their channel. First, let's consider strengths and weaknesses.

ORIENTATION STRENGTHS AND WEAKNESSES

For those of you who are supporting-giving (SG), it is terrific that you like to do something to benefit others. But do you get overinvolved in other person's problems and cannot say "No"? For those of you who are controlling-taking (CT), it is a big plus that you are quick to act and can handle a quick pace. But do you sacrifice thought for action and not pay attention to the needs of others? For those of you who are conserving-holding (CH), it is wonderful that you thoroughly examine the situation and look at all the options. But do you suffer from "analysis-paralysis," showing an inability to choose and act? For those of you who are adapting-dealing (AD), we love you for using the light touch and being flexible. But do you play around too much and switch sides too often in your attempt to keep everyone happy?

In almost every case, your weaknesses are your strengths overdone. For example, idealistic people become impractical (SG), persistent people become high pressure (CT), methodical people become plodding (CH), and flexible people become inconsistent (AD). Can you see that the problem is excess? By knowing your strengths and being alert to your excesses, you will work much more effectively with others.

How do you know when you are being excessive? This lesson was taught in grade school when your teacher told you to "Stop, look, and listen before you cross the street." Effective communicators stop and look at others—look at how they talk with their faces and bodies. They listen—listen to the sounds of their voices as well as their words. If SG people stop, look, and listen, they will know when they are being excessive because they will see that they make others feel guilty and get too emotional. CT people will know when they are being excessive because they will see that they overwhelm other people and are not tuned in to the needs of other people. CH people will see that they are tied down to rules and regulations and frustrate others by getting bogged down in details. AD people will see that others cannot figure them out and get impatient with their fooling around.

Thus, the first key to more effective interactions with others is to monitor yourself so that you will know when your strengths become weaknesses. Stop, look, and listen. When you can see by your effect on others that you are being excessive, it is time to take a breath and engage in some self-talk: "There I go again! It is time to back up and start again. Cool it." This is the time for supporting-giving people to become more practical, for controlling-taking people to back off, for conserving-holding people to get into gear, and for adapting-dealing people to get to work.

RECOGNIZING ORIENTATION OF OTHERS

Earlier, we identified a second key to relating effectively to others: diagnosing others so that you can interact on their channel. You can diagnose the Life Orientation of other people by listening to their words. Does this sound like any of your co-workers?

PERSON A
"Help is on the way."

"I believe in you."

"I'll back you up."

"My word is my bond."

"What would be the right thing to do?"

Which Life Orientation says these kinds of things? Person A is supporting-giving. Let's try to diagnose the other three styles:

PERSON B
"I hope we can see eye-to-eye."

"I'll play it by ear."

"I'm willing if you're willing."

"If you can't fight them, join them."

"Can we work it out?"

Life Orientation?

PERSON C
"An ounce of prevention is worth a pound of cure."

"Better safe than sorry."

"I'll believe it when I see it."

"Watch your step."

"What do the data say?"

Life Orientation?

PERSON D
"Full steam ahead."

"I've got to take a stab at it."

"Just try and stop me."

"This may be my last chance."

"What's the bottom line?"

Life Orientation?

Did you diagnose Person B as adapting-dealing, Person C as conserving-holding, and Person D as controlling-taking? Start listening to people you work with, and you will see that what they say will tip you off. Also, look at their offices. Controlling-taking people like to sit behind a big desk, but adapting-dealing people prefer to make their office more cozy. Conserving-holding people like to clear their desk top for one task at a time, but supporting-giving people prefer "random access piles" of material.

TAILORING INTERACTIONS

Once you have diagnosed someone else's Life Orientation style or styles, you can customize your interactions with them. For example, in presenting a proposal to a supporting-giving person, it might be more effective to emphasize that it offers the best approach. However, a controlling-taking person would be more influenced by the "not-to-be-missed" opportunity. On the other hand, a conserving-holding person would like to see the documentation. Finally, an adapting-dealing person would like to know that you have tested the waters.

Similarly, in suggesting some change in the organizational structure, each orientation will have a different concern.

SG: Will it benefit all?

CT: Who is in charge?

CH: Is this in our tradition?

AD: Will it gain acceptance?

Again, in trying to influence another person, the approach that is effective will vary because each person's orientation is different.

SG: Is the cause worthwhile?

CT: Is this an opportunity for advancement?

CH: Is this low risk?

AD: Will this increase togetherness?

Can you picture how you can tailor your dealings with subordinates? Supporting-giving staff are looking for recognition and opportunity to do what is right. Controlling-taking staff are looking for autonomy and an opportunity to advance. Conserving-holding staff are looking for fairness and an opportunity to be systematic. Adapting-dealing staff are looking for friendly relations and interactions. If you want to be an effective boss, each of these individuals deserves a different approach. On the other hand, can you picture how you can tailor your communication patterns with your superiors? SG bosses want your loyalty. CT bosses want your responsiveness. CH bosses want your attention to details. AD bosses want you to fit in.

LIFE ORIENTATION AND WORK ENVIRONMENTS

The Life Orientation approach also helps identify the most and least effective work environments for people. Which Life Orientations match the following work environments?

A. Unemotional ____________

B. Accepting ____________

C. Fast paced ____________

D. Reassuring ____________

If you matched conserving-holding to A, adapting-dealing to B, controlling-taking to C, and supporting-giving to D, you understand what people like in their work environment. What do they dislike? Conserving-holding people do not like a highly emotional setting. Adapting-dealing people do not like unfriendly workers. Controlling-taking people do not like boredom. Supporting-giving people do not like betrayal. Also, each Life Orientation responds differently to rules and regulations. CH people do not like constant change in the rules and regulations, whereas AD people do not like many rules and regulations at all. CT people do not like to feel restricted by rules and regulations, whereas SG people want to know that the rules and regulations are fair.

In summary, you will be a more effective manager if you can identify your own style and learn to build upon your strengths and to minimize your weaknesses by monitoring your excesses. In addition, you will become more effective if you pay attention to the preferences of other people and learn to address their needs and respond to their concerns.

REFERENCES

1. Olsen P. Forget the facilitating. Just vent! *The New York Times* June 14, 1998 (Business Section, p. 11).

2. Whitman NA. Comparison of seven training models which could be used to improve communication skills. *J Staff, Program, and Organizational Development* 1992;10(2):95–101.

3. Stewart TA. Escape from the cult of personality tests. *Fortune* March 16, 1998;80.

4. Atkins S. *The Name of Your Game,* 2d Ed. Beverly Hills, CA: Ellis and Stewart, 1991.

5. Atkins S. *Discovery Workbook.* Beverly Hills, CA: Stuart Atkins, 1973, 1978, 1990.

6. Atkins S, Katcher A. *LIFO Survey.* Beverly Hills, CA: Stuart Atkins, Inc. 1967, 1971.

ADDITIONAL READING

For additional information about Life Orientation training, contact:

Buckingham M, Clifton DO. *Now, Discover Your Strengths.* New York: Simon & Schuster, 2001. (Discussion of insights to help the manager. Marcus Buckingham is the first author of *First, Break all the Rules* [1999], mentioned in Chapter 7, and *The One Thing You Need to Know About Great Managing, Great Leading, and Sustained Individual Success* [2005]. Both of these, which are both also published by Simon & Schuster, are of interest in considering personality aspects of management.)

Business Consultants Network, Inc., LIFO Division at www.bcon-lifo.com/programs.htm. Accessed June 27, 2007.

Johnson S. *Who Moved My Cheese?* New York: G. P. Putnam's Sons, 1998. (A "classic" discussing how people of different personality types respond to change.)

Kroeger O, Thuesen JM, Rutledge H. *Type Talk at Work: How the 16 Personality Types Determine Your Success on the Job.* New York: Dell, 2002. (Additional discussion of Myers-Briggs applications at the worksite.)

O'Leary T. *Warriors, Workers, Whiners, and Weasels: The 4 Personality Types in Business.* Katonah, NY: Xephor Press, 2006. (Another approach to categorizing employee's personalities.)

CHAPTER

9

Team Building and Motivation

Health and safety challenges in the community, workplace, and government and regulatory environments are growing increasingly complex, and they frequently require a team effort to resolve. For example, as discussed in Chapter 12, we have found that conducting a comprehensive health and safety audit of a worksite requires occupational medicine, occupational health nursing, safety and ergonomics, and industrial hygiene representatives. Even if only a single health and safety discipline is involved in a program or project, the manager will frequently find it essential to instill a sense of cooperation and support among the professionals and staff involved in order to obtain quality results. Teams may be as small as two people or may involve a large number of individuals. The essential criteria of a team are that the members are working together on a common task, members have to interact and influence each other in order to accomplish the task, each member is essential to the effort, and the team effort is necessary for the satisfaction of individual needs [1–4]. Team members *contribute* to a team, but only the team as a whole *performs* [5]. Simply calling a group of people a *team* does not make them one, as has been demonstrated too often when a "group" effort was totally unsuccessful because a true team had never been formed [2]. Consideration of some basic aspects of team building may aid the manager in forming, maintaining, and motivating teams.

TEAM BUILDING

Typically, professional and staff members will have already been assigned to the health and safety unit, and the director has the responsibility of bringing these individuals together into a cohesive, supportive crew. As requirements change, the manager may need to form new teams, which may be subsets of the health and safety organization or include people from outside the health and safety group. The following discussion considers some aspects of building a team of health and safety professionals.

Communicating Effectively

Some health and safety directors have commented that effective communication is responsible for over 90 percent of their successes in team building. It is essential that all team members know what is to be done, when it is to be done, and what approach is to be used to get it done. It is absolutely critical from the outset that every member shares the same understanding of what is expected of the team [2]. Members should also have clear, concise knowledge of what is expected of them, as team members, with respect to individual responsibilities as well as standards of performance, quality, ethics, and other pertinent areas. One of our graduates experienced difficulties as a new director of a team of health and safety professionals, and she did not understand why the team was floundering until she asked the members what they saw as the purpose of their efforts. The diverse and disparate responses demonstrated to her that she had not been successful in articulating the focus of the team's efforts. Her team was not functioning effectively as a result of poor communications. When the goals, mission, standards, and similar factors were clarified through improved communication, her team became highly effective and was recognized for its quality work.

Employing a Participatory Approach

By definition, a team is composed of individuals working together in an interdependent mode. This is true whether the team is oriented along functional lines (e.g., industrial hygiene) or task (e.g., designing and implementing a comprehensive health and safety program). In order to be effective, team members have to be committed to the desired outcome(s), and a participatory approach can enhance their commitment. Typically, lack of involvement is the equivalent of lack of commitment. If members are brought together to receive their "marching orders" and they do not have any input into the content of those orders, the probability is that the outcomes of the team effort will be, at best, marginally effective. As previously mentioned, simply calling a group a *team* does not make it one. Conversely, a true participatory approach usually produces more effective plans, methods, and results than would ever have resulted from direction by a single person. An essential aspect of team formation is that each member is essential. Many of us have been on "teams" to which more than one physician, safety specialist, health promotion professional, or other specialist has been assigned. Without fail, team meetings tend to be scheduled at inopportune times due to last-minute or other conflicts. In such situations, many specialists on the team, such as physicians, believe they can skip the meeting because another physician on the team will fill in. However, if another physician believes likewise, it is possible that *no* medical specialist will be in attendance. The same factors apply to the other members of the team. The final outcome may be that team meetings are ineffective and the project fails.

Avoiding the Appearance of Favoritism

Usually, the director of a health and safety program is a professional in one of the disciplines represented in the health and safety group. As considered in Chapter 8 on personnel aspects of health and safety programs, a team leader must avoid the appearance of favoritism. If the individuals in the health and safety team believe, rightly or wrongly, that the team leader is favoring members in the same discipline as the leader, serious problems can result. The same outcome can occur in small health and safety units if the team members perceive the director is using inordinate amounts of travel funds, continuing education funds, or other unit resources. The significance of "perception" of favoritism must be emphasized. A manager may believe all professionals in the organization are treated equally with respect to vacation, on-call responsibilities, continuing education opportunities, advancement, and other similar issues. However, if the members of other disciplines *believe* that they are not being treated fairly, the result may be the same as if the director were consciously favoring one group. This situation is a clear illustration of the saying "Perception is reality." In one actual situation, the physician leader gave other physicians paid time off for continuing medical education (CME) even though no funds were provided for registration or other costs. The leader thought that he was treating all equally, as his policy stated that any member with CME requirements for licensing would be given up to 10 days off with pay. However, he denied paid time off for members of other disciplines, even though they stressed the needs they had for professional development courses. He was surprised when members of the other disciplines filed a grievance with the leader's supervisor over the unfair treatment. Awareness of the high sensitivity levels that may exist in this area can help the manager avoid major conflicts that could jeopardize the effectiveness of health and safety teams and programs. The team leader should anticipate complaints and conflicts if it appears that members of the same discipline as the leader benefit from favoritism. The astute manager will recognize this possibility and ensure that each member perceives treatment to be equal for all.

Providing Appropriate Feedback

It is as important to provide regular feedback to the team as to individual team members. As stressed in Chapter 7, praise and recognition should be as specific as possible and as close in time to the particular achievement as possible. Negative feedback must be specific and provided in a timely manner and should commence only after ascertaining how those involved perceive the situation. In considering negative feedback, it is particularly important to determine whether all team members clearly understand the desired outcome of the effort and the anticipated time for the project. If a PERT (Program Evaluation and Review Techniques) analysis or similar process is used to track the team's efforts, as discussed in Chapter 12, it is easier to ascertain whether a project is on track to provide the desired results. For example, in the health and safety audit described in Chapter 12, we had weekly meetings to review our status in meeting the PERT schedule and objectives. All team members received feedback on their own contributions to the effort as well as on the overall effort. This ongoing feedback was essential in achieving the successful outcome.

Delegating Effectively

Professional members of health and safety teams typically benefit from being given increased responsibility. Effective delegation is one means of providing this responsibility. However, it should be noted that the emphasis is on *effective* delegation. Telling someone what to do and also *how* to do it is not effective delegation. The delegation process must be a participatory one that involves the following steps:

1. *Decide on the anticipated outcome.* The manager should formulate a preliminary vision of the desired outcome. (Recall Covey's advice to begin with the end in mind [6].) The final picture of the outcome may change as discussions ensue, but the preliminary view will help in estimating time requirements, identifying potential people to do the project, and similar aspects.

2. *Select the appropriate person(s).* The choice may be obvious in some situations but more obscure in others. A potential pitfall is always picking the same individual because of past successes; this limits professional growth opportunities of others on the team and implies a lack of confidence in them. It is also important to avoid the trap of concluding "It would take more time to explain it than to do it, so I will do it myself." As discussed in Chapter 19 on time management, failure to delegate is a major management failure and can produce adverse impacts on the health and safety unit.

3. *Agree with the selected individual(s) on the desired outcome.* In some instances, the MBO (or similar) process in planning efforts may have helped define the desired outcomes. In other situations, outcomes may have to be developed during the delegation process. For example, is the desired outcome to accomplish a noise survey of a particular manufacturing facility within the next few weeks, or is it really to obtain and evaluate noise exposure data from all noise hazard areas and design an effective hearing conservation program? If the person responsible for the effort believes the former is the desired outcome but the manager expects the latter, all involved are going to be disappointed. Depending on the significance of the effort to be delegated, it may be necessary to initially

approach the activity through an MBO, problem-solving, or other structured process, as described in Chapters 5 and 14. However, the majority of delegated items are much simpler, and outcomes can be agreed upon with a few minutes of discussion.

4. *Agree on assignment of authority, time lines, and resource requirements.* The person who agrees to accomplish a delegated item must also be given the necessary authority to accomplish the work. (A manager's requiring that he or she approve every action or step is not delegating.) Similarly, it is essential to agree on timelines and necessary resources. Without these agreements, there is little probability of success.

5. *Agree on monitoring.* Even though the manager has delegated a project, the manager is still responsible to a supervisor for satisfactory completion of the project. Although that responsibility cannot be delegated, it is reasonable for the manager to monitor the progress of the effort. However, it is not reasonable to require the responsible professional to meet with the manager each morning for a "how goes it" discussion. This degree of monitoring is micro-management at its worst and can be expected to destroy the initiative and the enthusiasm of the person "given" the project.

Removing Obstacles

As a health and safety manager, one of your primary responsibilities is to provide the support needed to enable the team members to do their work. You provide the team the resources to do the job the way you want it done. In essence, you serve as a facilitator of the team's efforts. If you cannot provide needed resources or remove obstacles, you will have to modify your expectations and priorities for the team's outcomes.

Recognizing the Importance of Different Personality Types

All health and safety teams will be composed of individuals with different personalities. The manager has to recognize the strengths and weaknesses of the individuals on the team and work to have these capabilities complement one another rather than detract from the team efforts. In many situations, blending different personalities is a major challenge. Responding to different personality types is of such significance that it is discussed extensively in the preceding chapter.

Making Decisions

Health and safety managers and team leaders need to make decisions, as discussed in Chapters 11, 13, and 14. Few factors will have an impact as adverse on a team's enthusiasm and cohesiveness as having a team leader who delays making a decision indefinitely. All decisions involve risk, and we never have "all" the information we might want. A manager who accepts the extensive work and recommendations of a team and then takes no action on them should not be surprised to find a demoralized team with little interest in future efforts. An actual example of just such a situation is discussed in Chapter 13.

MOTIVATION OF TEAM MEMBERS

Health and safety professionals are not motivated by the same factors as those that motivate some other individuals. Certainly, wages and working conditions are important; however, a sense of responsibility and accomplishment are often valued much more. Positive, "real time" feedback and involvement in goal setting may be particularly gratifying to health and safety team members. The works of Maslow and Herzberg have focused on different motivational aspects applicable to members of a health and safety team [7,8]. Maslow developed a hierarchy of needs, ranging from basic physiologic and safety (security) needs to social (affiliation) needs and the highest needs of esteem and self-actualization. The lowest levels of needs need to be met before the higher ones can be realized. Wages and benefits primarily satisfy physiologic and security needs. Belonging to a professional health and safety team with satisfying goals and outcomes satisfy the highest needs. Being placed in a situation in which routine is the ruling principle and innovation is discouraged will be frustrating to the health and safety professional.

Herzberg observed that the "KITA" technique ("Kick in the ____," or, more euphemistically, abrupt application of force to the lower posterior of an individual) used in the past by many managers to motivate employees had poor results. Hygiene factors had to be provided before an employee reaches a state of no dissatisfaction. Persons had to be in a "not dissatisfied" state before they could be motivated. Hygiene factors include food, a safe place to work, the right tools for the job, and compensation. For example, inadequate compensation would lead to a state of job dissatisfaction that would compromise any efforts to motivate or enhance job satisfaction. Motivators, which can be applied once hygiene factors are in place, equate to Maslow's higher levels of development and include achievement, professional growth, task-relevant feedback, recognition, and self-actualization [1]. The theories of Maslow and Herzberg have been confirmed by surveys of employees. As noted by Dioro and Fallon, working conditions, job security, and self-actualization have been ranked ahead of compensation in importance [9]. However, other studies have indicated that money may be either a motivator or a hygiene factor [10]. Money may be a motivator if it is used as a form of recognition or feedback or is a measure of status, advancement, and achievement. Figure 9-1 illustrates how money may be a hygiene factor and compares the two motivational theories of Maslow and Herzberg.

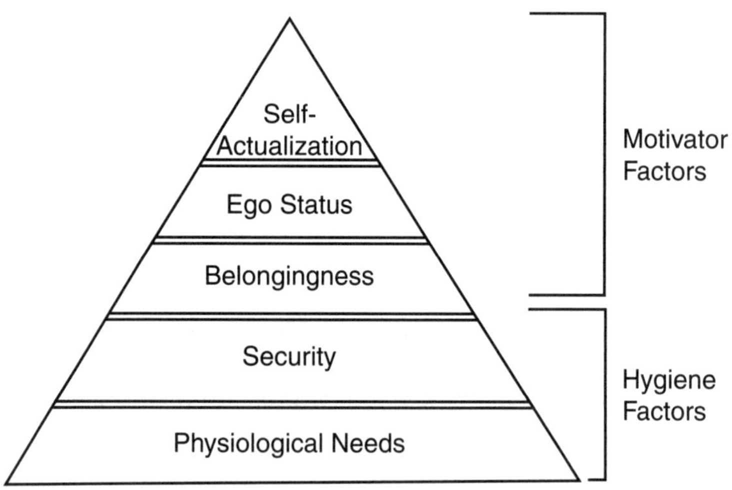

Figure 9-1. Theories of motivation. Both Maslow (left) and Herzberg (right) emphasize the importance of meeting basic needs and also providing personal recognition and satisfaction in order to motivate people.

Team members are specifically motivated by demonstrated interest of supervisors and managers. This was illustrated by Mayo's study of Western Electric's Hawthorne, Illinois, plant between 1927 and 1933 [1,11]. Investigators increased lighting levels to see if better conditions would improve productivity and were rewarded by a significant increase in output. They increased the lighting levels still further, and additional productivity increases occurred. However, the researchers were surprised to find that similar increases in productivity occurred in a control group setting, where lighting was not changed. (It is not clear from available data whether other factors, such as individual praise or increased pay, were altered in the control group.) Additional studies of changes in working conditions, e.g., more rest periods and company lunches, resulted in increased productivity. When the changes were rescinded, productivity continued to be high. The investigators came to believe that it was not the environmental changes that produced the improvements; rather, they concluded that the workers came to feel that management was concerned about them and their managers considered the workers important. This is known as the *Hawthorne effect*. The health and safety manager has to recognize the importance of this effect and the "people" aspects of his or her team.

In regard to the Hawthorne effect, which is described in numerous management and psychology texts, it should be noted that the concept of the effect may be valid but is based on poor science. Dr. Lee Ross, a psychology professor at Stanford University reviewed Mayo's study and found that only five workers took part in one portion of the study: Two of the five were replaced partway through the study because of gross insubordination and low output [12]. Another reviewer of Mayo's investigation found that participants had been paid a fixed hourly wage plus a piecework bonus based on total output of the department [13]. During the study, the pay of the piecework bonus for the five workers was based only on their collective output, so individual performance had a more significant impact. Additionally, the reviewer found that the results of the lighting portion of the study were published only as a brief summary in an engineering newsletter, and the original data had disappeared [13].

Regardless of the basis for, and validity of, the Hawthorne effect, Peters and Waterman observed that successful companies create in their employees the sense that their best efforts are essential to the success of the company and that they share in that success [14]. Demonstrated positive interest in people by managers was a major factor in increased productivity. This apparent confirmation of Mayo's studies is further illustrated by the ongoing emphasis on management personnels' walking (wandering) around worksites periodically [15]. This approach, commonly referred to by the acronym *MBWA* (see page 186), also demonstrates interest in the organization's people. Similarly, an "open-door" policy may show interest in people but only if people know that something will happen if they use the open door [15].

Those forming or leading teams need to recognize when a team is not functioning as desired and then determine and correct underlying causes. The need for effective communication among all parties has been previously discussed. Lencioni provides guidance on additional concerns in his book *The Five Dysfunctions of a Team: A Leadership Fable* [16]. He lists the five essential building blocks of a pyramid that are necessary for a team to be successful. The base of the pyramid is *trust.* Members are comfortable with each other and do not hesitate to make mistakes in attempting to achieve the goals of the team. Members can be completely open with each other in making comments or brainstorming. Once trust is established, *conflict* can occur, whereby members do not hesitate to challenge and disagree with each other in order to find the best solution or action. The next building block in the pyramid is *commitment,* meaning members readily "buy into" group decisions even though they may have initially opposed the action. Thus, all members feel that they have been given a fair hearing before the decision was made. Once these three building blocks are in place, members accept *accountability,* that is, they hold each other accountable to abide by decisions and standards established by the team and they convey this accountability to their peers. The peak of the pyramid is *results.* Members support what is best for the team without placing their own desires, egos, career plans, or other personal concerns ahead of the team's success. Problems with any of the five steps can lead to a dysfunctional team.

Unfortunately, many of us have been on teams that are missing one or more of the essential building blocks. In one instance, a team "member" would readily agree to any action or proposal developed by the team, never expressing any disagreement or conflict. He would then go to "his" group of professionals to complain bitterly about the team's results, noting the decision was made in spite of his strong objections. There were other similar instances, so trust was never established. As a result, the final outcome of the team's effort was so noncommittal it was useless. Recognizing the potential for team dysfunction may help team leaders and members achieve the necessary steps to accomplish an effective outcome.

In summary, the health and safety manager works through and with team members to achieve organizational goals and objectives. The manager's output is, in essence, the output of the team. By building and motivating effective teams, the manager helps members reach the highest level of self-actualization. Such an achievement pays dividends in increased productivity and personal satisfaction for the team members and the manager.

REFERENCES

1. Hersey P, Blanchard KH, Johnson DE. *Management of Organizational Behavior—Utilizing Human Resources,* 7th Ed. Upper Saddle River, NJ: Prentice Hall, 1996.

2. Marshall EM. *Transforming the Way We Work—The Power of the Collaborative Workplace.* New York: American Management Association, 1995.

3. Kanter RM. *On the Frontiers of Management.* Boston: Harvard Business Review Book, 1997.

4. Drucker PF. *The Effective Executive.* New York: HarperBusiness, 1966 (HarperBusiness ed 1993).

5. Drucker PF. *Managing in a Time of Great Change.* New York: Truman Talley Books/Dutton, 1995.

6. Covey SR. *The 7 Habits of Highly Effective People—Restoring the Character Ethic.* New York: Simon & Schuster (Fireside), 1989.

7. Maslow A. A theory of human motivation. *Psychol Rev* 1943;50:370–396.

8. Herzberg F. One more time: how do you motivate employees? *Harvard Bus Rev* 1968;46(1):56–58.

9. Diorio JA Jr, Fallon LF Jr. Compensation and benefits. In: Fallon LF Jr, Dickerson OB, Brandt-Rauf PW (eds). *State of Art Rev Occup Med* 1989;4(1).

10. Martin DD, Shell RL. *Management of Professionals—Insights for Maximizing Cooperation.* New York: Marcel Dekker ASQU Quality Press, 1988.

11. Mayo E. *The Human Relations of an Industrial Civilization.* Cambridge, MA: Harvard University Press, 1933.

12. Kolata G. Scientific myths that are too good to be true. *New York Times* December 6, 1998. Week in Review.

13. Rice B (Senior Editor, Psychology Today). The Hawthorne defect: persistence of a flawed theory. Available at: www.cs.unc.edu/~stotts/204/hohawth.html; accessed January 11, 2005.

14. Peters TJ, Waterman RH Jr. *In Search of Excellence.* New York: Warner Books, 1982.

15. Peters TJ, Austin NK. *A Passion for Excellence: The Leadership Difference.* New York: Warner Books, 1985.

16. Lencioni P. *The Five Dysfunctions of a Team: A Leadership Fable.* New York: John Wiley & Sons, 2002. (A fictional narrative describing the efforts of a new CEO to overcome five dysfunctions: lack of trust, fear of conflict, unwillingness to commit, avoidance of accountability, and inattention to results. See also Lencioni's *Overcoming the Five Dysfunctions of a Team: A Field Guide for Leaders, Managers, and Facilitators.* Published in 2005 by Jossey-Bass.)

ADDITIONAL READING

Blanchard K, Johnson S. *The One Minute Manager.* New York: HarperCollins, 2003.

Drucker PF. Management's new paradigms. *Forbes* 1998;162(7):152–177.

Harvard Busines Review (ed). *Motivating People for Improved Performance.* Boston: Harvard Business School Press (Results-Driven Management Series), 2005.

CHAPTER 10

Financial Planning and Budgeting Aspects of Health and Safety Activities

All health and safety managers have financial planning and budgeting responsibilities. Although there may be an administrator within the organization who can assist in the budgeting process, the manager still has the primary responsibility for allocating and monitoring resources. This is not to suggest that the manager must have extensive accounting capabilities. The manager should fulfill, or arrange for a consultant to fulfill, this role. The manager's primary financial responsibility is to be certain that all income and expenses are properly controlled and in accordance with projected income and expenses. The director also needs to know the major budget categories and allocations impacting the health and safety program. Managers who direct organizations with multimillion dollar budgets usually understand that their responsibilities include funding allocations, tracking income and expenses, and resources, among others. Resources that must be managed include personnel, facilities, equipment, time, and money. The following discussion focuses on financial planning and budgeting, but many of the same principles apply to managing other resources. However, a detailed consideration of economics, accounting, and finance is beyond the scope of this chapter. A series of *Occupational Medicine State of the Art Review* articles devoted to the management perspective provides excellent discussions of these areas as they pertain to health and safety activities [1–3].

PLANNING

General Considerations

Effective financial planning, oriented to aspects of interest in the near term as well as in future years, will assist the manager in dealing with projected financial changes that could benefit or impact the organization. Much of the information gathered during the health and safety planning process (discussed in Chapter 3), is directly applicable to financial planning. Forecasting and planning for changes in program requirements and activities over the coming years will help in projecting future financial requirements and opportunities for the health and safety unit. Plans to implement new programs or expand current activities will have a financial impact, and the manager will have to accommodate these needs in order to accomplish the planned efforts. Projections of new competition, potentials for "outsourcing" health and safety functions, and scheduled loss of a major revenue source are examples of financial concerns that the manager has to incorporate into "out year" financial planning. The need to incorporate such aspects applies whether or not the organization is part of a corporation, is hospital based, is a stand-alone clinic or has consultant status. Appropriate financial planning also facilitates preparation of realistic budgets.

In financial planning, it is desirable to develop *complete* information on all projected costs *and* revenues. Past experience can be of value in developing this information, but the manager has to apply appropriate modifications to reflect projected changes. Although complete information is the goal, all of us recognize that we will not be able to define projected costs and income precisely. However, managers usually find that consciously attempting to develop valid cost and income estimates is of great assistance in ascertaining what priority actions can be accomplished, given projected income during coming years. This information can then help the manager in providing effective, efficient services that meet the needs of the "customers."

Defining "real dollar" costs and income is only one aspect of health and safety financial planning. Another facet is attempting to quantify the "return" from the different aspects of the program. For example, estimating the "savings" from an injury reduction program, documenting cost reductions from effective case management, dollar benefits from an immunization program, and similar accounting activities help delineate the return on investment (ROI) of the health and safety unit. Figure 15-3, page 216, depicts some of the components of health and safety programs that can serve as a starting point in developing estimates of financial benefits of the program. (Although this figure refers to OEHS programs, the concept of multiple program aspects is applicable to any health and safety program.) This information can be of very real benefit if senior management of an organization decides to do a "cost comparison" of the health and safety function (in-house, contract, or consultant) with a competing group. Unless the costs and benefits of all desired aspects of the health and safety program are considered, comparisons may be misleading and result in inappropriate decisions. For example, comparing only the direct medical costs of treating typical injuries may suggest that a primary care clinic could provide the cheapest treatment with resultant savings, compared with the costs of an in-house clinic. However, when the savings that those with occupational and environmental health and safety (OEHS) expertise can provide through effective case management are included in the equation, the total cost to manage the various conditions may be markedly less for the in-house program. Similarly, consultant or OEHS specialty clinic total costs may be significantly below that of a non-OEHS function due to the latter's delays in recognizing the need for early treatment interventions, such as physical therapy, in order to get an employee back to work sooner than the "typical" treatment plan of non-OEHS providers [4]. The same principles apply in any other health and safety program. Standard guidance in these situations is to "avoid comparing apples and oranges." The key to effective comparisons is to compare costs and benefits of all aspects of the program, not just a selected portion. The information in Chapter 15 may be of assistance in defining the components of an OEHS program. Effective comparisons may demonstrate that the "cheaper" program does not even offer essential components, with regulatory compliance or design and implementation of monitoring and surveillance programs being only two examples of components that may be missing. In addition to material in Chapter 15, recent reports provide additional examples of defining return components and developing return on investment (ROI) information [5–10]. Again, the same approaches are beneficial in determining the actual ROI of any other health and safety program.

Business Plans

In accomplishing financial planning, information obtained through the process of creating a business plan often aids in developing a more comprehensive picture of financial opportunities and concerns that will impact the health and safety program. If a manager wishes to develop a new health and safety program or accomplish a major expansion of a current program, constructing a business plan may be helpful in designing the program and in obtaining funding support. Another important benefit of using a business plan approach is that it may reveal significant gaps in knowledge that could lead to errors in structuring health and safety activities. These errors may, in turn, have severe impacts on the overall program or even jeopardize its existence. For example, the Academic Health Sciences Center recently decided to purchase a group of primary care clinics. However, an apparent failure to comprehensively analyze the clinics' competition and lack of recognition of the impact of pending changes in reimbursement procedures or policies almost resulted in bankruptcy of the clinic program. Conversely, a manager who recognizes the need for more information or the potential for future problems can take actions to prevent deleterious effects and also improve current and projected health and safety activities. More detailed information on developing a business plan is provided in Chapter 11.

THE BUDGET CYCLE

In most organizations, financial budgets are prepared and approved on a cyclical basis. Usually, the cycle period is one year but does not necessarily start with January. For example, the federal government's fiscal year begins October 1, whereas the budget period at some companies begins July 1.

Regardless of the start date, the manager will typically begin developing the health and safety program budget months ahead of the beginning of the fiscal year. If new to an organization, the manager may find that the budget already in place provides a useful starting point for developing a proposed new budget. However, it is important to ensure that using the current budget as a basis to develop the new one is appropriate and will promote the accomplishment of desired goals and objectives. Developing a budget in the "usual" format because "it has always been done that way" can lead to a mediocre program with few innovations or significant accomplishments.

THE BUDGET PROCESS

Identifying Income, Costs, and Required Budget Categories

An initial step in both financial planning and the budgeting process is to identify all projected income and expenses. Although it appears obvious, this step has been a source of consternation and problems for many health and safety program directors.

Failure to identify all projected income and to accurately estimate the dollar amounts can cause priority aspects of the program to fail, aspects that otherwise could have produced important benefits. Because of lead times involved in obtaining funding for major projects, it may not be possible to utilize funds effectively if their availability is not identified until late in the fiscal year. Having excess funds on hand at the end of a fiscal year, with important portions of a program still unaccomplished due to "lack of funds," is not the mark of a good manager. In some organizations, the "excess" funds may be taken back by the company or funding agency. This is not uncommon in government organizations. Further, since, apparently, the health and safety manager "could not use" and "did not need" the money, the next health and safety program allocation may be reduced by an amount equaling the "excess." Expending excess monies by last-minute purchases of cartons of paper clips, tablets, and other items to support the paperwork mill will, likewise, not impress senior managers. Even if all funds are used, a 10-year supply of staples and paper clips (actually observed at one activity) does not bode well for the manager who authorized the purchase. Indeed, the manager may be subsequently encouraged to find another position in a different organization where his talents might be more "appreciated."

Similarly, failing to identify all expense categories will also create problems. For example, one new manager developed a comprehensive budget to reflect optimal use of accurately estimated, projected resources. However, late in the review process, it was noted that the manager had not included personnel fringe benefits (health insurance, retirement funding, life insurance, and similar items). Since the fringe benefits were approximately 35 percent of employees' salaries, major reprogramming had to be done quickly. As a result, both the budget and the program's effectiveness were adversely impacted that year.

Similar problems have occurred because the original budget omitted travel, utilities, rent, or some other item. Omissions often occur because of a misconception that expenditures would be covered by a central fund, or in some other manner, or had been taken care of at a new director's prior location. Carefully reviewing the current budget, and spending some time discussing proposed budgets with an organization's budget officials, should help a new manager identify the costs that program income will have to cover. Specifically, overhead costs, such as the cost of building insurance, utilities, maintenance agreements, and similar items, can be identified so that they can be included in the budget. In addition, particularly if estimated budgets for future years have to be submitted, the planner can ascertain the inflation factor to be used for the future, or "out," years. It also should be determined whether or not the organization uses "cost centers" to help identify specific requirements for various program functions. This will help avoid budgeting omissions that could jeopardize a part of the program.

Appendix B provides examples of some typical budget categories that are often used in health and safety activities.

Establishing Priorities and Defining Budget Allocations

As noted previously, it is important that management define priorities for program efforts. Delineating priorities facilitates resource allocation. It is highly unlikely that any health and safety program will ever have all the resources it desires; every action involves a resource cost. Consequently, the effective manager must ensure that the limited resources available are allocated to accomplish the highest priority components of the program. Lower priority items can then be addressed as funding permits. Use of MBO or a similar process will aid in this activity, since defining resource costs for each project is a specific required part of these processes. The manager can then more easily decide on the best use of available resources.

In developing budget funding requests, the health and safety director must recognize constraints that are established by the corporation or organization. For instance, there may be a limitation on the "head count" for the unit. Adding a safety and ergonomics professional, for example, may necessitate balancing the action by eliminating another position. "Hiring freezes" can impact budget efforts in a similar manner. Being aware of such restrictions facilitates both budget planning and the process of explaining to other members of the organization why staff cannot be expanded to desired levels.

In evaluating needs and establishing priorities for limited funds, the director may have to decide whether to "make or buy" a product or service. That is, should an effort or project be accomplished "in house," within the organization, or on a contractual basis, with an outside group? Deciding on one of these alternatives may require extensive cost-effectiveness analyses. As with the comparison of sources of health and safety support and activities described above, it is important to define the total costs and projected benefits of the options. Carefully defining all the costs involved in obtaining the same results through different approaches can make the choice relatively obvious. Some companies often require cost-effectiveness analyses for major budget proposals, that is, comparisons of the costs of different approaches to providing the desired outcomes.

Obviously, the program manager should not set priorities in a vacuum. The other professionals in the unit should have input into the process, and the advisory committee (discussed in Chapter 3) can also furnish valuable assistance. Further, the health and safety director's supervisor(s) may provide direct inputs, guidance, and suggestions. If the CEO has emphatically stated that the organization will have a health promotion disease prevention (HPDP) program, it is obviously unwise to withhold funding arbitrarily from the HPDP activity to support "higher priority" items. At the very least, the health and safety manager should explain why funding limitations preclude implementing HPDP activities. Such a discussion may result in allocation of additional resources or revision of either the CEO's or the health and safety manager's priorities.

BUDGET APPROVAL

Once the manager has developed a budget, the review process, which should finally lead to approval and adoption, begins. The layers of review will vary with the size, organizational structure, and philosophy of the institution as well as regulatory requirements and other factors unique to a particular organization.

Tying health and safety budget requests to organizational goals and objectives increases the probability of approval. As mentioned in Chapter 3, incorporating the health and safety unit's long-range plan and projected funding needs into the overall plans and goals of the company, government agency, clinic, or other organization also facilitates approval and helps avoid subsequent day-to-day struggles over funding. Similarly, delineating expected benefits and using appropriate cost-effectiveness or other analyses can provide important support for the budget. As suggested in previous chapters, the health and safety director usually does not have sufficient authority or power to acquire extensive new or additional resources. In such situations, it is necessary to "borrow" influence from others; therefore, line managers, board members, legislators, local or state government officials, or others must be cultivated to help champion health and safety requests. The previously mentioned advisory committee may have laid the groundwork needed to obtain such support. The discussion of political aspects of leadership in Chapter 13 provides additional information on obtaining support for health and safety programs.

In many organizations, it is unusual for a manager's budget to be approved as originally submitted. Revisions usually are required. In some instances, the manager may have developed a budget without identifying income sources for all projected expenditures. If the manager has not established funding priorities, reviewers may arbitrarily "balance" the budget by making cuts that become apparent only after the program receives the new budget. In other instances, the reviewers may reduce projected health and safety allocations because of perceived higher priority needs within the organization. Numerous other factors may result in reduced allocations. It is desirable that the health and safety director have the opportunity to defend or be an advocate for the budget during the review process. This advocacy is often permitted in many organizations with health and safety programs. In such negotiations, if the manager knows precisely what he or she wants, it is more likely to be approved. Effective planning (discussed in Chapter 3) helps delineate what is needed and the priorities for each item.

Defending the requested funding and budget allocations requires that the health and safety manager use the same skills and techniques utilized by other organizational managers. The other managers also believe in the critical importance of their programs and can be expected to do a superb job of justifying financial and other support. If the health and safety director attempts to justify the program's funding request only by noting the funds are "needed to maintain and improve worker health," the chances for success are limited. Such "motherhood" justifications do not work well in today's corporate environment. Limited resources exist at all levels of an organization, and requests for allocation of resources to health and safety activities must be justified in the same manner as that for other nonhealth functions.

BUDGET MONITORING AND EVALUATION

Once the health and safety budget is approved and awarded, it cannot be overemphasized that the manager must monitor both the expenditures and the progress of the activities that the expenditures were intended to cover. For example, if monitoring reveals that 50 percent of available funds have been spent to conduct health surveys but only 25 percent of the surveys have been done, adjustments are necessary. Procedures like those discussed in Chapter 12 can assist the manager in conducting such monitoring.

An effective budget monitoring and evaluation system is meant to provide timely information so that discrepancies can be detected and effective adjustments can be made. Unfortunately, many institutional administrators cannot provide current expenditure information, since records cannot be compiled until several weeks to months after the expenditure. Accordingly, the manager may have to develop a method to access needed information on a more "real time" basis. For example, our university budget monitoring reports typically arrive one to two months after the expenditures or fund receipts. As a result, we arranged to hire a person to track and monitor our expenditures and income much more closely. This monitoring helped assure that we were obtaining the maximum benefit from our resources. These "real time" methods do not necessarily have to be pinpoint in their accuracy. They need to provide only enough information for the director to obtain a realistic picture of progress and expenditures.

An important aspect of budget monitoring is to ensure that procedures are in place to prevent "diversion" of funds by an employee. Finance members of an organization or a consultant accountant can assist in designing processes, such as requiring persons receiving checks be separate from the person(s) making deposits, with cross checks to confirm all income is properly managed. Similar procedures for expenditures also help prevent inappropriate financial transactions. Unusual delays in payments or other similar irregularities are often clues that there is a problem. For example, in one department, members were told that it had "always" taken six months to get reimbursed for travel expenses. When one member was finally reimbursed, she noted that the payment was listed as reimbursement for a computer purchase rather than travel. Another member found that the amounts in the account used to deposit his consultation and other outside income was less than he had calculated. He was initially advised that the shortfall was due to the institution's delay in processing checks, but he was not satisfied with the explanation and requested an audit. The audit confirmed that a trusted employee had been embezzling funds, an action facilitated by lack of appropriate financial controls in the department.

In another recent incident, a secretary stole $1 million from a school district foundation. She wrote checks to her own charge card accounts. Those reviewing expenditures had checked only the check numbers to see that all had been processed and therefore had accepted the secretary's statements as to what the expenditures were for [11]. If something does not seem "quite right" in the financial arena, it is wise to have an outside evaluation of financial transactions and processing.

PREPARING FOR "FALLOUT"

In contrast to environmental fallout with the attendant health concerns, *budget fallout* refers to additional funds that may unexpectedly become available. Budget fallout may result from delays or cancellation of a program, revenues greater than projected, or similar events. Departments within an organization may then be queried regarding possible use of the unexpected additional funds. Approved funding using the additional revenues is frequently granted with the stipulation that they be used quickly, usually by the end of a fiscal year. The manager who has planned ahead may be able to capitalize on these situations. If major health and safety expenditures have been identified in the budgeting process, even though current funds are unavailable to cover them, the health and safety manager may be able to convince senior managers to use additional resources that become available to meet the identified needs. The senior managers are more likely to give their approval if the manager has carefully justified the expenditure in the initial budget (action on which had to be deferred due to lack of monies).

While budget parameters may specify that projected expenditures must not exceed available resources, there is usually no harm in identifying other, unfunded, requirements. During initial budget preparations, the manager may have contacted vendors to ascertain actual costs as well as the time needed for delivery of a product. By demonstrating that the extra funds can be used for a needed project within the budgeting time constraints, the health and safety manager may be in a much better position than his colleagues who have not developed similar detailed options. The senior managers who allocate the additional funds would rather see money spent on previously identified, well-justified projects than on supplies per se. For example, a parent organization realized that a particular unit had grossly overestimated expenditures for the year and thus additional funds were available. The situation was recognized in September, the month at the end of the organization's fiscal year. The health and safety director had previously justified the need for a cardiac catheterization replacement unit and had arranged for rapid delivery if funds became available. The organization transferred several hundred thousand dollars to the health and safety unit, and the catheterization equipment was in place by the end of September.

REFERENCES

1. Diorio JA Jr, Fallon LF Jr. Compensation and benefits. In Fallon LF Jr, Dickerson OB, Brandt-Rauf PW (eds). *State Art Rev Occup Med* 1989;4(1).

2. Fallon JB. The science of accounting. In: Fallon LF Jr, Dickerson OB, Brandt-Rauf PW (eds). *State Art Rev Occup Med* 1989;4(1).

3. Fallon JB. The essentials of finance. In: Fallon LF Jr, Dickerson OB, Brandt-Rauf PW (eds). *State Art Rev Occup Med* 1989;4(1).

4. McCunney RJ, Anstadt G, Burton WN, et al. The competitive advantage of a healthy work force (editorial). *J Occup Environ Med* 1997;39:611–613.

5. Anstadt GW, Lester DL, Powell BH, et al. The business planning process applied to an in-house corporate occupational medicine unit. *J Occup Med* 1991;33:354–357.

6. Anstadt GW. Contracting for occupational health service—an insider's view. *J Occup Med* 1994;36:443–446.

7. Kalina CM, Fitko J, Barthel CW. Business process design—using information technology systems in the occupational health service business plan. *Am Assoc Occup Health Nurses Journal* 1997;45:337–341.

8. Greenberg PE, Finklestein SN, Berndt ER. Economic consequences of illness in the workplace. *Sloan Manage Rev* 1995;36:26.

9. Conti DJ, Burton WN. The economic impact of depression in a workplace. *J Occup Med* 1994;36:983–988.

10. Brady W, Bass J, Moser R Jr, et al. Defining total corporate health and safety costs—significance and impact. *J Occup Environ Med* 1997;39:224–231.

11. Lyon J. Embezzler says audit gaps eased theft. *Salt Lake Tribune* June 28, 2007. A-1, A-9.

ADDITIONAL READING

Berman K, Case J, Knight J. *Financial Intelligence: A Manager's Guide to Knowing What the Numbers Really Mean.* Boston: Harvard Business School Publishing, 2006. (Designed to enable managers to understand what goes into a financial statement and read and understand income statements, balance sheets, and other financial data.)

Brigham EF, Houston JF. *Fundamentals of Financial Management* (Concise Edition with Thomson ONE-Business School Edition and TMV Card.) 5th Ed. Cincinnati: South-Western, 2006.

Finance for Managers (Harvard Business Essentials Series): *Your Guide and Mentor to Doing Business Effectively.* Boston: Harvard Business School Publishing, 2003. (A primer for corporate and small business managers to enhance their ability to impact financial planning, budgeting, and forecasting.)

Financial Management: Principles and Applications. 10th Ed. New York: Prentice Hall, 2004. (A "classic" that includes chapters on forecasting, planning, and budgeting, international business finance, and multiple other topics. Good resource for the manager with major financial responsibilities.)

CHAPTER 11

Entrepreneurship: Implementing or Expanding Health and Safety Programs

with Karl Blake and Don Marano

RECOGNIZING CHALLENGES INVOLVED IN DEVELOPING HEALTH AND SAFETY INITIATIVES

Although this book focuses primarily on management activities of in-place health and safety programs, some professionals may decide to propose new start-up programs as they go through the steps described in Chapter 3 on planning. Others may decide to create a full-time consulting program, incorporating the principles and activities described by Dr. McCunney in Chapter 18. If new programs are to be successful, it is essential that the individuals planning such efforts recognize the requirements and challenges that entrepreneurial initiatives involve.

A student in a graduate health and safety program was asked what he planned to do after graduation, approximately two weeks away. He responded that he planned to work in a part-time position that he had secured and also start formulating a health and safety consulting program, with a goal of working as a consultant on a full-time basis in two or three years. When asked how he would fund his consultant program, he replied that he really did not need funding because he already had a computer with on-line capabilities, fax equipment, and a cellular phone he could use for consultant calls. He agreed to visit a successful local health and safety consultant group to discuss his plans. After the visit, he stated how much he had learned and how surprised he was that starting a new business was so complicated.

Failure to recognize the requirements of starting a new program, whether as an expansion of a current activity or as a start-up effort, is one reason that over 90 percent of new programs fail within the first two years. Other reasons include under-capitalization, inadequate cash flow, and an inability to determine whether efforts will be profitable.

A new manager of a health and safety program was surprised to find that the occupational medicine clinic was losing money. The clinic director planned to conduct a marketing program to increase the number of patients in order to reverse the unfortunate financial situation. The manager reviewed the clinic's financial statements and found it was not readily apparent why the clinic was losing money. He then decided to meet with the billing secretary to determine expenses and profit derived from a typical patient. Since most companies prefer "single-point" billing, in which they pay one inclusive bill rather than multiple bills for evaluations and clinical procedures over a period of months, the clinic billed the company for the patients' evaluations and hospital procedures, then reimbursed the hospital. The manager went through billing for a hypothetical patient and found that the basic physician charge for each patient barely covered the overhead cost and was significantly less than other clinics' charges. The secretary reported that the clinic director had maintained the fee in order to "be competitive." In addition, the clinic had not increased charges for hospital procedures (radiographs, pulmonary function tests, laboratory studies, and similar procedures) for several years—also in order to "be competitive." Review of the clinic charges versus hospital charges revealed that the clinic was consistently charging less, in some instances markedly less, than the hospital's charges, which had increased over the years. In essence, the clinic lost money on each patient! Attempting to make up the shortfall by volume was a sure path to bankruptcy!

Avoiding such problems requires that those managing or implementing health and safety programs obtain excellent, "real-time" financial information. Many of us have not had any bookkeeping or other financial training and have had little experience with interpreting financial statements, cash flow statements, and similar information. In such situations, it is imperative that the health and safety unit hire a person with appropriate accounting or financial background or obtain necessary financial expertise by other means. Relying on a parent organization's financial support may be inadequate for a number of reasons. For example, in one organization, accounts were reconciled several months after transactions had occurred. In addition to the delayed reconciliations, periodic "roll-ups" demonstrating excellent positive balances were provided. Unfortunately, the roll-up information did not include known obligations against the accounts, such as salaries, resulting in the accounts actually having zero balances at the end of the year. If the organization had attempted to use the erroneous large positive balances shown on the roll-up statements to fund shortfalls of other units, the health and safety organization would have been seriously compromised. As is true in other situations, this example underscores the truism "misinformation is worse than no information."

OBTAINING ASSISTANCE IN PLANNING A NEW EFFORT

Health and safety managers may have little experience in planning major new or expanded programs. The situation is analogous to the discussion of accounting and other financial management aspects discussed in the last chapter. The student's "abrupt onset of reality" episode described at the beginning of this chapter highlights the limited experience and knowledge many of us have in this arena. As with financial planning, obtaining assistance from those with expertise in planning start-ups or other major efforts can be invaluable. A colleague recommended discussing the topic of this chapter with Mr. Blake—one of the co-authors of this chapter—because he had saved my colleague and his group time and money in developing a start-up health and safety program. My associate emphasized that Mr. Blake had made their entire effort much more effective and better managed than it would have been without his guidance.

In some instances, those with necessary expertise may provide some aid at no cost. However, if the effort will involve significant time, it is likely there will be charges. Quoting my colleague, "It was the best money we spent."

One approach to planning a new effort is to develop a business plan. Obtaining and considering the information required for a business plan can be of great value in defining opportunities and avoiding costly mistakes. The next section summarizes some aspects of developing a business plan.

CREATING A BUSINESS PLAN

In planning any *major* program, including start-ups, creating a business plan is important. If the endeavor requires significant additional funding from a parent organization or other sources, it is almost always necessary to prepare a business plan. Attempting to implement a major effort without a business plan can lead to disaster. Table 11-1 summarizes the content of such a plan. The following information on business plans is provided to assist the reader in appreciating the steps, content, and complexity of a plan. Recalling the concerns about financial expertise mentioned previously, it may be appropriate to obtain professional assistance in developing a business plan.

The essential components of a typical business plan are described in the following discussion.

Statement of Confidentiality

Because of the considerable time and effort involved in developing a business plan, it is essential that the contents be protected. Standard components of a statement of confidentiality are the requirement that any party privy to the business plan must have written permission to disclose it to others and a statement that because any disclosure may cause serious harm to the owner, legal action may be taken if a disclosure is not approved.

Executive Summary

This portion of the business plan may "make or break" a proposal, particularly if funding is being sought. Not surprisingly, funding

Table 11-1. Typical Components of a Business Plan

Typical Components of a Business Plan
Statement of Confidentiality
Executive Summary
New business activities or services
Mission statement
Objectives
Keys to success
Financial overview
Company Summary
Overview
Balance sheet—2 or 3 years
Services To Be Offered, or Services Delivery Plan
Description of services to be offered
Budget for required personnel, equipment, etc.
Market Analysis
Current, projected demand
Targeted marketing strategy—what population groups or segments are targeted
Professional capabilities and requirements, e.g., licensing, certifications
Competition and buying patterns, with emphasis on competition analysis
Strategy and Implementation Summary
Competitive edge
Marketing strategy—*how* will the program be marketed to the targeted groups
Sales (or revenue) forecast, including costs of producing sales (or revenue)
Management Summary
Financial Plan
Important assumptions
"Break even" analysis
Pro forma profit and loss statement
Projected cash flow
Projected balance sheet
Business ratios

sources receive multiple applications for support, and a weak executive summary may result in rapid discarding of the proposal by potential funders. Typically, an executive summary includes *proposed new business activities, or services,* such as increasing accessibility to patients or availability in underserved areas, as well as the rationale for them. The aim is to succinctly state why there is a demand for the services, what their value is to consumers, and why consumers would be willing to pay for the services or even pay a premium for the new offerings. This should be followed by a *mission statement,* then a statement of *objectives,* such as increasing the number of patient visits by 50 percent. Next, there should be a section describing *keys to success,* such as professionalism, personalized care, and repeat business and referrals, with appropriate succinct descriptions of each. Emphasis could be placed on how to most efficiently and effectively reach the customer, or consumer, and the time line for making customer contacts.

Since the plan has to "sell" the banker, venture capitalist, CEO, or other decision maker, it is important to "cut to the chase" and provide a *financial overview* that clearly demonstrates sufficient payback over time. The importance of reaching and convincing decision makers must be recognized, and the guidelines in Chapter 6 on organizing health and safety activities can assist the person proposing the business plan in determining the decision levels involved.

A technique that may be helpful in creating an executive summary is to begin only by writing portions of the plan or ideas and finish the summary after the rest of the plan is developed. This strategy has proved very helpful in preparing grants and similar funding applications, as the material in the rest of the document will facilitate deciding what aspects to emphasize in the summary.

Company Summary

An executive summary should usually be followed by a company summary, including the ownership and an overview of the organization's recent activities and financial information, with bar graphs or similar depictions of past performance regarding patients seen, gross income, and net income as well as accompanying balance sheet figures for the past two or three years.

Services To Be Offered, or Services Delivery Plan

A description of the services to be offered or services delivery plan should provide a description of proposed services. It should also include a budget for personnel ("head count"), equipment, and other costs associated with delivery of services. Cost projections should cover increased staff and equipment to respond to projected increases in the number of patients or customers and for new equipment required to stay competitive.

Market Analysis

The services section should be followed by a market analysis or "needs assessment" (see page 24), to justify the projected increase in patients seen, with a breakdown of different categories of patients. For example, the breakdown might include current and projected numbers of acute care patients, hazardous materials examinations, independent medical evaluations, consultations in-house and at plant worksites, or similar categories. Projections of increases in each category would then permit estimates of categorical and total increases in number of patients and resultant income for three to five years. A *target market segment strategy* should be included in the market analysis. This section permits descriptions of how clients come to the clinic or other service agency—of population segments that will be "targeted" for the new services. Are they self-referred, referred through contracts with companies, through workers' compensation policies, or by other means?

A crucial aspect of a market analysis is a discussion of professional, licensing, and other requirements of those who conduct similar programs, followed by a section on *competition and buying patterns*. Overall, since other health and safety programs provide similar services, a key element in purchase decisions by companies and insurers (a health and safety program's customers) is trust in the professional reputation, skill, performance, and knowledge of the health and safety professionals involved. Costs are a concern, but an important aspect is whether the users (customers) are happy with their initial and follow-up experiences. In one new health and safety program, a patient's medical record was confused with that of another patient of the same name, even though checks were supposed to prevent such a situation. The confusion came to light only when the patient presented a prescription and the pharmacist noticed that the stamped birth date on the prescription did not match that of the patient. The patient notified the company using the clinic, and the company immediately canceled the contract with the clinic.

In analyzing the competition, other aspects besides patient charges must be examined. What type of services do competitors offer? Can the new program enhance the offerings? How do competitors promote their services? Is the target market for the new program the same as the proposed program and thus subject to significant competition? What have competitors done well and what have they done poorly? For example, one new program planned to use a typical voice-mail system to route calls until developers found that another, established competitor prided itself on the fact that a "real human" would answer any call within four rings. The competitor's "human touch" could potentially overcome a number of advantages of the proposed new program due to the fact that "please listen to all options as our options have changed" irritates many potential customers to the point they hang up and try another competing organization. Aspects to consider in evaluating the competition are listed in the Table 11-2.

Strategy and Implementation Summary

The next aspect of many business plans is a strategy and implementation summary. This provides an opportunity to iterate the rationale for the program and the likelihood of attracting and keeping new users, specifically the *competitive edge* of the program. If the thrust of the new program is caring, compassionate, high-quality, personalized care of the highest professional standards that is readily available to people in the targeted geographic area, these emphases can be stressed. If all practitioners are board certified, while competitors employ providers without such qualifications, this distinction may also provide a competitive edge. In

Table 11-2. Example of Analysis of Competition

List competitor(s) and evaluate/compare them, using the following criteria. (Note if the evaluation is based upon perceptions or documentation.)

- Image (perceived)
- Caliber and ability of personnel
- Facility(ies) and location(s)
- Scope of services provided
- Capacity (growth potential), availability, and access (includes space, personnel, and supporting services)
- Market share
- Price/charge competitiveness
- Advertising/promotional effectiveness
- Financial and clinical outcomes
- Summary of advantages relative to competitor(s)
- Summary of disadvantages relative to competitor(s)

essence, the competitive-edge section is devoted to defining what makes the new program different from competitors.

The next portion of the strategy and implementation summary should address the *marketing strategy*. How will the new program be marketed to the targeted groups? Although advertisements in newspapers may be of assistance, advances in health care are such that relying only on these may be ineffectual. Other media advertisements may be prohibitive in cost, but the new program should provide, at the least, such aspects of marketing as open houses, visits to potential clients, brochures to leave during visits, follow-up calls, and effective Web pages. Visits to potential clients for health and safety services are particularly important, and the Pareto principle, as discussed in Chapter 3 on planning and prioritization, is applicable to such visits. It is essential to prioritize visits so that they are scheduled with the organizations that are likely to provide the greatest return. Leaving brochures and website URLs usually facilitates follow-ups. Some health and safety professionals may have the capabilities to design Web pages, but use of marketing and information technology (IT) professionals with extensive experience in Web design may be more cost effective. As with the need for the expertise required to perform pro forma and other financial balance sheet activities, it is very important that the program's Web page be one that will attract viewers and display the primary advantages of the new program.

Obviously, the cost of marketing has to be included in the budget for the operation. In essence, the marketing strategy has to attract patients and retain them. The next section of the strategy and implementation summary should be a *sales (or revenue) forecast*. The point in this section is to present a realistic run-down on forecasted sales. The figures below provide an example of a revenue forecast:

	2007	*2008*	*2009*
Walk-in acute care	$20,000	$25,000	$30,000
Contract patients	$25,000	$35,000	$50,000
Hazardous duty examinations	$10,000	$20,000	$25,000
Independent medical evaluations	$5,000	$15,000	$22,000

The narrative must provide the rationale for the projected increases in revenue (sales), ensuring that the information is based on reasonable estimates. Projected expenses of producing the revenue, including equipment, consumables, rent, utilities, insurance and insurance billing, and similar expenses should follow the sales projections.

Management Summary

A management summary provides the opportunity for the owners to discuss their management experience and expertise. Demonstration of a proven track record greatly aids potential funding groups in evaluating the risk of a proposed new program and helps instill confidence in reviewers that the business plan will be successful.

Financial Plan

The final section of the business plan is the financial plan. The plan should include the latest financial data available prior to the implementation of the new program and the projected profit or losses for the years after implementation. The plan may begin with *important assumptions,* such as a continued strong economy that will result in maintenance of current employment levels and new businesses that will require health care. Other assumptions could be that there will be no changes to current medical insurance payment scales, short- and long-term interest on any loans will remain stable, utilities will not increase beyond estimated inflation rates, accounts receivable will continue to be paid in full within sixty days, and similar projections.

Any new program may run a deficit for some time until new business is generated. This potential requires a *break-even analysis* as part of the business plan to show a monthly break-even amount—that is, a demonstration of how much business income is necessary to break even and begin earning a profit. A *pro forma profit-and-loss statement* that includes information for the prior two years or so and projections for out-years should follow this information. The next item is *projected cash flow.* This section will be scrutinized closely by any potential funding source, since inadequate cash flow remains one of the leading causes of program failures. Bar graphs are useful in depicting current and projected monthly net cash flow and cash balances.

The next portion of the financial plan is a *projected balance sheet.* This is developed on a yearly basis, with current and recent years followed by projections for the new effort. It includes:

Assets	*Liabilities and Capital*
Current assets	Accounts payable
Accounts receivable	Current borrowing
Other current assets	Other current liabilities
Total current assets	Long-term liabilities
Long-term assets	Paid-in capital
Accumulated depreciation	Retained earnings
Total assets	Earnings
	Total capital
	Total liabilities and capital
	Net worth

The last section of the financial plan is the *business ratios.* This includes anticipated ratios for profitability, risk, and return. Examples of entries are gross margins; selling, general, and administrative expenses as percentages of sales; profit before interest and taxes; total debt to total assets; and similar information.

This overview of a business plan is provided to demonstrate the scope of information required for an effective plan. As previously mentioned, it is often advisable to obtain assistance from a consultant or a colleague who has previously prepared successful plans.

FUNDING NEW PROGRAMS

If the health and safety unit is part of a larger organization, the manager will have to advocate funding for an expanded program. In this situation, it will be important to have the support of other senior managers, since there are always demands for additional funding from multiple groups. The new proposal must "stand out" from other requests and present a well-conceived rationale for the expanded program. As noted elsewhere in the text, there is a difference between simple "approval" by a senior manager and "support." As one student commented, "Now I understand what happened." He had proposed a new effort that would be funded by the parent organization. His immediate supervisor approved the plan, and the student began contacting personnel and making other arrangements to start the program. He was surprised when his proposal was put in the "approved but not funded" category. He realized during the class session that his supervisor's "approval" was only the first step. His supervisor had other proposed projects under consideration that were competing for limited funding from the organization. The supervisor strongly supported two other projects that took the available funds, so the student's project could not be started.

If the new initiative is a "stand alone" situation, the manager may have to obtain funding from other sources. Personal savings may be used, but it is essential that sufficient funds be available to cover business expenses until the effort begins to generate profits. Some health and safety start-ups have pooled their funds with those from a limited number of partners to obtain needed amounts. The manager may consider loans from family and friends, but such loans should be carefully structured to avoid "misunderstandings." Such aspects as repayment schedules, rate of interest charged, collateral (if any), and similar aspects should be clearly defined. Banks may be hesitant to loan funds to a new venture, so it may be necessary to turn to venture capital sources. Venture capitalists usually want part ownership in the new program. It may be possible to sell stock in the company, but those planning such an effort must ensure they comply with securities regulations and the company is properly formed or incorporated under the laws of the state where the unit is located.

One pitfall that can quickly cause cash flow problems for consultant organizations is the failure to calculate an appropriate hourly rate to charge those using the organization's services. For example, if a person is paid at the rate of $20.00 per hour, some believe that a charge of twice the hourly rate is appropriate. However, the following calculations, based on experience of an actual consulting group, demonstrate that such a "rule of thumb" approach is not correct:

Total direct payroll (DP)	$945,935	
Total indirect payroll (Costs that cannot be billed, e.g., receptionists, accountant, manager and others when not working on project)	$1,143,935	121.1% of DP
Employee benefits (Medical insurance, life insurance, bonuses, pension, payroll taxes)	$560,800	59.3% of DP
Overhead (Insurance, rent, utilities, telephone equipment rental, professional development, office supplies)	$613,800	64.9% of DP
	Total	245.3% of DP
Hourly rate for an employee earning $42,000 per year:		
Average annual base pay:	$42,000	
Direct hourly pay (42,000/2,080 hours per year)	$20.19	
Total burden (@ 245.3% of DP)	$49.53	
Subtotal	$69.72	
Plus profit at 10%	$6.97	
Total hourly rate	$76.69	

This simplified calculation demonstrates that the health and safety unit's charging twice the base hourly rate will result in losses for the company. The data also illustrate the importance of "utilization rates" in consultant organizations. Employees will not be able to work on projects and thus not be able to submit billable hours 100 percent of the time. The non-billable time contributes to the indirect payroll costs. In some groups, the utilization rate approximates 50 percent, so it is necessary to receive $2.00 for each salary dollar to cover this cost, at a minimum.

Similar adverse financial impacts may occur if there is uncertainty as to whether or not various activities conducted by the new effort will be profitable. The example of the occupational medicine clinic at the beginning of the chapter illustrates this. In another instance, a company thought that each of its services was profitable. The company planned to expand two services since they met with such high customer approval. It was only when a consultant analyzed the services that it became apparent that the two services targeted for expansion were actually losing money. Expanding these would have had disastrous effects on the company.

SUMMARY

The primary purpose of this chapter is to illustrate some of the aspects involved in starting a major new health and safety program. Small efforts may not require such detailed analyses and documentation, but recognizing, and accomplishing, requirements for larger efforts will increase the probability of gaining necessary support and developing an effective program that has a sound financial base. Despite the considerable work involved, a well-constructed business plan serves as the basis for a good strategic plan for a health and safety unit. (See the discussion of planning and prioritizing in Chapter 3.) Updating the strategic plan every three to six months will help the plan "live" and keep the organization on track to accomplish the goals and objectives delineated in the plan.

ADDITIONAL READING

Two books by Michael Gerber are particularly helpful for anyone starting a new business:

Gerber ME. *The E-Myth Revisited—Why Most Small Businesses Don't Work and What To Do About It.* New York: HarperCollins, 1995.

Gerber ME. *E-Myth Mastery—The Seven Essential Disciplines for Building a World Class Company.* New York: HarperCollins, 2007.

Other resources include:

Abrams R. *The Successful Business Plan—Secrets and Strategies,* 4th Ed. Palo Alto, CA: The Planning Shop, 2003.

Anstadt GW, Lester DL, Powell BH, et al. The business planning process applied to an in-house corporate occupational medicine unit. *J Occup Med* 1991;33:354–357.

Czisny K. Developing an occupational and environmental health services business plan. Presentation at the American Occupational Health Conference, Boston, MA, April 26–29, 1998.

Kalina CM, Fitko J, Barthel CW. Business process design-using information technology systems in the occupational health service business plan. *Am Assoc Occup Health Nurses J* 1997;45:337–341.

Leone F. Marketing occupational medicine effectively. Presentation at the State-of-the-Art Conference, Phoenix, AZ, October 19–21, 1998.

Leone FH. The physician working with the business community. In: McCunney RJ, *A Practical Approach to Occupational and Environmental Medicine,* Philadelphia: Lippincott, Williams & Wilkins, 2003.

Marcoso JR. *How to Write a Winning Business Plan.* New York: Simon & Schuster (Fireside), 1985.

McKeever M (Lisa Goldottas, ed). *How to Write a Business Plan,* 4th Ed. Berkeley, CA: NOLO Press, 1995.

CHAPTER 12

Evaluation of Health and Safety Programs and Projects

with Richard E. Johns, Jr.

Establishing goals, designing methods to accomplish them, and implementing new programs can be professionally satisfying. Similarly, defining the true nature of a problem, gathering data about it, developing alternative solutions, selecting the best solution, and implementing it can be both intellectually stimulating and satisfying.

Unfortunately, a critical component is missing from the above—that is *evaluation.* This omission is a major cause of program compromise or failure. New projects or programs can generate significant enthusiasm. However, taking the steps necessary to recognize that a program is not meeting established criteria, is behind schedule, or is not producing desired results is too often given a lower priority or is even omitted from the implementation plans. If the director does not detect the problem(s) early enough to institute appropriate corrective actions, serious difficulties may ensue.

A few years ago a health and safety program in a major metropolitan area had enjoyed a series of "outstanding" annual evaluations for a number of years. Less than a year after its final outstanding evaluation, it was described as "totally useless and a money-loser." It was not producing "desired results." The program was abruptly terminated as a result of that rating. There had been no change in the program director, health and safety staff, or program components. Patients and clients had continued to express appreciation for the high-quality service provided. Obviously, something changed, and in this instance it was a change in the organization's CEO. The CEO was responsible for evaluating the health and safety program, and the new incumbent used different evaluation criteria than had his predecessor. Unfortunately, the health and safety director had not ascertained the new evaluator's standards. Rather, the director assumed that the same evaluation criteria used previously would continue to be used. By the time the changed situation was recognized, it was too late, and all health and safety personnel had to find positions in other organizations.

If a health and safety manager does not include evaluation processes in the unit's efforts, or if the manager and supervisors do not agree on evaluation criteria, the results can be as disastrous as those described above. Even if the evaluation criteria are specified, problems may still occur. Shortly after I arrived as the new director of a U.S. Air Force hospital, I reviewed the most recent Health Management Inspection (HMI) for the hospital. I determined that the unit had done well but identified a significant shortcoming: The medications in the hospital's ambulances were outdated. I rapidly developed a checklist that the primary driver of ambulances was to complete each month. The checklist specifically required that the driver check each medication to ensure it was current. A few months later, I learned that the hospital would be undergoing a repeat HMI. A major concern of any unit undergoing an HMI was to avoid any "repeat discrepancy," since that could indicate persistent compromise of the care of ill or injured patients. I reviewed all actions taken since the last HMI and felt confident that all had been corrected. However, shortly before the HMI was to begin, I decided to check to be certain that the ambulances contained no outdated medications. I was surprised to find that the longest that any ambulance driver had used the checklists was approximately two months! Multiple outdated medications were present in the ambulances. In developing the new procedure, I had failed to recognize that the drivers were replaced by new drivers every four to eight weeks as part of a cross-training program. As a result of the failure to adequately evaluate the components of the "solution," the same problem existed. Fortunately, the later evaluation permitted correction of the situation before the HMI occurred. Of more importance, transported patients would be assured of receiving current medications, if needed, as a result of new, more effective procedures, which were subsequently evaluated monthly.

EVALUATORS OF HEALTH AND SAFETY PROGRAMS

The health and safety manager must recognize that the program will be evaluated by different groups or individuals. Although only one individual or committee may make a decision about the effectiveness and funding of a health and safety organization, that decision may well be influenced significantly by other evaluators. In considering such evaluators, it may be helpful to use a "triangle" approach. This method is somewhat similar to the "Host, Agent, Vector" aspects of the epidemiologic triangle. However, the evaluation triangle concept includes two triangles, as depicted in Figures 12-1, 12-2, and 12-3.

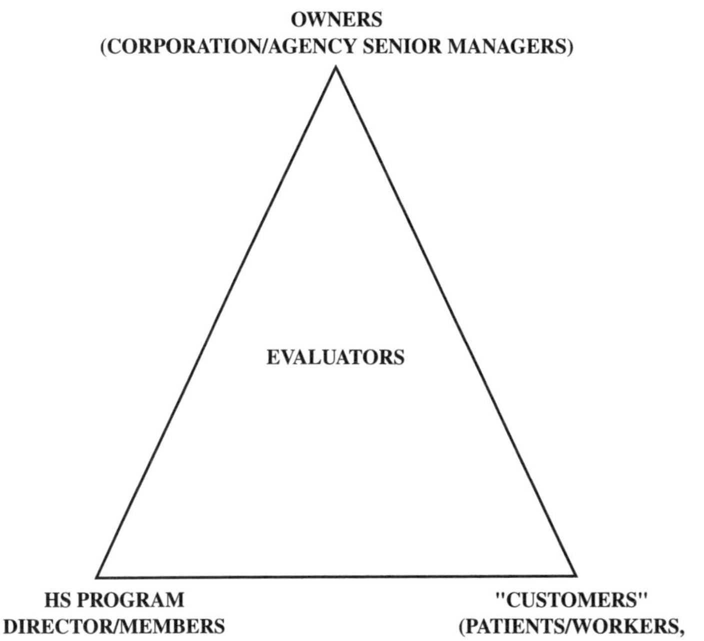

Figure 12-1. "Management Triangle" depicting some of those who evaluate health and safety programs.

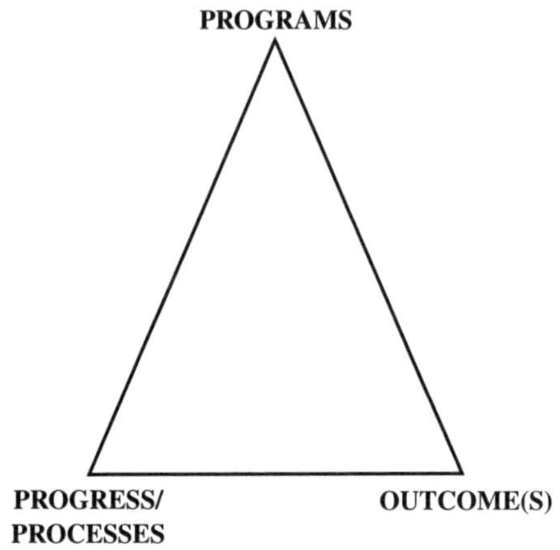

Figure 12-2. "Management Triangle" depicting some aspects of health and safety activities that are evaluated by evaluators shown in Figure 8-1.

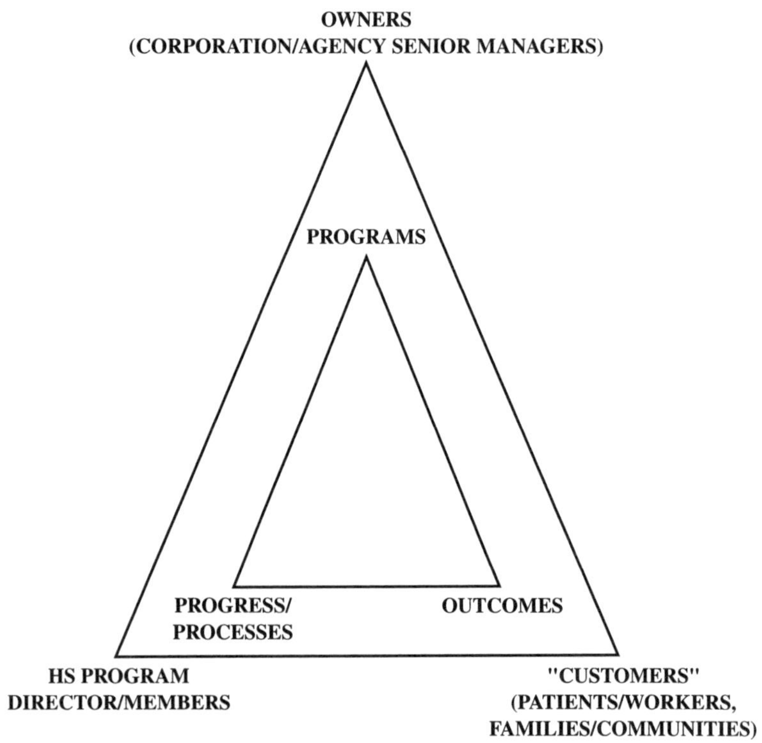

Figure 12-3. "Management Triangles" depicting inter-relationships between evaluators (larger triangle) and aspects evaluated (smaller triangle).

"Owners" of the Health and Safety Program

The first triangle, Figure 12-1, focuses on the evaluators. As noted, these include the "owners" of the program, as manifested by the health and safety director's superiors or oversight committee. In this regard, and as illustrated by the summary of the "useless" health and safety program, it is essential that the health and safety manager and the supervisors responsible for the evaluation agree on the criteria to be used. The director also needs to recognize that a significant change in the owner group may necessitate a change in the evaluation standards. The health and safety manager must ascertain the need for such changes as soon as possible after the owner composition changes.

Health and Safety Program Director and Members

The second group of evaluators includes the director and members of the health and safety organization. It is, of course, appropriate for the director to evaluate the program on an ongoing basis. However, the director may not fully recognize that each health and safety team member is also evaluating the program in some manner. Evaluation by program members can have a significant influence on the other evaluators, depending on how members' evaluations are communicated. A sense of pride and enthusiasm on the part of those working in a health and safety group is conveyed verbally and nonverbally to all of those using or interacting with the organization. If opposite sentiments pervade the health and safety unit, they will usually be transmitted, unfortunately, to an even wider group of people. It should be noted that the wise director will recognize that evaluations by the health and safety staff or support members often have as much, if not more, impact than those by health and safety professionals.

Customers

The final, and particularly important, group of evaluators depicted in Figure 12-1 are the customers, or users, of the program. In many health and safety programs, users include all the worksite employees, groups of citizens, or other populations served by the health or safety components of the program. The users may also include those with illnesses or injuries who receive care as part of the health and safety program. Typical terms used to identify the ill or injured individuals include *consumers, family units, covered lives, clients*, or, as preferred by this author even though the term may be considered outmoded, *patients*. The term *patients* has usefulness if health promotion or other services are provided to "well" individuals, since its use facilitates divisions between those requiring clinical care and those requiring other services. In addition to these "users," there may be other company managers, particularly line managers, whose personnel obtain support from, or interact with, the health and safety unit. Political, legislative, academic, and public leaders in the communities or regions served by the unit may also evaluate the effectiveness and competence of the health and safety group, particularly if environmental or health promotion and disease prevention activities are prominent in its health and safety activities.

Judicial and Government Administration Members

If the health and safety program is concerned with workers' compensation, return to work, recommendations under the American with Disability Act, or similar administrative or legal activities, members of the judicial and/or government administrative system will likely also evaluate the health and safety program. The process charts developed by Dr. Johns (Figure 12-4) are intended to assist in developing similar processes to assure reviewers and evaluators that proper processes were followed in developing recommendations or taking actions, ensuring that decisions are based on a logical process that applies equally to each person and are not made capriciously. The first flow chart is a typical one for managing ADA (Americans with Disabilities Act) and fitness-for-duty concerns, the second focuses on workers with performance issues, and the third is used by Workability Centers, LLC.

EVALUATION COMPONENTS

In addition to acknowledging the diverse evaluators of the health and safety program, the director should also be aware of the areas used for evaluations. Some typical areas are depicted in Figure 12-2 as follows.

The Health and Safety Program Itself

The overall health and safety program is often evaluated with respect to its goals, structure, member expertise, and program

Process I: ADA Disability and Fitness-for-Duty Case Management Model

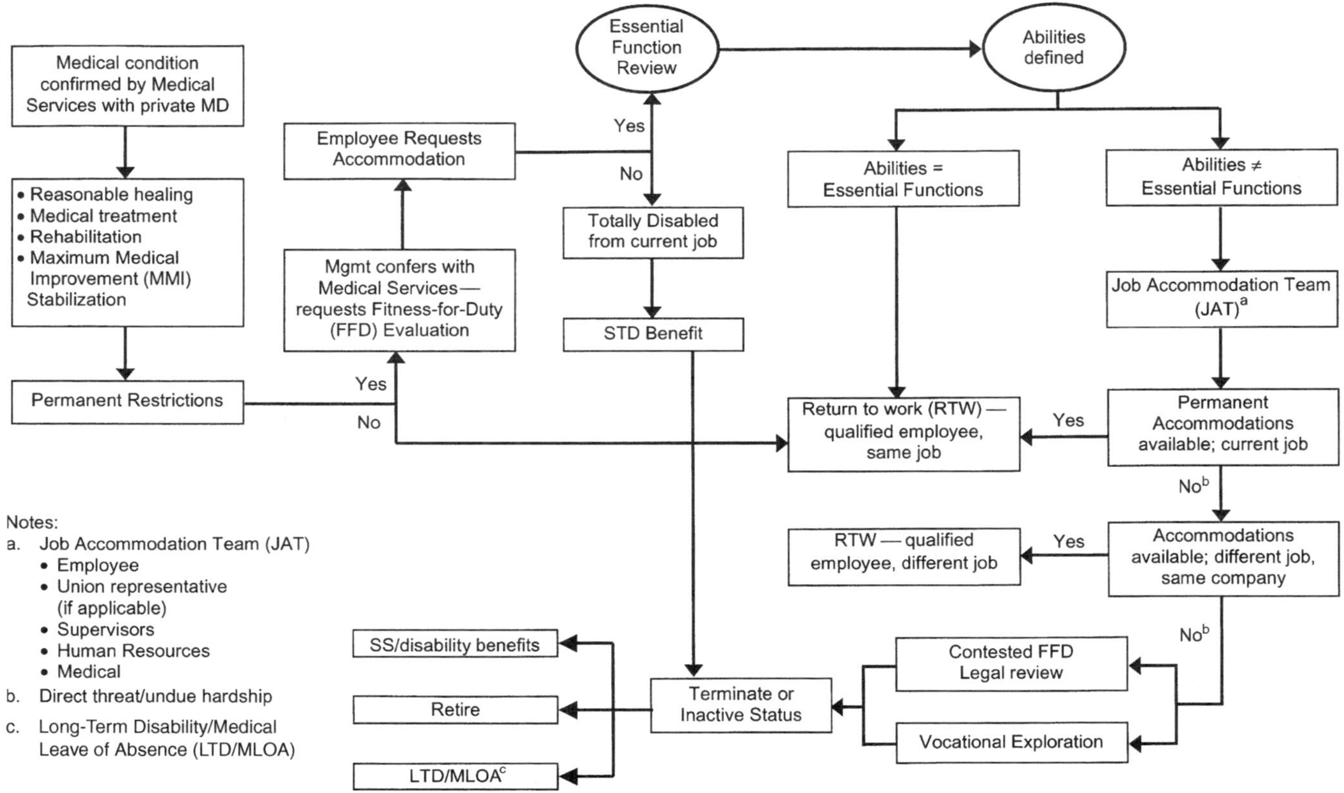

Figure 12-4. Process flow charts to assist in uniform management of Americans with Disabilities and return to work (RTW) considerations. (Courtesy of Richard E. Johns, Jr.)

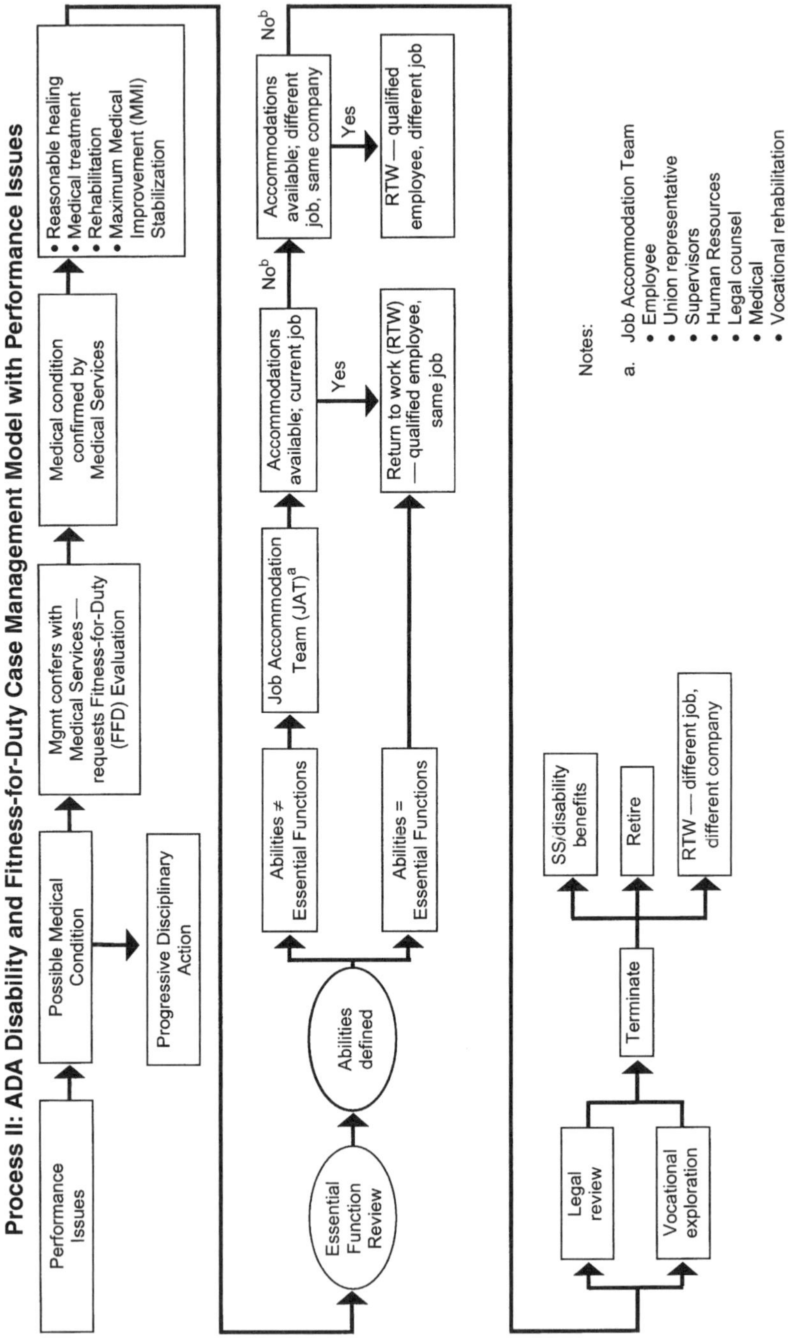

Figure 12-4. *Continued*

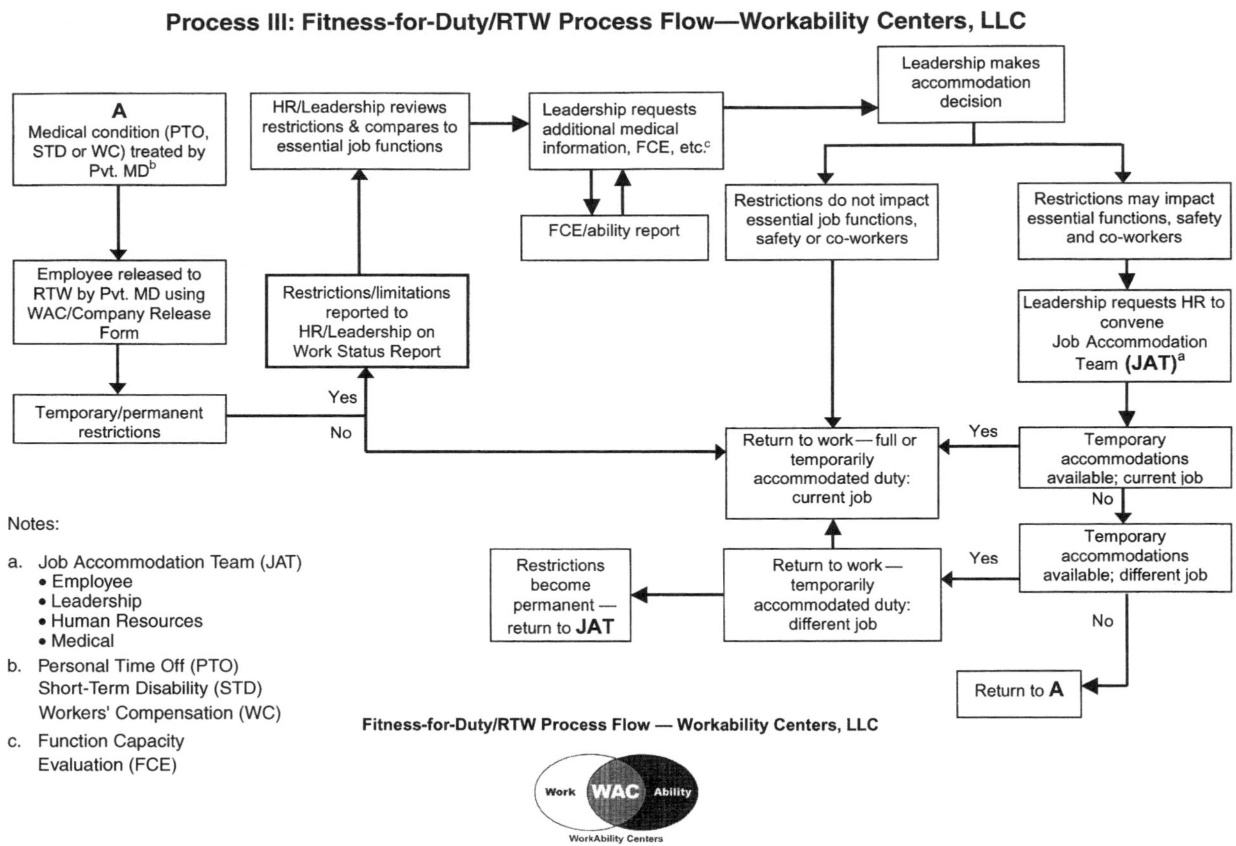

Figure 12-4. *Continued*

content. The scope and perceived quality of health and safety activities are also typically considered. The director may find it necessary to take a "marketing" approach to help ensure that the diverse components of a comprehensive program are appreciated by evaluators. It has not been unusual, as a result of such an effort, to find that others have a new appreciation for activities stipulated by regulatory or legal requirements that the evaluators previously considered as only a "nice to have" aspect of the health and safety program. Similarly, the director should take advantage of opportunities to show the relationship of the program's components to the vision, mission, and goals of the organization.

Progress and Processes

Another area of evaluation is the progress in accomplishing a program or project as well as the appropriateness of the processes involved. Evaluators typically downgrade a program if the products or results are consistently late or behind schedule, even if the final result is a high-quality outcome. Difficulties in prioritization, as well as poor personal and staff time management, are often major contributors to delayed results. Poor organization may also be a factor. The final portion of this chapter describes one method, the Program Evaluation and Review Technique (PERT), that many have found useful in tracking project progress, finding out early if problems are occurring so that the manager can take action to get the effort back on schedule.

Outcomes

Outcomes are a major, and often the most important, evaluation factor. Do the results achieve the desired objectives? As noted by McCunney, one important measure of a health and safety program's success is its impact on the rate of work-related illnesses and injuries [1]. In some health and safety programs, outcomes may be measured by patient or other customer satisfaction scores on surveys, absence of medical complications, responses to emergency situations (see Chapter 4), or other aspects of the programs. Additional outcome metrics may include cost per service provided; return on investment; absence of OSHA, Medicare, or other regulatory citations; perceived effect on employee or other population morale or turnover; reduction in accident rates; and influence on health or safety-related litigation costs. In many instances, the health and safety director is expected to quantify the benefits of the program, providing approximate dollar values for claimed reductions in workplace illnesses or injuries, cost benefits from childhood immunization programs, impacts on workers' compensation or health care insurance costs, or other aspects. As noted in Chapter 15, such quantification is particularly significant if an organization is attempting to determine whether it is more cost effective to provide services in-house or by contracting with an outside organization. In such situations, it is essential that the manager delineate all the benefits, as well as costs, of the different program approaches. (See Chapters 10, 15, and 16.)

Besides outcomes that lend themselves to quantification, there are other outcomes that are more difficult to quantify. These include areas such as quality and patient (user) satisfaction. Indirect indicators, such as low levels of complaints by users, may provide some useful information. Techniques to evaluate different aspects of the health and safety program are discussed in the next section.

Figure 12-3 depicts the interrelationships between the evaluators and the criteria used for evaluation. Any group of evaluators may use any combination of evaluation indicators to establish the group's overall evaluation of the health and safety activities of its members. The important aspect of this consideration of evaluators and what is evaluated is for the manager to be certain to recognize the significance of what the different evaluators are evaluating and convey appropriate evaluation information to each of those doing the evaluations. Failure to accomplish this may result in the unfortunate outcome described at the beginning of this chapter.

SELECTED EVALUATION TECHNIQUES

Audit

An approach that some health and safety groups, and particularly occupational health and safety groups, have taken in evaluating their programs is to have an "audit" of the program by an in-house or consultant organization. This is different from a typical financial audit, although financial management may be considered in the audit process. The auditors look at "indicators" or "criteria" to evaluate how well the health and safety program is accomplishing its functions. An example of audit criteria is provided in Table 12-1. Examples of actual audit instruments used to evaluate some occupational and environmental health and safety (OEHS) components are provided in Appendix C. Development of criteria, or indicators, for such an audit may best be accomplished as a joint effort of the health and safety organization and the one doing the audit. In some circumstances, it may be decided that standards that have been independently established may be most useful. Ideally, all will agree before the initial audit that the results will serve as a baseline for improvements to be evaluated by follow-up audits. If the initial audit is to be used for disciplinary or other adverse per-

Table 12-1. Example of Health and Safety Audit Criteria (selected portions only—see Appendix C—for complete examples)

Program Audited: HazWopper (29 CFR 1910.120)

Program Criteria	Status C/NC/O/NA[a]	Comments
1. Specific hazardous substances (HS) identified by Safety/Industrial Hygiene		
2. Medical surveillance inclusion criteria defined:		
a. All employees exposed to HS at or above specific HS PEL's; or		
b. Above PEL's for substances, without regard to respirators, for 30 days or more per year; or		
c. All employees who wear respirators for 30 days or more or as required by CFR 1910.134		
3. Content of the medical examination made available to employee as determined by examining physician		
Overall compliance rate: (#C minus #NC divided by #C) = #%		
Recommendations:		

Signature of medical auditor | Date

Signature of plant medical services representative | Date

[a]C = compliant; NC = non-compliant; O = optional; NA = not applicable.

sonnel actions, it should not be surprising if the validity and objectiveness of responses are jeopardized. The concept of baseline and periodic audits can be used in any health and safety program, although the components of the audit instruments will be changed to document the particular activities of that organization.

Surveys

In evaluating health and safety programs, surveys may be illuminating. The health and safety director should recognize, however, that if a scale is used in a survey, it may provide biased or misleading information. This holds true even if the survey is professionally designed. It is not possible to consider all aspects of survey design in this chapter; however, if a "basic, low-cost" survey is used, the chance of bias is particularly significant. For example, providing a five-item rating scale (unacceptable, far below average, good, excellent, outstanding) tends to "force" a good or better rating. Past experience suggests that few evaluators rate any health or safety service as unacceptable and many will hesitate to rate "their" organization as far below average. The difficulty can be compounded if the report is "lumped" by noting that "__% rated the health and safety services good, excellent, or outstanding." Ignoring the problems in attempting to define the difference between excellent and outstanding "lumping" the three top ratings can create the impression that "many" gave the services excellent or outstanding ratings even if the majority of respondents provided only a "good" rating. In spite of these concerns, properly designed and utilized surveys can provide information that may indicate both levels of satisfaction and trends.

Program Evaluation and Review Technique

One method that has proved to be particularly beneficial in evaluating the progress, or status, of program projects is the *Program Evaluation and Review Technique (PERT)* [2,3]. Although the actual PERT process requires certain specific steps, the term is often used generically to indicate one of many tracking processes to evaluate progress. Other terms, such as *Gantt charting* or *milestoning,* may also be used to denote the use of an evaluation process based on PERT principles. Similarly, *project management* may be used to indicate application of a tracking system similar to PERT.

PERT Components. In the PERT process, the manager breaks a project or major objective into the specific tasks needed to accomplish the project. Time needed for each task is estimated. The tasks are then placed in a logical, sequential network over time so that steps dependent on earlier actions can be identified. These efforts to define every necessary step and estimate time requirements are some of the most valuable aspects of the PERT process [3]. A disadvantage is that a typical PERT process does not identify the resources required for each step, but an earlier MBO approach to developing the overall objective alleviates this limitation somewhat [2] (see Chapter 5). In depicting sequencing, "looping" must be avoided—that is, no down-stream (later) event should be necessary to initiate an up-stream (earlier) step or action. A PERT, or similar approach, will define a "critical path." This path is usually defined as the series of tasks or activities that take the longest time from a project's beginning to end. It may also be considered the path which identifies target times that must be met in order to complete the project on time.

Table 12-2 illustrates the tasks that were necessary in accomplishing a health and safety audit of a microelectronics facility. It is based on recent experience in conducting such audits. This information can be used in conjunction with the symbols shown in Figure 12-5 to develop a PERT chart, as depicted in Figure 12-6. For the sake of clarity, connections between activities and events are not shown in Figure 12-6. Rather, the relationships are depicted schematically in Figure 12-7, which represents a more typical PERT chart.

A number of computer programs are available to assist in constructing event and time schedules. Once the appropriate chart has been developed manually or by computer, it gives the project director a method to track progress in accomplishing the effort. If one activity slips, the overall impact on the project is evident. For example, if contract negotiations to begin the audit depicted in Table 12-2 and Figure 12-6 slip, it will either be necessary to compress other activities or extend the projected completion date by a comparable amount. Similarly, if an activity, such as review of the OSHA logs, does not begin on time, the effort may still be com-

Table 12-2. Actions Needed to Accomplish a Health and Safety Audit

Initial meeting with contractors to define scope of audit, time frame, reports required, and similar items
Develop proposal for audit
Obtain proposal approval by contractor
Accomplish literature search
Accomplish in-depth review of pertinent literature
Review plant OEHS data
Accomplish IH portion
  Accomplish plant surveys (2)
  Evaluate sampling procedures
  Obtain any additional samples required
  Analyze results of plant survey, sampling, and literature review
  Prepare IH report
Accomplish OM/OHN portions
  Accomplish plant survey
  Review medical data, OSHA 200's, etc.
  Evaluate and analyze incidence/severity data
  Evaluate emergency response capabilities
  Evaluate worker and community right-to-know programs
  Analyze data and results of literature review
  Prepare OM report
Accomplish Safety and Ergonomics portion
  Accomplish plant survey
  Evaluate safety policies
  Evaluate accident reports/trends
  Delineate safety hazards
  Define ergonomic concerns
  Analyze data and results of literature review
  Prepare Safety and Ergonomics report
Prepare final report incorporating information in discipline-specific reports

pleted on time if the review is done faster than initially planned. This will still permit the information from the OSHA log review to be available at the originally projected time.

PERT Applications. As time passes, the manager can document both progress and problems in meeting the initial schedule. Figure 12-8 depicts input to the PERT chart as the actual project proceeded; it shows that slippage in obtaining proposal approval caused slippage in other areas, such as beginning the plant surveys. Nevertheless, the two industrial hygiene (IH) and one occupational medicine/occupational health nursing (OM/OHN) plant surveys

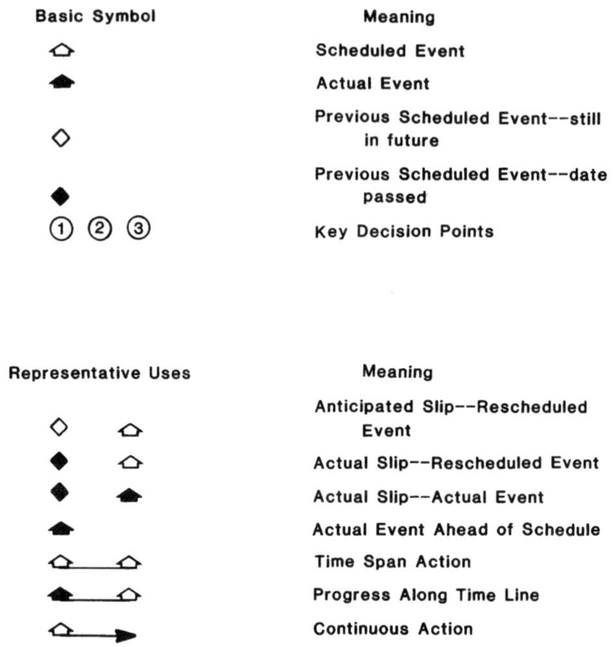

Figure 12-5. Examples of symbols used to indicate schedules for activities. These symbols can be used to develop a PERT chart, as shown in Figure 12-6.

were completed on time. A slippage in beginning sampling procedures also did not preclude finishing that effort on time. In these situations, identifying the problems early allowed the health and safety team members to adjust their schedules so that it was still possible to complete the tasks on time. Although preparation of the OM/OHN report began on time, slippage in completing it delayed the onset of preparation of the final report. Thus, the time to write the final report had to be shortened if the original completion date specified in the proposal was to be met. Significant extra effort was consequently required to complete the final report by the projected due date.

As illustrated in the prior discussion and in Figure 12-8, a particularly important advantage of tracking progress by PERT or a similar procedure is that it helps identify problems early enough to take corrective steps to put the effort back on schedule. Thus, if a particular project requires a fixed number of physical examina-

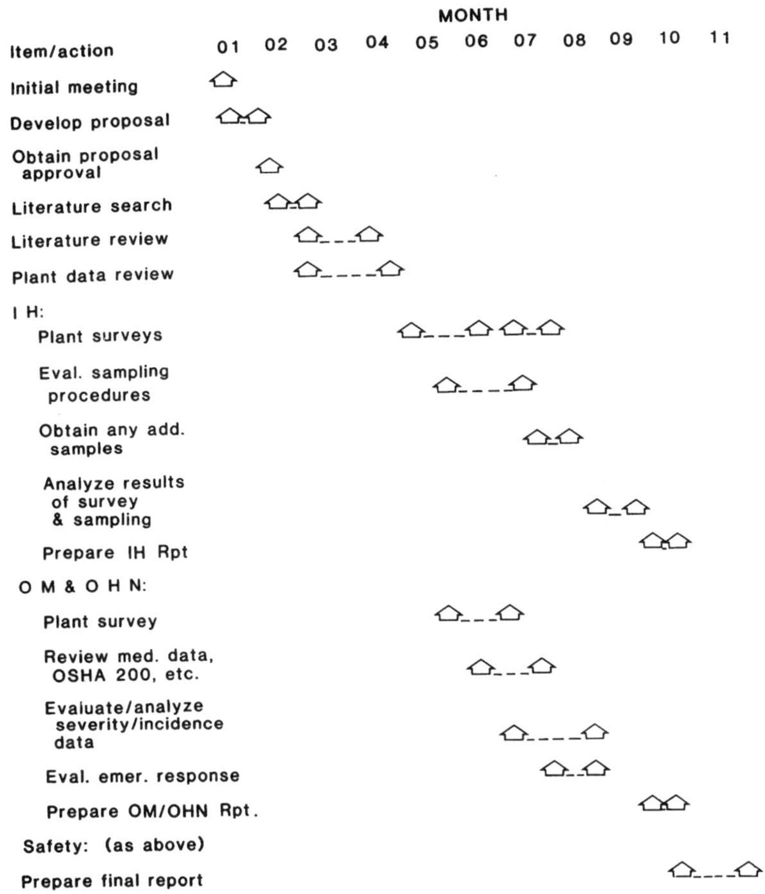

Figure 12-6. PERT (milestone) example. This figure depicts the projected start and completion dates for the components of a health and safety audit of a manufacturing facility. It delineates desired progress and establishes times that activities must be completed if the total effort is to be accomplished on time.

tions, interviews, or safety surveys each week, the manager will be alerted to problems in achieving the necessary numbers as they arise rather than be faced with the problem shortly before the results are due.

As with MBO, PERT takes time and effort. Although it is a logical follow-up to the goal planning and MBO processes, because of

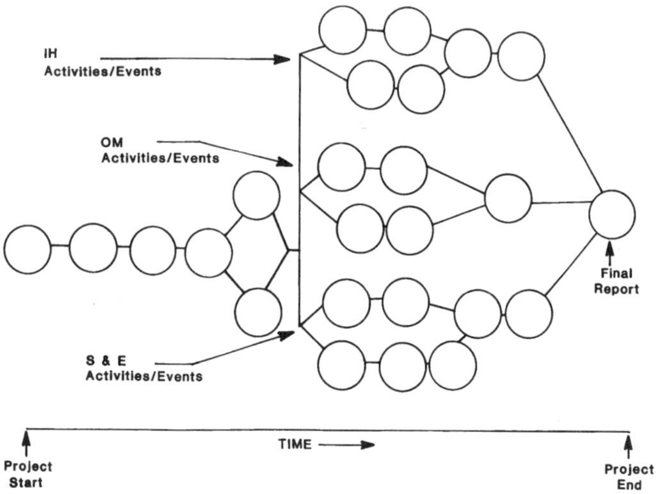

Figure 12-7. Schematic PERT chart of activities and events depicted in Table 12-2 and Figure 12-6.

the time required to develop a PERT chart, it obviously is not appropriate to use the process for every management activity. However, it can be quite valuable for major efforts and can help the manager get a project done on time and achieve quality results. At our center, use of a modified PERT approach has been essential in preparing major competing renewal applications, such as the one for the NIOSH Education and Research Center grant. The application submitted in July 2006 was over 3,500 pages. The application received highly favorable reviews, in part because faculty scheduled section completion times so that (1) the contents of the entire grant could be properly coordinated and (2) discussions and comments throughout the grant complemented each other. Such accomplishments add credibility to the entire health and safety program, which may help the manager obtain continued support for current and future activities.

OTHER EVALUATION APPROACHES

There are numerous other evaluation techniques that may be company- or organization-specific, but it is absolutely essential

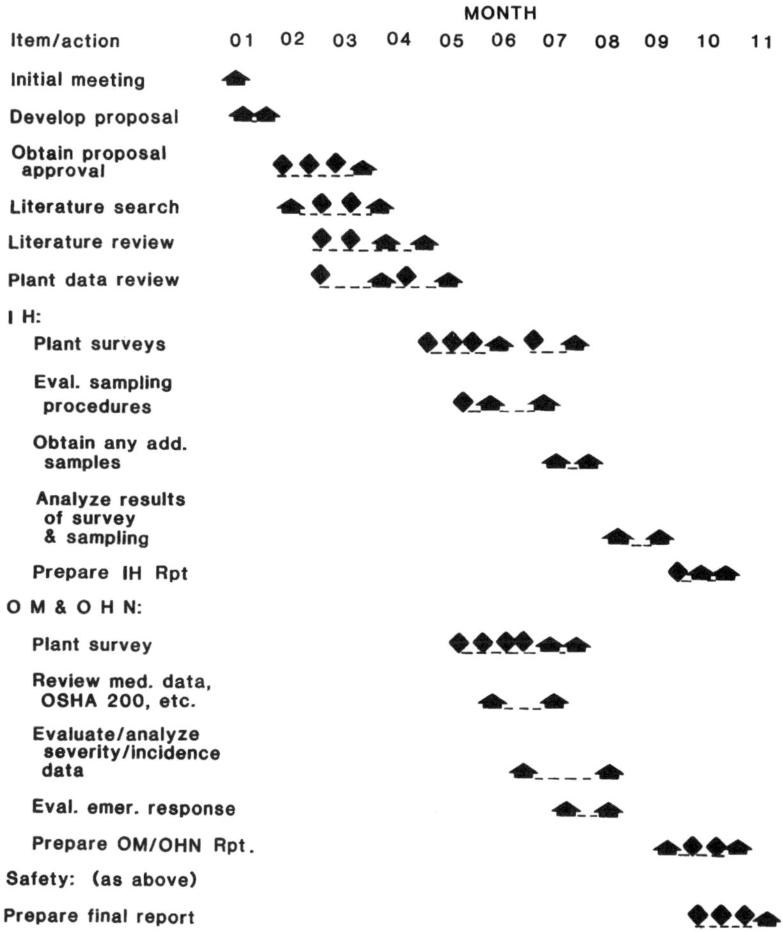

Figure 12-8. PERT (milestone) example (actual progress). As items are accomplished, symbols are filled in. Diamonds indicate actual slippages from projected times. Arrows indicate actual start and stop times.

that the health and safety director and the director's supervisors agree on the evaluation methods and content. The director should remain in close communication with evaluators to ascertain any changes in evaluation criteria in time to make any necessary program adjustments. Such interactions help avoid surprises that could jeopardize support or even continuation of the health and safety program.

REFERENCES

1. McCunney RJ (ed). *A Manager's Guide to Occupational Health Services.* Beverly, MA: OEM Press, 1996.

2. Fallon LF Jr, Brandt-Rauf PW. An overview of management. In: Fallon LF Jr, Dickerson OB, Brandt-Rauf PW (eds). *State of Art Rev Occup Med* 1989;4(1).

3. Drucker PF. *The Effective Executive.* New York: HarperBusiness, 1966. (HarperBusiness edition published 1993).

ADDITIONAL READING

Lewis JP. *Project Planning, Scheduling, and Control: A Hands-On Guide to Bringing Projects in On Time and On Budget,* 4th Ed. New York: McGraw-Hill, 2005.

CHAPTER 13

Leadership: What Works and What Doesn't

with Kurt Hegmann

LEADER, MANAGER, OR BOTH?

In addition to having management responsibilities, many managers are or will be in leadership positions. Managers who assume leadership roles will likely incorporate their experiences in working for other leaders, provided they profited from seeing what worked and what didn't work for these leaders. Multiple texts that focus on leadership are available, and these can be of assistance to one who assumes a leadership role. Although the qualities of a good leader are complex and difficult to define, an "I know a good leader when I see one" approach to understanding leadership is insufficient. Comprehension of basic leadership principles and common traits of successful leaders is essential.

First, it must be recognized that there are differences between being a leader and being a manager. A leader leads the organization, provides the vision for the group, promotes group enthusiasm to accomplish goals and objectives, and focuses on the "big picture." A manager implements plans, allocates resources and people to accomplish goals and objectives, coordinates activities, and tracks progress. Ideally, an individual in a management position is both a good leader and a good manager, but difficulties may occur if an individual in a leadership position is good in one area and not the other, particularly when the individual does not recognize the deficiency(ies). An excellent leader who is unusually successful in motivating employees but has no management interest or skills may experience total program failure unless others accomplish management requirements for the leader. This situation has been the underlying cause of some new companies' failure in spite of superior products and dedicated work forces. Although some leaders may not be effective managers, there is increasing emphasis on the need for managers to also be leaders. Recent management publications, such as those by Deming [1], Covey [2], Collins [3], and others stress the need for good leadership if the organization is to be effective.

Deming, the author of the "total quality management" concept, notes that the job of a manager is to provide leadership [1]. In their book *A Passion for Excellence: The Leadership Difference* Peters and Austin similarly stress the importance of leadership in making corporations, or activities within businesses, successful [4]. In a later publication, Covey notes that, whereas managers work within a system or paradigm, leaders focus on new systems, efforts, and paradigms [5]. A leader is concerned with deciding what needs to be done and then doing the right things, whereas a manager focuses on doing things right. Coates [6] observes that a good leader never loses sight of the big picture and has the commitment necessary to grasp sufficient details, create and communicate the leader's vision or purpose, and instill a sense of urgency in all members of the organization to achieve its purpose. Coates, along with Drucker [7] and many others, also emphasizes the importance of motivating and supporting workers as they turn visions into realities. A health and safety manager's failure to provide leadership can result in a stagnant organization that just continues to do what has always been done, causing a reduction in the relevance and effectiveness of its efforts. In time, such an organization may become totally superfluous, with neither those served by the health and safety program nor the parent organization realizing the benefits that an effective health and safety program *could* provide.

DEFINING LEADERSHIP CAPABILITIES

Various studies have underscored the difficulties in defining what makes a good leader. For example, the military services have conducted extensive studies to try to define the qualities that enable a pilot to be a good leader in combat flying. Although criteria could be established that would predict success in completing a flight training program, the investigators were unable to define the qualities that would consistently result in leadership success. Some leaders who demonstrated bravado and risk-taking proved to be good leaders, but much more conservative flyers also became highly successful leaders. Other individuals, with traits that were apparently identical to those of proven leaders, were failures as leaders.

Compounding the difficulties in defining leadership capabilities is the fact that an effective leader in one setting may not be able to succeed in another one. In World War II, General George Patton was a highly effective leader in combat. He reportedly let nothing stand in the way of getting the job done, whether it was the enemy, the U.S. military bureaucracy, or senior members of the Allied Forces. However, it is doubtful that his style of leadership and management would have served him well in a peacetime position, such as a corporate chief executive officer (CEO) or a president of a university. Certainly, Patton had charisma. However, Drucker emphasizes that charisma is not a requirement for leadership, noting that other highly effective leaders, such as Dwight Eisenhower, George Marshall, and Harry Truman did not possess any more charisma than a "dead mackerel" [7].

LEADERSHIP STYLES

Some have attempted to define leadership capabilities by focusing on different styles of leadership, according to the following categories:

1. *Autocratic:* You know where you stand with such a leader, also known as a "my way or the highway" type of person. There are very rare (typically brief) situations in which this style may work, involving resolving prior ineffectual management disasters, provided the autocrat has good solutions. There are situations in which all managers will be required to make immediate decisions, e.g., the actions required to solve an urgent problem, such as a workplace violence issue; however, the overall management style for a successful manager should not be

autocratic outside of these kinds of brief interventions. This style works poorly for health and safety and other professionals, since it stifles creativity and leads to poor morale.

2. *Bureaucratic:* This type of leader remains removed from hard decisions and exhibits a focus on procedures, resistance to change, and inflexibility. A professional will find it difficult to work for such a leader unless the individual can circumvent or manipulate the bureaucracy. At one Strategic Air Command (SAC) base where I (*RM*) served as the health and safety manager, flight surgeons had been repeatedly denied permission to fly in the primary aircraft, with the excuse there were no extra seats available, even though it was known that seats were often unoccupied. I met with the chief of operations to again ask to schedule flight surgeons to fly, since such experience was essential in establishing the rapport with crew members needed to provide appropriate medical support. When my request was again denied, I replied that I understood but would have to report that we were in violation of a SAC supplement that required such flights. When asked, I provided the directive that I "happened" to have with me, upon which the operations chief gave his permission for the requested flights, with the comment "Oh, geeze Doc (or words to that effect), you got me!" The bureaucratic style of leadership is never successful in occupational health and safety units and should be avoided. Occasionally, in large organizations, it is "successful" for a time—that is, to "tread water" and merely attempt to survive.

3. *Laissez-faire:* After agreement on desired outcome, this leader gives large amounts of independence to the organization's professional members. This style works well with self-starters who can work independently and with many academics. This leadership style fails if: (1) the organization needs a change in direction, (2) the supervisee(s) require directives, and/or (3) there are problems with individuals completing tasks. In any of these settings, direct managerial intervention is required.

4. *Democratic:* This leader develops mutual trust among the organization's professionals. This type of leader understands and embraces the increasing emphasis on participatory approaches

to management. However, this orientation can be quite time consuming, as it requires numerous formal and informal meetings. In addition, it may fail if there is a need for direction and the supervisees are thoroughly split in their opinions, or they are headed in the wrong direction.

SOME LEADERSHIP REQUIREMENTS

Although it is difficult to define specific capabilities that a leader must have, a number of studies have suggested common threads in the characteristics of effective leaders as well as certain skills that are particularly beneficial to leaders. The presence of these do not ensure a leader's success or failure, but they appear to help. Some of these traits and skills are listed in Table 13-1.

The following discussion considers some important leadership capabilities in more detail.

Providing a Vision

As discussed in Chapter 3, providing a vision of where the organization will be in five or more years is an important motivating

Table 13-1. Leadership Skills and Characteristics [2,4,8]

Conceptual skills—thinks creatively and strategically, works toward visions of what should be
Communication skills—clearly communicates what is to be done (relevant vision, goals, and outcomes) and what standards are expected; tactful; persuasive
Human skills—able to motivate colleagues and team members to a common purpose; delegates effectively
Facilitator—provides resources required by others; removes obstacles; strongly supports associates and subordinates
Confident—demonstrates personal confidence as well as confidence in team to get job done
Focused—focuses on first things first; doing right things right
Decisive—makes decisions with imperfect data without delay
Knowledgeable—possesses knowledge of the business and primary activities of the organization
Driven, committed—never gives up, strongly committed to successful outcomes, sets work example
Organized—personal as well as business activities
Tolerant to stress—adapts to continually changing demands
Assumes responsibility—accepts responsibility for outcomes but shares successes
Trustworthy—reliable, consistent, manifests honesty and integrity in all dealings

factor for the health and safety group. It is essential to have all the members of the health and safety unit involved in establishing the organization's vision. However, the leader needs to demonstrate conceptual ability in guiding the unit in developing a vision that will be attainable, generate enthusiasm among the organization's members, and be accepted and supported by senior managers supervising the health and safety leader. Ideally, the vision should also generate enthusiasm and respect among the health and safety unit's "customers" and others concerned with its activities. A wise leader recognizes that not only must the vision process be participatory but also that sufficient time must be devoted to the process to underscore its importance.

Communicating Effectively

Regardless of the leadership skills and qualities an individual leader may have, it is essential that the leader recognize the importance of effective communication. Successful leaders invariably have stressed that such communication was responsible for 90% or more of their success. As discussed in Chapter 9, a graduate of our academic program who was leading a health and safety program reported that her organization did not seem to be functioning as well as she would like, even though she had told the group what her vision and priorities were. Remembering the stress on communication in her graduate program, she had a meeting of her professionals to ask them what they saw as their primary emphases and how they should be proceeding. She was surprised to find multiple answers, many that were quite different from the thrusts she thought that she had communicated. After agreement on priorities and efforts, she commented that her group was functioning like a "well-oiled machine." As Blanchard and Johnson stressed in *The One Minute Manager,* people must know what is expected and what good performance is [8]. Communication is discussed more extensively in Chapters 20–22.

Setting Standards

An effective leader must also set standards for the organization. The manager should participate in establishing performance and ethical standards for that activity. This aspect involves setting examples for others. Accepting poor-quality work, such as poorly organized reports or letters with typographical errors, will give the appearance to other members of the unit that such work is acceptable, and those dealing with the organization will likely believe that the poor work reflects the low quality of its professional activities. Permitting some members to perform at less than expected levels (assuming there is no underlying health or other problem involved), will result in a mediocre organization. Conversely, establishing appropriate expectations for quality and performance will engender the respect of those the health and safety organization serves as well as the members of the organization.

The need for quality has been underscored by Deming and others. Deming's observations, which are credited with revitalizing Japanese industry after World War II, have led to the concepts *total quality control* and *total quality management* [1]. The strategies inherent in these concepts are designed to detect problems at their source, rather than when the customer uses an item, and to provide rapid responses to anything that may jeopardize customer satisfaction. Peters and Waterman have similarly documented the importance of quality in making companies successful [9]. Recent emphasis on the *Six Sigma* process reflects the importance of quality in an organization's activities [10]. How well the health and safety director satisfies program "users," or "customers," by providing quality products, can determine how successful the health and safety program is. (See Chapter 25.)

Establishing Integrity Standards

Integrity is of paramount importance. "Your word is your bond" in highly effective companies, as Peters notes [4,9]. Once trust is lost, it is usually impossible to regain it. In one organization, the leader announced that a subordinate unit was to receive a major budget cut due to a new budgeting process imposed by the senior administration of the parent company. He noted that he had no choice because of directions he had received, and the new budgeting process was a "black box" that he could not access. Subsequently, the director of the subordinate unit met with the senior administration of the parent organization and was told that the budgeting formula was readily available and that the formula would have allocated significantly more funds to the unit than the leader had designated. Additionally, he discovered that the leader who had announced the cut had total flexibility in allocating funds to the various subordinate units in the organization. This experience, combined with similar other events, resulted in the subordinate unit director's distrust of the leader and his staff and had a demoralizing impact not only on the members of the leader's organization but also on others who learned of these events.

As noted in Chapter 7, it is important for a leader to "begin with the learner" in ascertaining *why* a problem is occurring rather than to "automatically" blame the individual involved. Otherwise, the perception of the leader's fairness and integrity may suffer. A leader's shifting the blame rather than admitting a mistake poses a greater threat to this perception. Even relatively minor incidents can demonstrate to an organization's members that the leader does not have integrity. For example, an administrative assistant had repeatedly asked her supervisor if he wanted her to print some pictures that she knew would be needed for an approaching event. When the supervisor's wife called him to find out why she still did not have the photos he had promised, he replied that the administrative assistant had been "slacking off." The assistant, who happened to be standing in back of the supervisor during the conversation, immediately resigned. Thus, the organization lost a valuable member due to the supervisor's lack of integrity. An effective leader recognizes the importance of admitting "I was wrong" or "I made a mistake" when the admission is warranted.

A director, or other member of the unit's team, who compromises personal or professional integrity standards may jeopardize both program activities and the health and safety of those the organization serves. Compromises in integrity "to get the job done" can suggest that expedient actions that could jeopardize health or safety are acceptable. The tragic results of the apparent initial cover-up of asbestos hazards dramatically illustrate the consequences of not maintaining appropriate ethical standards. Conversely, early reporting of adverse events can benefit countless workers even if there is a short-term negative impact.

Making Decisions

Decision making is another essential leadership capability, as described more fully in Chapter 14. A decision-making process that did not work is illustrated by the example of a manager who responded to repeated expressions of concern about administration of his organization by appointing an "administrative task force" to develop recommendations for improving the organization's administrative processes and support. The task force, which was given a near-term deadline, was headed by a respected leader who used a total quality management approach to the effort. Task force members met frequently during the few months allocated to the process and produced a series of significant recommendations by the due date. The manager, after several weeks, referred the recommendations to a group of senior members of the department. That group did not have any additional recommendations in its report to the manager; the manager then referred the recommendations to a management committee. After more than a year from the time the report of the task force was submitted, no action had been taken on any of the recommendations. Discussions commenced and spread throughout the organization's senior managers, eliciting increasing vituperation on their part. The manager's and the management committee's delays and inaction on this "priority" effort were demoralizing to all who had spent so much time and effort in developing the recommendations. Additional aspects of team building and motivation are considered in Chapter 9.

Providing Feedback

In addition to the points made in the discussion of feedback in Chapters 7 and 9, it is important for leaders to provide frequent feedback. As Blanchard and Johnson comment, "catch someone doing something right" and then tell the person [8]. Feedback should occur as close as possible to the event, although if negative feedback is necessary, it may be advisable to first have a "cooling off" day. Similarly, it is also wise to avoid immediate responses to e-mails that are sent in an emotionally charged environment or situation or that could otherwise be interpreted as negative, since they are invariably misinterpreted in such circumstances. *Specific* favorable feedback will convey a genuine interest in the members of the organization and encourage them to continue with quality performance. By contrast, the "Atta-boy/girl" letter, in which the name of the recipient is typed in a different font than the rest of the letter and the comments are so general that they could apply to anyone, often generates a response that is the opposite of that which the leader desired. Appropriate feedback will also demonstrate support for those one leads. Members of an organization will typically expend their best efforts if they believe that they have or will receive the strong support and recognition of those leading them.

Management by "Walking/Wandering Around" (MBWA)
It is difficult to provide positive feedback if the leader spends his or her entire day in the office, especially if the door is closed. The emphasis on MBWA underscores the need for a leader to know what is going on in the organization and what the "climate" is—whether people seem to be committed to their work or, instead, to just be doing their jobs [4]. It is most gratifying for a leader to see an employee smile and stand or sit a little straighter when the leader provides a specific compliment while walking around the workplace.

MBWA reaps other benefits as well. When I (*RM*) directed hospitals, I visited every patient every weekday morning. I had learned this from another leader that impressed my wife by such visits. The purpose of my visits was not to check on the treatment, as we had very competent heath providers to provide care, but to be certain that we were providing the best support possible during a trying situation. One Monday, I was making rounds with the chief nurse and came to the last room at the end of the hall. I entered a female patient's room and saw an old-fashioned school bell on the nightstand. I commented that I had a bell just like it and that mine had been used by my mother when she taught school. I then asked if she was a school teacher. She replied, "No, that's in case I need to call a nurse!" Apparently, the room's call system had failed during the weekend, and someone had substituted the bell because a technician to fix the system was not available over the weekend. I had always prided myself that the care we gave in Air Force hospitals was equal to, or better than, that in surrounding community hospitals, and using a school bell to call a nurse did not meet my patient-care standards. The chief nurse, who was as embarrassed as I was, found that the problem had been reported and follow-up action had been taken that morning, but the repair part was on backorder and might not be available for several days or weeks. Although I was initially advised not to go outside of Air Force supply channels, we purchased the repair part at a near-by electronics store so the system could be made operable that day. (I offered to make up the $1.40 difference if necessary.) Had I not been visiting the hospital rooms, the situation might have existed for an extended period of time, which would have been unacceptable and possibly a threat to patient safety.

It is difficult to attribute a problem to any one action, but it is possible that the debacle at Walter Reed Hospital, where outpatients were housed in rooms contaminated by mold as well as insects and rodents, might have been prevented if the leader had personally visited the rooms in Building 18. As General Weighman commented, "We did not see where some of these soldier-patients were living, and we should have" [11].

LEADERSHIP LESSONS FROM A JANITOR

Moschgat describes an experience in leadership he had while at the U.S. Air Force Academy [12]. A Mr. Crawford was his squadron's long-time janitor. Few of the cadets gave him much notice, offering little more than a nod or curt greeting as they rushed to and from the dormitory. Mr. Crawford did his work so well despite his advanced age that there was never cause for complaint. A humble man, he seldom spoke unless he was addressed by a cadet. In reading about World War II, Moschgat discovered that Mr. Crawford had been awarded the Congressional Medal of Honor, the nation's highest award for bravery. Upon learning this, the attitude of all the cadets toward Mr. Crawford changed, and Moschgat, after ruminating on this experience, developed the following leadership lessons [12]:

1. Be cautious of labels. (Mr. Crawford was not "just a janitor.")

2. All deserve respect. (A particular application of this in making a hiring decision is taking a potential hire to dinner at a fine restaurant to observe the way in which the job applicant treats the wait-staff.)

3. Courtesy makes a difference.

4. Take time to know your people.

5. Anyone can be a hero. (Demonstrated by the recent action of a bystander in New York City when an ill person fell between subway tracks.)

6. Be humble. (In the book *Good to Great,* the authors found that many of the great companies had leaders that were humble and did not take credit for themselves, in contrast to some flamboyant leaders who may produce short-term results but at the long-term expense of the company.)

7. Life won't always hand you what you think you deserve.

8. Don't pursue glory; pursue excellence. No job is beneath a leader.

9. Pursue excellence—whatever the job, be the best you can be.

10. Life is a leadership laboratory—be ready to learn from others.

POLITICAL ASPECTS OF LEADERSHIP, INCLUDING COMMUNITY RELATIONS

As noted in Chapter 12, a health and safety manager will have many evaluators. In addition to organizational supervisors, the leader may need to interact with multiple other organizations and people, including local, state, and federal government members; the public at large; the media; local and other medical communities; health care facilities; professional organizations; and labor representatives and other groups. Depending on the circumstances, the leader may need to gain support from one or more of these groups in order to have an effective health and safety program. In gaining support, the health and safety leader must recognize that, just as with direct supervisors, these other individuals may have little knowledge of health and safety activities. The leaders must provide appropriate justification for the health and safety program or initiative. "Mother/Father-hood" statements, such as "We must do this program, or the health of all of our citizens will be harmed," will generate little enthusiasm. In addition, projected costs and benefits must be discussed in language that all concerned can understand. In discussing planned programs, as professionals it is easy to slip into using familiar jargon. Stating to a lay decision-making group that the significance of a study will be evaluated with a "two-tailed t test" will likely result in blank stares and little or no support for the program(s).

The wise leader establishes relationships with those who may be affected by or needed to support a program well in advance of the event. In this regard, community relations and interactions may play vital roles in supporting and improving a health and safety program. For example, Utah's Rocky Mountain Center for Occupational and Environmental Health had spent years interacting with businesses who hired our graduates or who benefited from consultations or other support. The center's advisory board included a state senator, the head counsel of the largest workers' compensation insurer in the state, a member of the labor commission, the head of a major union, industry representatives, and other interested parties in senior positions in the state. When the center was faced with a major reduction in state funding, all these parties came together to support a bill, developed by the workers' compensation fund legal counsel, to permit workers' compensation underwriters to donate a small percentage of premiums written to the center for a dollar-for-dollar credit on the tax underwriters paid the state. Numerous companies and other stakeholders testified on behalf of the center, wrote letters of support, and generally made the bill a priority for their lobbyists. This strong political support resulted in overwhelming majorities for the bill in both the Utah House and Senate. That the bill passed a few months after it was proposed is a direct result of the relationships that had been built over the years.

A FINAL THOUGHT ON LEADERSHIP

The writer Charles Edward Montague (1867–1928) stated, "There is no limit to what a man can do so long as he does not care a straw who gets the credit for it." Otherwise stated, you can accomplish great things if you do not care who gets the credit!

REFERENCES

1. Walton M. *The Deming Management Method.* New York: Putman, 1986.

2. Covey SR. *The 7 Habits of Highly Effective People.* New York: Simon & Schuster (Fireside), 1989.

3. Collins J. *Good to Great—Why Some Companies Make the Leap and Others Don't.* New York: HarperBusiness, 2001.

4. Peters TJ, Austin N. *A Passion for Excellence: The Leadership Difference.* New York: Warner Books, 1985.

5. Covey FR, Merrill AR, Merrill RR. *First Things First: To Live, To Love, To Learn, To Leave a Legacy.* New York: Simon & Schuster, 1994.

6. Coates C. *The Total Manager: Break Out of Your Department and Manage the Whole Business.* London: Pitman, 1995.

7. Drucker PF. *Managing for the Future: The 1990s and Beyond.* New York: Truman Talley Books, 1992.

8. Blanchard K, Johnson S. *The One Minute Manager.* New York: HarperCollins, 2003.

9. Peters TJ, Waterman, RH Jr. *In Search of Excellence.* New York: Warner Books, 1982.

10. Brue G. *Six Sigma for Managers.* New York: McGraw-Hill, 2002.

11. Philpott T. Exclusive interview: Maj. Gen. George W. Weighman, March 14, 2007. Available at: www.military.com/features; accessed March 30, 2007.

12. Moschgat J. Be Cautious of Labels—A Janitor's 10 Lessons in Leadership. E-mail communication, January 18, 2002.

ADDITIONAL READING

Blanchard K. *Leading at a Higher Level—Blanchard on Leadership and Creating High Performance Organizations.* Upper Saddle River, NJ: Prentice Hall, 2007.

George B, Sims PE. *True North—Discover Your Authentic Leadership.* New York: John Wiley & Sons, 2007.

Harvard Business Review on Leadership (Collection of pertinent Harvard Business Review articles). Boston: Harvard Business School Press, 1990–1998.

Iacocca L. *Where Have All the Leaders Gone?* New York: Scribner, 2007.

Zaleznik A. HBR classic—managers and leaders: are they different? *Harvard Business Review* 1992;70(2):126–135.

CHAPTER 14

Decision Making and Problem Solving

with Jeffrey S. Lee

> The man who insists upon seeing with perfect clearness before he decides never decides.
>
> *H. F. Amiel*

In developing programs, responding to problems, and accomplishing day-to-day activities, managers have to make decisions and solve problems [1,2]. Many decisions and problem-solving efforts will be relatively easy, but others will tax the skills and capabilities of the manager. It is possible to delegate these functions, or obtain recommendations from a committee, but the manager is ultimately responsible for the final actions.

NEED FOR DECISION MAKING

In their book, In *Search of Excellence,* Peters and Waterman document the significance of making decisions by noting that successful companies have a bias for action and are not paralyzed by the decision-making process [3]. The same principle applies to service organizations, such as health and safety units.

One manager for whom I worked commented that he frequently made wrong decisions but that at least he made decisions. He considered himself successful if the majority of his decisions, and particularly the most important ones, were correct the majority of the time. One might argue with this supposition, but my personal experience has been that the worst managers are those who delay making decisions indefinitely. Both the organization and its members suffer from such stagnation. Supervisors must recognize that "not deciding now" is a decision and is often the worst possible one.

Ordinarily, it is not possible to have *all* of the desired information needed to make a decision. Certainly, a health and safety director wants to obtain and evaluate as much information as reasonably possible before making an important decision, but repeatedly deferring action until "one more study" has been done can severely impact an organization. It is essential that the manager recognize the fine line between reasoned deliberation and "paralysis by analysis," or procrastination. Delaying the decision can readily lead to being "overtaken by events (or OBE'd)," to the detriment of all involved. If the pending decision is a complex one, it is common to think that insufficient information is available to make the decision. However, it is very rare for a manager to ever have "sufficient" information to ensure that only the "correct" decision is made. It should be recognized and accepted that risk increases with uncertainty. It is also typical of complex decisions that there may not be a clear-cut "right" or "wrong" decision. The wise manager recognizes that these circumstances make it easy for others to criticize the final decision, since those critiquing the action can always hypothesize that another course of action would have been better. The health and safety director who accepts these constraints and still makes a decision can be expected to reap benefits for the organization much more than an individual who fails to make a decision for fear of being wrong or criticized.

In most, but admittedly not all, instances, even the most significant decision does not have to be a final decision. If a decision does not produce the expected results, in most instances the decided-upon course of action can be re-evaluated and a new decision made if necessary. If a wrong decision could cause complete failure of an activity or an organization, it is obviously appropriate to make every effort to make the right decision. On the other hand, in many situations, a decision that proves to be wrong will not have a disastrous impact, and it may be relatively easy to make corrective changes. Certainly, an organization will not benefit from constant changes in direction, but recognition that one course of action is not working can help in refining subsequent decisions to the advantage of the overall program. The point is that a manager must consider viable alternatives, then decide on a course of action and ensure that the results are monitored to evaluate the success of the decision. Then, the manager needs to *move on* to work on other needs of the organization, not delay action to the point that the organization is crippled or fatally impacted.

Note, however, the requirement to weigh alternatives appropriately before deciding on a course of action. In considering various courses of action, it is essential the manager ensure that a potential solution can be implemented—that is, that it is *viable*. For example, a manager advised the director of a subordinate unit of a major program change that would have a severe negative impact on the unit, and advised the director that the change was directed by the parent organization and "beyond the manager's control." The director then found that the manager's statement was incorrect, and the manager had total control over the situation leading to the change and could have modified the negative effect. When this situation was brought to the manager's attention, the manager then decided to remove the unit director and announced his decision. There was no question the removal was in the manager's purview. However, it was only after announcing the action that the manager found that he did not have another member with the expertise necessary to direct a project that brought his organization over $1 million per year. In view of the fact that the contract funding would cease without a qualified director in place, the manager had to rescind his action, with obvious negative impacts on the perception of his capabilities as a manager throughout the organization.

PARTICIPATORY APPROACHES TO DECISION MAKING

Since health and safety professionals are usually in a staff position, they have to "sell" many of their decisions to local government officials, state or federal legislators, line managers, workers, and the public. The professional enjoys a certain level of "authority," even as a staff member, by virtue of expertise, knowledge, and position in the organization. However, this authority may not be sufficient in obtaining "buy-in" by public or political decision makers; it is essential that the other decision-making managers trust the health and safety manager. Such trust is generally developed over a relatively long period of time, but it can be lost rapidly through a series of bad decisions—hence, the significance of a *sound* decision-making process. One approach, mentioned in Chapter 5 on planning and prioritization, that may be of assistance in making some decision is forming an advisory committee. Since involvement tends to equal commitment, having those affected by the outcome involved in the decision-making process may greatly facilitate the manager's "selling" the decision. Of course, this participatory approach should also extend to the other professionals in the health and safety organization. Their involvement in the decision-making process will strengthen their commitment to implementation of the decision.

In asking colleagues to participate in a decision-making process, it is essential that the "ground rules" be clear to all involved [1]. It may be that the manager wants to reach a consensus on the decision, with each participant's "vote" having equal weight. All those deliberating must recognize that a consensus may not be reached, in which case everyone will vote again to reach a final decision. In many situations, some participants will vote for a decision other than that selected by the majority. Regardless, all should be expected to support the action finally approved.

In other instances, the director may advise those considering the situation that their input will be appreciated and valued. However, the director will make, and be responsible for, the final decision. This approach was summarized well by a staff member in an office where I once worked. This individual observed, after a long discussion of possible decisions by voting members, that the result was "a typical democratic process. The vote was eight to one, and the one won." As might be expected, the one winning vote was cast by the senior officer in charge. The important aspect of this example is that the rest of us knew from the start that our input in an advisory capacity had been considered, and all subsequently strove to make the decision work. (It should perhaps be noted that it subsequently became clear that the decision by the most experienced person was the best possible one!)

The manager may delegate decision making. As Drucker notes, effective executives do not make large numbers of decisions; instead, they concentrate on making the really important ones [2]. If the manager delegates decision making, the delegator must have trust that the individual(s) assigned the task will make a carefully considered, reasoned decision. The delegator must also be committed to accepting, acting on, and supporting the final determination. Responding otherwise can rapidly produce loss of trust and commitment on the part of the individual performing the task and also by other members of the organization.

One technique that can be of value in "brainstorming" and similar events is a "forced" *analysis*. Typically, brainstorming sessions end with enumeration of multiple options for a health and safety unit, and while all may be of value, limitation of resources precludes incorporating all of the options into the goals of the organization. In such situations, it has been found helpful to ask each member to rate the options with a "1" vote (equal to 5 points), a "2" vote (equal to 3 points) or a "3" vote (equal to 1 point). (Participants cannot combine their votes to divide the total of 6 points to equal 2 points each.) Tallying the count usually clearly identifies the top three priorities of the group. Of equal significance, this approach helps ensure that the group feels that the members have "had their say" and believes the rankings are the best possible outcomes.

PROBLEM-SOLVING TECHNIQUES

Another approach that may be of value in making decisions is to structure the process in a *systems analysis format* [4]. This technique emphasizes a systematic approach to defining the problem, gathering information, and formulating a solution. It is an outgrowth of problem-solving techniques used to make decisions regarding specific problems. As with the MBO and PERT processes described in prior chapters, a formal problem-solving process is not necessary for every problem or decision. Many day-to-day problems can be solved relatively easily through an informal approach. Other problems are more complex, and a systematic approach can be particularly effective for such concerns. As Covey notes, an ideal outcome of negotiation, problem solving, and other efforts is a "win-win" result, where all involved gain through the process [5].

In some instances, and particularly with complex problems, there may be inherent conflicts in potential outcomes that preclude the "ideal" solution. In these situations, a more formal process can help achieve the "best possible" solution. One step-wise process that many have found useful is the following:

1. Define the problem.
2. As appropriate, obtain upper management support for the problem-solving effort.
3. Ascertain factors bearing on the problem:
    a. Facts
    b. Assumptions
    c. Definitions
    d. Criteria, including constraints and objectives
4. Gather, organize, and evaluate data and factors bearing on the problem.
5. Develop possible solutions, ideally with representatives of groups involved or affected by the problem or possible solutions.
6. Test possible solutions.
7. Recommend the best possible solution.
8. Delineate the action(s) needed to implement the solution.
9. Specify evaluation procedures.

Define the Problem

As with deciding what to do during the planning process, this step in the problem-solving process may look simpler than it is. Too often, considerable effort is expended to solve a problem only to realize later that the actual problem was not the one that it was thought to be. As a result, the "solution" did not solve the actual problem. For example, a health and safety member at one plant became concerned over the failure of workers to use safety glasses in eye hazard areas. A health and safety group was appointed to study and solve the problem. Considerable effort went into issuing memoranda pointing out that company directives required use of safety glasses, obtaining new posters to urge workers to use eye protection, and asking senior management to brief workers, during periodic meetings, on the importance of the requirement. These efforts produced no increase in the usage of safety glasses, and the real problem was only detected when a first-line supervisor was asked why protective glasses were not being used. The response was that the glasses were unavailable from supply. A quick check revealed a problem with the manufacturer. By switching to another supplier, glasses were obtained and the usage rate rose dramatically. This incident not only illustrates the difficulties that can occur if problems are not properly defined but also underscores the need for input from the "customer" in the problem-solving process.

Another difficulty in defining problems is stating, or defining, a problem too broadly. For example, "Reduce substance abuse in our plant" appears to delineate a problem, but the statement does not define the specific problem. "Reduce marijuana use in employees age 25 or less" facilitates a more focused approach because it more concisely defines the problem. Defining problems has some of the same requirements as establishing objectives in a management by objectives process or a similar approach. In some instances, attempts to define the problem more specifically will lead to the conclusion that the original problem can best be managed if it is broken into several problems that should be handled separately.

Finally, problem definition may be hampered by hidden agendas. We were asked, as an independent, "outside" group, to do a health and safety audit of a manufacturing facility to confirm that the exposure measurements obtained by the plant industrial hygiene group were valid. Management believed that concern over the validity of the measurements was leading to significant unrest among the employees. In attempting to define the problem, we included input from the workers in the plant. It soon became apparent that a very real problem was the fact that the workers did not trust management. One underlying aspect was that the workers could detect the odor of chemicals even at levels well within OSHA standards. If we had simply done split-sample analyses to confirm the accuracy of the measurements by the plant industrial hygienist and then submitted a report confirming the sample validity to management, it is doubtful that the underlying problem would have been solved. A more comprehensive approach was developed to evaluate the extent and health significance of any exposures. We then advised both management and the workers that we would report our findings to both groups simultaneously. Fortunately, we were able to confirm the accuracy of the measurements and also substantiate the high quality of the plant's health and safety program. In addition, we discussed the fact that olfactory sensitivity is such that odors may be detected for some agents at levels far below those that the most sensitive instruments can measure and much below exposure limits established by OSHA. Follow-up six weeks later documented that workers' complaints and other evidence of unrest had become essentially nonexistent.

Obtain Upper Management Support for the Problem-Solving Process

If a defined problem has significant ramifications outside the health and safety unit, it is advisable to obtain senior management's support of the problem-solving process before extensive effort is expended. Significant costs are involved in the process, and a check with supervisors may reveal their overriding concern with another problem area. Discussion may result in the allocation of available time and other resources to a different, higher priority problem area and deferring action on the original problem. Of course, it is essential that such a decision implies no immediate health or safety threat as a result of the deferred action. Obtaining management support also facilitates implementation of the final solution recommended by health and safety members.

Ascertain Factors Bearing on the Problem

Factors that have bearing on the problem are:

1. Facts
2. Assumptions
3. Definitions
4. Criteria, including constraints and objectives

The first three factors (facts, assumptions, and definitions) are relatively straightforward; however, it is essential to ensure that all factors bearing on the problem are properly categorized. Stating something is a "fact" because "everyone" knows it to be true may actually be accepting an assumption as a fact. Carefully defining the factors can help avoid such confusion. The fourth factor (criteria) should be specified to help ensure possible solutions will meet the specified requirement(s) or constraint(s). For example, there may be a financial constraint, such as the proposed solution must not exceed $6,000 in total costs. Another criterion involving a requirement or constraint might be to develop a solution that can be implemented within six weeks after approval. Delineating appropriate criteria is important not only in identifying possible solutions but also in evaluating the implemented solution.

Gather, Organize, and Evaluate Data

Often, gathering data about the problem and the factors bearing on the problem will result in redefining the problem as well as suggesting solutions. It is important to evaluate the quality as well as the content of the information. If the information is valid, it can be useful in developing plans and solutions. If the data are not valid, time and effort may be wasted. The need for valid information is illustrated further in Chapter 15 on developing corporate health and safety risk management programs. In the efforts discussed in that chapter as well as in other problem-solving efforts, the adage quoted in Chapter 11, "Misinformation is worse than no information" holds true.

Develop Possible Solutions

Usually, there are several different approaches to solving a problem, and it is important to consider alternatives rather that concentrating on just one solution. Involving representatives of those impacted by the problem or affected by the solution in the problem-solving process can be especially valuable. This type of participation would have prevented the misdirected efforts of the health and safety members in the eye-protection example described above.

Test Possible Solutions

The criteria previously established can be used to test possible solutions to see if they meet the criteria. In applying criteria, it may be useful to prioritize the requirements. Typically, a particular solution may not meet all the criteria established, and the manager and others involved may have to decide which of the possible solutions will meet the largest number of the most important, highest priority criteria.

Recommend the Best Possible Solution

It is usually most effective for the manager or problem-solving group to recommend only one solution. However, it may be appropriate to list other solutions considered, perhaps including reasons that each was discarded. The manager responsible for developing the solution will not gain credibility if a series of possible solutions are presented without a recommendation regarding a specific solution. Such an approach will be recognized by most senior managers as an attempt to avoid responsibility, and it can have deleterious effects on the manager long after the problem-solving effort is completed.

Delineate Action(s) Needed to Implement the Solution

In addition to recommending a single solution, it is usually appropriate to prepare "action" correspondence necessary to implement the solution. If the senior official concurs with the selected solution and proposed actions, the signed action correspondence will then activate the solution. Delaying this step until the proposed solution is approved can result in wasted time for those implementing the solution as well as for those approving the recommended course of action.

Specify and Implement Evaluation Procedures

It is essential that the effectiveness of the solution be evaluated after it is implemented. Too frequently, in the press of other requirements, the evaluation step is omitted. As a result, a solution either may not fix the problem or, after a few weeks, may no longer be effective. Finding that one is faced with a repeat of the same problem that was thought to have been resolved will discourage the director and other members of the health and safety unit. This situation will also cause concern among senior executives and others dependent on, or affected by, the health and safety function. The end result of failing to evaluate how well the solution really solved the problem may be loss of credibility, a loss that may have prolonged adverse effects. A classic example of the mistake of not including follow-up evaluation in "solving" a problem is that which I made at a medical facility when outdated medications were found in crash ambulances (see page 158).

SUMMARY

The problem-solving process depicted above is only one of many systematic approaches that the manager may use in making decisions. For example, one method that works well for some managers is to try to define how the problem under study could be caused. By defining all the factors needed to cause the problem, methods to solve or prevent the problem may become evident.

Regardless of the method used, it is important that, once significant problems are identified, the problem-solving process be conducted as expediently as possible without compromising the quality of the effort. Failure to proceed rapidly compromises morale, demeans the importance of the problem, and may open the door for liability for the organization.

REFERENCES

1. Hersey P, Blanchard KH, Johnson DE. *Management of Organizational Behavior—Utilizing Human Resources,* 7th Ed. Upper Saddle River, NJ: Prentice Hall, 1996.

2. Drucker PF. *The Effective Executive.* New York: HarperBusiness, 1966 (HarperBusiness edition 1993).

3. Peters T, Waterman RH Jr. *In Search of Excellence: The Leadership Difference.* New York: Warner Books, 1985.

4. Martin DD, Shell RL. *Management of Professionals—Insights for Maximizing Co-operation.* New York: Marcel Dekker ASQC Quality Press, 1988.

5. Covey SR. *The 7 Habits of Highly Effective People—Restoring the Character Ethic.* New York: Simon & Schuster (Fireside), 1989.

ADDITIONAL READING

Gentile MC, Goodpaster KE, Piper TR (eds). *Managerial Decision Making and Ethical Values.* Boston: Harvard Business School Publishing, 1989.

Groopman J. *How Doctors Think.* Boston: Houghton Mifflin, 2007.

Hammond JS, Keeney RL, Raiffa H. *Smart Choices: A Practical Guide to Making Better Decisions.* Boston: Harvard Business School Press, 1999. (A concise discussion of decision making that uses examples to illustrate underlying principles. Accolades include "best book I know" on the subject and "exceptional resource.")

Harvard Business Review on Decision Making. (Collection of pertinent HBR articles.) Boston: Harvard Business School Press, 1965–2001.

CHAPTER

15

Containing Health and Safety Costs: Health and Safety Risk Management Programs

Most senior managers in corporate, government, and other settings are well attuned to risk management programs structured to reduce legal liability exposure. The concept of health and safety risk management, as provided by a comprehensive health and safety program, is not so widely acknowledged or understood by senior executives. However, calculations using Health Care Financing Agency data indicate that total health care costs for occupational and nonoccupational illnesses and injuries for U.S. workers at the end of the last decade exceeded $1.25 trillion annually [1]. This translates into health care expenditures of over $10,000 per worker, considering both direct and indirect costs. The costs for only work-related illnesses and injuries were estimated at $171 billion [2]. The National Coalition on Health Care reviewed multiple studies and reported that total national health expenditures rose 6.9 percent, two times the rate of inflation, in 2005. Total health care spending was $2 trillion, an amount representing 16 percent of the gross domestic product [3]. Schulte reviewed the literature from 1990 to 2005 and noted that estimates of the costs related to occupational disease and injury ranged from $128 billion to $155 billion per year [4]. If successful in reducing corporate health cost risks by melding a number of diverse activities into a comprehensive corporate health and safety risk management program, the health and safety professional not only contributes to the corporation's profitability but also improves worker health, safety, and productivity. The program will probably increase the corporation's competitiveness as well. The same principles apply in developing and implementing health and safety risk-reduction programs as those applicable in other, noncorporate settings.

DEVELOPING A PROGRAM FOR CORPORATE HEALTH AND SAFETY RISK MANAGEMENT

For purposes of discussion and illustration, this chapter focuses on methods that can be used to identify and reduce the costs of work-related illnesses and injuries—in essence, how to reduce the corporate health risk. These techniques are also applicable in other settings. Whether consultant, contractor, or corporate employee, the health and safety manager can consider developing a corporate health and safety risk management program an appropriate health and safety effort. In many settings, developing such a plan will be established as a health and safety manager's responsibility during initial interactions with supervisors, as discussed in Chapter 3 and Appendix A. In other instances, however, it may be appropriate to obtain specific approval from supervisors or senior executives prior to developing a risk management proposal. The health and safety manager must judge the necessity for obtaining approval before the effort. If it is deemed necessary to obtain approval, and support, for the effort, some managers have found the approach discussed in the final portion of this chapter—Advocating a Health and Safety Program: The Costs of Not Having a Health and Safety Risk Management Program—useful in delineating the benefits of a comprehensive program.

As in other health and safety efforts, it is often beneficial to include representatives from other appropriate sections of the company in developing the risk management program. (See the discussion of an advisory committee in Chapter 3). As noted by Belk and colleagues, candidates to consider for membership on a risk management team may come from corporate benefits, employee and labor relations, financial, legal, management information systems, and/or management groups [5]. A participatory approach is usually optimal for obtaining support for the program from line managers and workers.

Some of the steps involved in developing a health and safety risk management program are listed in Table 15-1.

Table 15-1. Steps in Developing a Risk Management Program for Corporate Health Costs

1. Define present health and safety costs, including direct and indirect costs.
2. Ensure validity of the health and safety cost data.
3. Estimate potential health and safety costs using recent experience.
4. Identify high cost/high risk aspects.
5. Establish objectives for the program.
6. Develop risk management plan, including:
   a. Reactive, traditional, "above the line" portions.
   b. Proactive, prevention and promotion, "below the line" portions.
   c. Financial plan for both portions.
   d. Evaluation procedures for both portions.
7. Obtain management approval and support.
8. Implement the program.
9. Evaluate and revise the program and repeat appropriate steps.
10. Ensure appropriate communication at each step and on an on-going basis.

The following discussion considers the steps in Table 15-1 in more detail.

DEFINE PRESENT HEALTH AND SAFETY COSTS

In considering methods to handle increased health and safety costs, it is useful to begin by defining the corporation's health and safety risks. The *corporate health and safety risk* can be defined as the corporation's exposure to costs associated with health or safety problems. Health and safety costs are made up of both direct and indirect costs. Direct costs include items such as clinic or hospital charges, provider charges, costs of medications, costs of destroyed plant equipment, fees for laboratory services and procedures, disability and insurance payments, and custodial or home health care expenses. Indirect costs include those related to absenteeism, presenteeism (present but not fully effective due to illness or other factors), lessened productivity, loss of training investments, costs of training replacements, administrative costs, legal fees related to health problems and injuries, and fines due to OSHA or other regulatory agency citations, to name a few. Tables 15-2, 15-3, and 15-4 list some of the direct and indirect cost factors [1].

Table 15-2 identifies many of the direct costs related to a specific illness or injury; Table 15-3, indirect costs related to a specific

Table 15-2. Direct Costs Related to Specific Illness or Injury

Medical Care
Physician, other provider services
Inpatient
Outpatient
Clinic, hospital services
Inpatient
Outpatient
Ancillary diagnostic services, e.g.,
Laboratory
Radiology
Electrocardiography
Electromyography
Other
Patient-specific medical supplies, equipment
Medications, pharmacy services
Rehabilitation, occupational/physical therapy
Employee assistance counseling
Other (to be specified for specific situation)
Other Injury- or Illness-Specific Requirements
Workers' compensation payments (as applicable)
Sick pay (as applicable)
Other benefits (as applicable)
Compliance with OSHA, other reporting required by the illness or injury
Costs of temporary employee to accomplish tasks of ill or injured worker
Case management costs
Vocational rehabilitation counseling
Case-specific litigation costs
Case-specific human resources or personnel costs
Specific accommodations required by the Americans with Disabilities Act
Other (to be specified for specific situation)

Reprinted with permission, American College of Occupational and Environmental Medicine.

illness or injury; and Table 15-4, indirect costs that are not related to a specific illness or injury. It should be noted that many identified costs are applicable to non-work-related illnesses and injuries in both workers and family members covered through organization benefit plans. However, some costs are specific to work-related illnesses or injuries or to programs designed to prevent such illnesses or injuries, such as compliance with OSHA requirements. Still others are related to methods to accommodate individuals with illnesses or injuries that may be related either to work

Table 15-3. Indirect Costs Related to Specific Illness or Injury

Reduced Productivity

- Absence of injured, ill employee
- Shift in activities of coworker to accomplish absent employee's work
- Increased supervisor effort to cope with absence of employee
- Temporary or long-term absence of corporate memory possessed by the ill or injured employee
- Start up, training time for replacement employee
- Start up, training time for returned ill or injured employee
- Development of limited work position for ill or injured worker, as appropriate
- Reduced effectiveness of "near-by" coworkers
- Overtime pay

Impacts on Competitiveness

- Potential for reduced customer satisfaction due to absent employee and effects of employee's absence
- Effect of greater-than-projected medical costs
- Increased risk of illness or injury by temporary, replacement employee due to limited time for hazard, safety training, or other factors
- Increased insurance premiums
- Increased overtime costs
- Increased training, retraining costs
- Increased legal costs, including class action defense, coordination of new policies to respond to event or prevent recurrence, and similar costs
- Loss of senior management time as required to respond to event
- Reduced performance/effectiveness of returning ill or injured person
- Effects on labor relations, including requests for hazardous pay, new safety programs or equipment, strike potential
- Potential for adverse media coverage
- Effect on worker morale (also impacts productivity)
- Requirements for increased quality control efforts, as required, for replacement or returning employee
- Increased human resources, personnel department costs associated with efforts to replace ill or injured worker
- Medical, industrial hygiene, safety costs involved in investigation of accident or exposure site
- Risk-management activities involved with respect to investigation of accident or exposure site or other activities

conditions or to nonoccupational etiologies, such as efforts required by the Americans with Disabilities Act (ADA).

In general, it is easier to define direct costs more precisely than indirect costs. The current financial picture may indicate the existing

Table 15-4. Indirect Costs Related to Other Health and Safety Requirements, Not to a Specific Illness or Injury

Organizational Health and Safety Program (In-House, Consultant, Contract) Costs
Staffing to provide care for ill or injured employee on-site (or costs to arrange such capability off-site)
Regulatory compliance, including medical/industrial hygiene monitoring, surveillance programs
Development, maintenance of capabilities for case management
Employee assistance program, other than costs for specific illnesses or injuries
Development of health and safety policies
Data processing and data management costs
Research expenditures, as appropriate
Wellness, health promotion, immunizations
Health and safety committees
Evaluation of options for provision of services
Program evaluations
Interactions with other organization departments and managers, including risk managers
Drug and alcohol testing programs
Pre-placement, periodic examinations, evaluations
Other, organization specific
Other Organizational Costs Related to Health and Safety Activities—Other Than Those Involving Specific Illness or Injury (see Tables 15-2 and 15-3)
Human resources, personnel
Benefits
Legal
Labor relations, unions
Management (other than health and safety)
Other, organization specific

Reprinted with permission, American College of Occupational and Environmental Medicine.

direct cost of health risk exposure, or "dollars at risk." However, the health and safety manager must ensure that even direct cost data are as complete and accurate as possible. On occasion, I, as well as colleagues, have found that direct health cost information from a human resources or safety department either does not include all health care cost categories or the cost data have significant errors, with resultant understating of total costs. In addition, senior managers often are not cognizant of the concept of indirect costs. It is important to consider indirect costs because, as illustrated in Tables 15-3 and 15-4, multiple aspects may be involved with significant financial impacts. The 2003 ACOEM Health-related Productivity Roundtable noted that studies had indicated employers pay out $2.00 to $3.00 (and up to $10.00 in some industries) for every dollar of direct medical costs [6].

If all costs are not recognized and reflected in the financial plan for a new risk management program, there may be an apparent increase in health and safety costs at the end of the first year because, if the financial plan had been properly designed, it would have identified all costs, including those not previously calculated. Current, *valid* information is crucial in developing health and safety risk management programs. If health and safety directors suspect that information is invalid or incomplete, they must work to improve the quality of information or at least identify the deficiencies and develop reasonable estimates to replace the questionable items. It is essential that supporting data reflect actual current injury and illness occurrences and costs. A risk management program based on inaccurate data can result in subsequent failure of the program and loss of credibility with senior management.

In collecting and evaluating data, it is often helpful to compare the information with the standards in similar industries. The Bureau of Labor Statistics compiles illness and injury incidence and severity information, and groups it into different manufacturing categories. This type of information can be useful in identifying excess health problems in a particular plant or a specific operation within a plant [4,7]. Such normative analyses, comparing observed costs with expected costs, can be used to evaluate the incidence of health problems, their severity, and both direct and indirect costs.

ESTIMATE POTENTIAL COSTS

Past experience can help the planner estimate future trends in both direct and indirect costs. This information aids in defining the objectives for the program, since it highlights areas that significantly impact the corporation or the health and safety of groups of workers or other populations.

In estimating potential risk, important data include anticipated length of employment by workers; projected age, sex, education, and other changes in work force composition; projected changes in job activities that may increase or decrease risk of illness or injury; and similar factors. Obviously, it is essential to work closely with senior management to obtain the information. This exchange will not only help gain essential data but also may prove particularly valuable in getting managerial support for the risk management program.

Health risk information is also obtainable through a Health Risk Assessment (HRA) instrument. This device may be used to gather preliminary information about the health risks of individual workers; this information can then be used in developing a corporate risk management program. In obtaining data to estimate risk, it may only be necessary to apply the HRA to a representative sample of the work force. The information from the HRA survey can then be used to develop an appropriate program for the entire work force. The HRA instrument can also be used as an important component of the total risk management program and periodically administered to each worker in the program. Such usage facilitates tracking the progress of the individual worker in reducing his or her health risks. In applying HRA instruments to samples or whole groups of workers, the restrictions on obtaining health information stipulated in the ADA and the Health Insurance Portability and Accountability Act (HIPAA) must be met (see Chapter 23). Whenever used, the HRA instrument should be simple and provide information on both current health risks and projected longevity or other benefits if risk factors are reduced or eliminated. (See also the discussion of HRA instruments in Chapter 17.)

A helpful aid in estimating total costs is the Blueprint for Health, developed by the Health as Human Capital Foundation, the American College of Occupational and Environmental Medicine, and the National Business Coalition on Health [8]. The Blueprint is free on-line and enables the user to estimate absence and productivity costs as well as total costs of employee health. It is based on analytical methods and metrics developed from a large dataset of U.S. employers. The user enters data on a particular work force, and cost estimates are then calculated. Work force data include gender, age, marital status, geographic location, number of employees and percent annual salary versus hourly salary, salary ranges, and health insurance deductible levels. "What if" scenarios can be entered to determine the impact of various changes in the health program.

IDENTIFY HIGH-COST/HIGH-RISK AREAS

Once data have been obtained and analyzed, activities can be ranked with respect to total costs, average cost per worker in an activity, and costs per worker who is injured or contracts an occupation-related illness. The Pareto approach discussed in Chapters 3 and 25 can be of particular benefit in this endeavor. The underlying concept of the Pareto technique is to focus your efforts where you obtain the greatest results. Figures 15-1 and 15-2 illustrate this approach. The rankings in Figure 15-1, depicting costs of injuries and illnesses, define areas where health and safety risk-reduction efforts can significantly improve health and also help control or reduce costs. In this example, focusing interventions on prevention of back injuries could produce the greatest benefits. Figure 15-2 indicates that if all areas in which back injuries occur cannot be addressed, it would be most appropriate to concentrate on the warehouse workers. As another example, it may be found that one particular department has high total health care costs. However, if the department has many workers, the average cost per worker or incident may be lower than that of another department. As illustrated in Figures 15-1 and 15-2, a

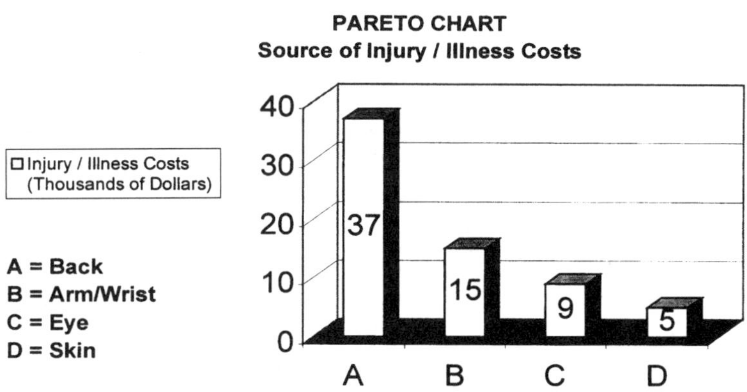

Figure 15-1. This Pareto chart identifies back conditions as a major source of health care costs. Figure 15-2 depicts causes of back problems.

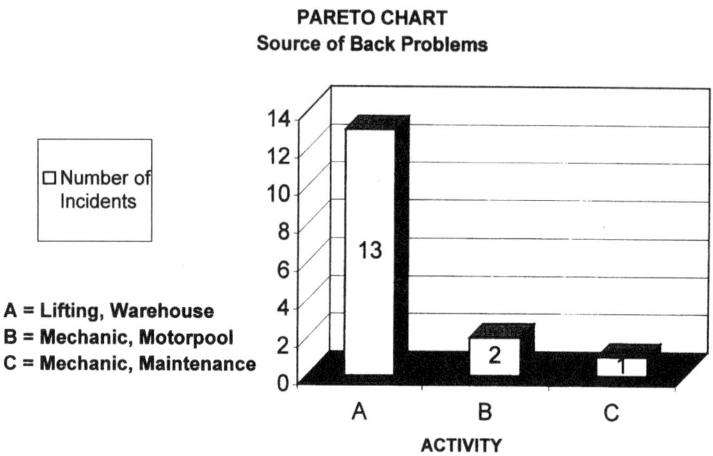

Figure 15-2. This Pareto chart depicts the major source of back injury and indicates where focused interventions could produce the greatest benefits.

relatively few workers in a small department had had back injuries that produced very large costs per injured worker. In such instances, it may be appropriate to direct the limited risk-management resources toward this department with few workers, where interventions can have more impact, than toward departments with more workers with milder, routine illnesses less amenable to intervention.

The program planner should remember, however, that a corporate risk-reduction program has to meet the needs of the corporate managers responsible for approval and support of the program. If the managers direct their concerns to one group or specific health promotion effort, the health and safety planner must factor these requirements into data evaluation and program design. For example, the pilots at a military base may represent only a small fraction of the base population. They are generally healthy and seldom have health problems. However, they are a critical factor to the accomplishment of the mission of the base. Further, the majority of senior managers at a base are pilots. Thus, it is essential that planners recognize this small group's significance and include their concerns in any risk management program. Similar principles apply in other corporate or community settings.

ESTABLISH OBJECTIVES FOR THE PROGRAM

Objectives for the risk management program can be defined when delineating the high-risk areas. Defining objectives should include review of any existing objectives as well as development of new objectives. If the team described in the first section of this chapter (page 204) has not been involved up to this point, it is important that others now *be* included. Delineating objectives should be a group effort involving members indicated previously.

Once risk management program objectives have been defined, in-place programs and activities can be evaluated and used, or modified, to meet the objectives. The health and safety planner can then determine whether additional or new measures are required.

DEVELOP THE RISK MANAGEMENT PLAN

Approaches

As current and potential corporate health and safety risks and program objectives are being defined, the planners can begin to develop the health risk management program. Two approaches have been found useful in developing plans.

Reactive Approach. The first approach targets the sick employee and is mainly designed to reduce health care costs; it may even result in improved quality of care. This approach may be thought of as *reactive, traditional, or nonpreventive.* Examples of the applicability of this approach are:

Designing health insurance and benefits programs, including co-payments and amount of deductibles.

Using the least costly provider, often through assignment to a specific Health Maintenance Organization, Preferred Provider Organization, or other managed care entity.

Requiring second opinions prior to selected health care activities.

Increasing use of out-patient facilities rather than hospitalization.

Requiring application of practice guidelines.

Requiring that an individual who is ill for more than a specified period of time be visited by a company nurse or other member of the company health department.

Establishing pre-admission certification requirements.

Obviously, this approach can compromise health care availability and quality unless it is carefully managed. In some instances, requirements for chart review, second opinions, and use of specific groups of providers may result in overall improvement in the quality of care a group of workers receives. Studies of managed care initiatives suggest that this approach, in particular, will become more prevalent in the coming years [9–11]. Similarly, increasing use of practice guidelines and "best practices" can be expected in the future [12,13].

The traditional approach can also result in significant savings. For example, the corporate medical director of an airline hired a nurse at one of the company's major facilities to visit all employees who were absent for more than three days. This process helped ensure employees were receiving appropriate care for their illnesses or injuries. The program also resulted in a reduction in the average number of days lost for the more common illnesses. The savings exceeded the cost of the nurse by several thousand dollars. This approach has been expanded in some settings by establishing a *case management* or *patient assistance,* function. These activities are frequently staffed by experienced occupational health nurses. These professionals review illnesses and injuries, especially those that are high cost, and help make certain that proper, well-coordinated care is provided expeditiously. In addition to controlling costs, these efforts may reveal additional sources of assistance for patients or their families that benefit all concerned.

Proactive Approach. The second approach, directed toward the well employee, is oriented to improving the health, safety, and productivity of the worker. Sometimes, services in this portion of the risk management program may also be available to the workers' families. This approach includes health and safety promotion and disease and injury prevention programs and may be thought of as *proactive* or *prevention and promotion oriented*. Examples of this approach's activities are:

Providing focused interventions to prevent illnesses or injuries.

Using health and safety audits to define areas that could be improved to prevent potential problems and losses.

Accomplishing risk screening.

Using validated health promotion and disease prevention techniques.

Evaluating new processes or materials to ascertain need for new surveillance or monitoring programs.

Chapter 17 provides details on the health promotion and disease prevention portion of a proactive approach. In some instances, a proactive health and safety program has produced reductions in projected health care cost increases, whereas in other situations, actual overall health care expenses have decreased.

In general, risk management programs that include both approaches appear to offer the best potential for significantly reducing costs while also improving the health, safety, and well-being of the employee. Figure 15-3 depicts the structure of a program combining the two approaches.

Financial Plan

The financial plan for the program can be developed while other processes are being completed. This plan must include all projected costs, including those for in-place programs or facilities. Similarly, projected cost reductions or savings should be estimated. It may be beneficial to provide projected costs and savings for various portions, or modules, of the program. In this way, part of the program might be implemented if funding precludes adopting the whole program at once. Depending on the scope of the program, it may be appropriate to develop a formal business plan. Further details on developing business and financial plans for program components are provided in Chapters 10 and 11 and their References and

("ABOVE THE LINE"—REACTIVE)

Coordinate/ Provide Illness- Injury Care	Accomplish Case Management	Evaluate/ Reduce W.C./ Health Ins. Costs	Accomplish Accident Investigations	Evaluate/ Control Health Care Costs	Respond to Regulatory Citations/ Queries (OSHA ADA, CAA, Etc.)

("BELOW THE LINE"—PROACTIVE)

Define Major Components of Direct/Indirect H&S Costs	Design Focused Interventions For "High Cost" Areas	Accomplish H&S Audits to Define Program Strengths/ Weaknesses	Respond to Media, Employee Concerns	Evaluate Hazards, Other H&S Concerns of New Processes, Chemicals, Etc.	Develop Focused HP/DP Programs as Warranted, Inc. H&S Info. Programs
Ascertain Need for/Implement Surveillance, Monitoring, Sampling Programs	Monitor Regulatory Actions; Establish Compliance Priorities	Orient H&S Programs to Improve H&S and Also Support Organizational Mission/Goals			

Figure 15-3. Components of corporate health and safety risk-management programs.

Additional Reading, but it must be emphasized that a well-constructed business or financial plan is vitally important in obtaining management approval and support for the effort.

Evaluation

In developing the program and the financial plan, it is important to determine methods that will be used to evaluate the corporate risk-management program. Many of the costs and benefits of the reactive portion can be defined in dollar amounts. However, it is more difficult to measure results of any attempt made to improve quality through this approach. Similarly, it may be possible to define costs of the proactive portion, but measuring results can be much more difficult. Information gained in determining baseline costs and use of the components listed in Tables 15-2, 15-3, and 15-4 can be of assistance in developing cost and benefit estimates. The ACOEM Roundtable noted the need to develop metrics that "can be tracked longitudinally to define, quantify, and benchmark the 'health' of a . . . population." In addition, productivity metrics are needed to calculate the return on investment (ROI) from programs that are designed to improve health of a group of workers or other individuals [6]. The NIOSH publication *Guide to Evaluating the Effectiveness of Strategies for Preventing Work Injuries* provides information to assist in program evaluation [14]. A recent benchmarking study echoed the need for measurement of program effectiveness [15]. Chapter 16 provides detailed information on evaluation processes and resources. Chapter 12 provides details on program evaluation approaches, techniques, and constraints.

Information systems can be of significant value in gathering and analyzing data, often on a "real time" basis. Timely data permits adjustments of interventions or other efforts to control risks or costs in midstream, helping to preclude unwelcome surprises at the end of the fiscal year. It must be recognized, however, that current and evolving information systems can set the stage for data overload that may stifle a program [1]. Trend and key data points can be valuable, but micro-reporting by means of detailed analyses can rapidly reach the point of diminishing returns in terms of useful information provided. These concerns should be incorporated into the planned evaluation processes.

Obtaining Management Approval *and* Support

After developing the projected program and the financial plan, the program director has to get both management approval and support. Communication with senior managers during the planning process should facilitate the process. Not only will these interactions provide necessary information, but they will also help identify concerns or problems. The planner then has the opportunity to resolve problems before the plan is formally presented.

In presenting the plan, the director must convey the importance of the plan in helping other managers accomplish their jobs more effectively. Those responsible for approval usually expect the director to present the proposal in the same manner and detail that would be required of any other senior manager. In presenting the program, past experience has demonstrated that use of a summary depiction of the program, such as that in Figure 15-3, greatly assists in conveying the concept of the program. With proper preparation and effective communication skills, the planner should be able to obtain project approval. Involving the planning team or advisory committee members may be of great assistance to the program director in obtaining the necessary approval. Chapters 3 and 17 provide additional suggestions on methods to obtain both approval, and, particularly important, support, for the program.

Implementation and Follow-up

Once the project is approved and implemented, the director must ensure that the agreed-upon evaluations are performed. Good and bad results, as well as progress and problems in the program, should be reported appropriately in order to maintain credibility with senior management. The reports should be accompanied by recommendations on how to overcome difficulties or further enhance the effectiveness of the program. Through ongoing evaluations and communication, the program director should be able to ensure that the risk management program supports the goals and objectives of the organization and simultaneously improves the health, safety, and productivity of the workers.

Communicating Effectively

As in other management activities, effective communication is important during all facets of the planning and implementation of the health and safety risk management program. Ongoing, regular communication is crucial to the program's success. Chapters 20 and 21 provide suggestions to facilitate appropriate, effective communication.

Advocating a Health and Safety Program: The Costs of Not Having a Health and Safety Program

Although many senior corporate managers recognize the need for a comprehensive health and safety program, others do not. Some believe it is only necessary to provide "some kind" of care for illnesses or injuries. Others relegate health and safety to a low priority, manifested by the place of health and safety activities in the organization and by minimal funding levels. In discussing health and safety programs with managers who do not appreciate the value that a health and safety program can add to a company or organization, it is sometimes helpful to focus on the cost of *not* having a health and safety program. Some of the components of such an approach may be emphases on the following:

Failure to Define Total Costs of Illnesses and Injuries. Tables 15-2, 15-3, and 15-4 summarize the results of an extensive review of a particular company's direct and indirect health care costs. The review documented the previously mentioned limitations present in many cost figures for injuries and illnesses. This was particularly the situation in indirect cost estimates but also existed to some extent in defining the direct portion of costs. The tables list the numerous factors involved in calculating such costs. If there is not a structured approach to defining the costs, as provided by a comprehensive health and safety program, costs will typically be understated by significant amounts. This situation can mean that the impacts of the costs on the profitability and productivity of the organization will not be recognized. Rather, they may be included in personnel or "other" categories, with little attention paid to the real significance of health and safety expenditures. Without definition of costs, there is no opportunity to design focused implementations for high-cost areas, with expected high rates of return from such interventions based on a Pareto approach.

Failure to Implement Surveillance or Monitoring Procedures. Although the Code of Federal Regulations (CFR) specifies monitoring, surveillance, and examination requirements for many industrial activities, a number of the requirements are complex. For example, a health and safety manager who is not familiar with the CFR's requirements may consult only the general duty requirements and fail to recognize that additional requirements exist for the construction industry that the person is supporting and that the introduction of hundreds of new industrial chemicals and processes each year can lead to inappropriate exposures if the manager does not evaluate the need for new monitoring or surveillance programs. The end result could be unnecessary illnesses or injuries that would have a major financial impact on the company.

Failure to Meet Other Regulatory Requirements. It is not appropriate to use "scare" tactics when discussing the need for a health and safety program to address health and safety risks. Senior managers may have little concern about OSHA citations, noting that few inspectors are available, and, even when fines are imposed, they are often negotiated to much lower levels. Some would even consider the periodic fine as "a cost of doing business" and a much lower cost than having to establish a comprehensive health and safety program. However, it is appropriate for the manager to point out that OSHA and other regulatory requirements have been established to protect the health and safety of workers, that the costs associated with illnesses or injuries due to failure to comply with requirements are significant, and that the costs can be avoided in the overwhelming majority of instances through effective health and safety programs. It also is appropriate to observe objectively that more recent legislative and regulatory agencies' actions have increased the impacts of failure to comply. For example, mentioning (without inappropriate emotion) that the Clean Air Act not only specifies company fines for violations but also specifies fines and possibly jail time for senior managers, may increase interest in health and safety efforts.

Failure to Have Health and Safety Programs that Support a Company's Mission and Goals. Some managers still believe that a health and safety program will benefit only workers and will be a detriment to the company's efforts to be productive and profitable. However, recent publications emphasize the competitive advantages of a healthy and safe work force [16]. Figure 15-3, which depicts the results of an analysis of several companies' health and safety programs, underscores the contributions that a comprehensive health and safety program can provide a company. Figure 16-2 (page 229) also illustrates benefits of an effective health and safety program. In addition to improving the health and safety of workers, coordination of care, case management, demonstrated concern for the worker through interventions and health promotion programs, visible on-site monitoring of working conditions, interactions with unions and with management, information programs to allay workers' concerns, and other aspects of comprehensive health and safety programs can enhance productivity and morale while reducing sick leave, turnover, and loss of experienced personnel [17]. The result is increased profitability and competitiveness of the organization. (See additional discussion of benefits of a healthy work force in Chapters 16 and 17.)

Emphasizing the major adverse impacts that may ensue due to the absence of a health and safety program, and depicting the advantages accruing if a program is established, can help the program director to obtain the support needed to implement or expand health and safety efforts to benefit workers and the organization.

REFERENCES

1. Brady W, Bass J, Moser, R Jr, et al. Defining total corporate health and safety costs—significance and impact. *J Occup Environ Med* 1997;39:224–231.

2. Leigh JP, Markowitz SB, Fahs M, et al. Occupational injury and illness in the United States. Estimates of costs, morbidity, and mortality. *Arch Intern Med* 1997;14:1557–1568.

3. National Coalition on Health Care. Health insurance cost—Facts on the cost of health care. Available at: www.nchc.org/facts/cost.shtml; accessed April 30, 2007.

4. Schulte PA. Characterizing the burden of occupational injury and disease. *J Occup Environ Med* 2005;47:607–622.

5. Belk HD, Harris JS, Wood LW. A strategy for health care value management. *J Occup Med* 1991;33:386–389.

6. The health of the workforce and its impact on business. ACOEM Health-related Productivity Roundtable, March 24–25, 2003.

7. Leigh JP, Miller TR. Ranking occupations based upon the costs of job-related injuries and diseases. *J Occup Environ Med* 1997;39:1170–1182.

8. Blueprint for Health—A Framework for Total Cost Impact. Available at: blueprint.hhcfoundation.org; accessed June 30, 2007

9. Sparks PJ, Feldstein A. The success of the Washington Department of Labor and Industries Managed Care Pilot Project: the occupational medicine-based delivery model. *J Occup Environ Med* 1997;39:1068–1073.

10. Green-McKenzie J, Parkerson J, Bernacki J. Comparison of workers' compensation costs for two cohorts of injured workers before and after the introduction of managed care. *J Occup Environ Med* 1998;40:568–572.

11. Bernacki EJ, Guidera JA. The effect of managed care on surgical rates among individuals filing for workers' compensation. *J Occup Environ Med* 1998;40:623–631.

12. American College of Occupational and Environmental Medicine. Glass LS (editor). *Occupational Medicine Practice Guidelines—Evaluation and Management of Common Health Problems and Functional Recovery of Workers,* 2nd Ed. Beverly, MA: OEM Press, 2004.

13. Rogers B, Randolph SA, Mastroianni K. *Occupational Health Nursing Guidelines for Primary Clinical Conditions,* 3rd Ed. Beverly Farms, MA: OEM Press, 2003.

14. Robson LS, Shannon HS, Goldenhar LM, et al. *Guide to Evaluating the Effectiveness of Strategies for Preventing Work Injuries.* National Institute for Occupational Safety and Health, 2001. Pub. 2001-119. Available at: www.cdc.gov/niosh/docs/2001-119.

15. Goetzel RZ, Schechter D, Ozminkowski RJ, et al. Promising practices in employer health and productivity management efforts: findings from a benchmarking study. *J Occup Environ Med* 2007;49:111–130.

16. McCunney RJ, Anstadt G, Burton WN. The competitive advantage of a healthy work force: opportunities for occupational medicine (editorial). *J Occup Environ Med* 1997;39:611–613.

17. Smith KJ, Everly GS Jr, Haight GT. An empirical analysis of a corporate health promotion intervention on reported sick leave absenteeism. *Benefits Q* 1990;6:37–45.

ADDITIONAL READING

American College of Occupational and Environmental Medicine. Health and Productivity Toolkit, 2006. Available at: www.acoem.org/publication.aspx?id=48; accessed November 27, 2007.

Emmett EA. What is the strategic value of occupational and environmental medicine? Observations from the United States and Australia. *J Occup Environ Med* 1996;38: 1124–1134.

Fielding JF. Corporate health cost management. In: Fallon LF Jr, Dickerson OB, Brandt-Rauf PW (eds), The management perspective. *State Art Rev Occ Med* 1988; 4(1):121–144.

Goetzel RZ, Jacobson BH, Aldana SG, et al. Health care costs of worksite health promotion participants and non-participants. *J Occup Environ Med* 1998;40:341–346.

Harris JS. Economics of occupational medicine. In: McCunney RJ, *A Practical Approach to Occupational and Environmental Medicine,* 3rd Ed. Philadelphia: Lippincott Williams & Wilkins, 2003.

Harris JS, Loeppke RR. *Integrated Health Management. The Key Role of Occupational Medicine in Managed Care, Disability Management, Productivity, Prevention, and Integrated Delivery Systems.* Beverly Farms, MA: OEM Press, 1998.

Herzlinger RE. The failed revolution in health care—the role of management. *Harvard Bus Rev* 1989;67(2):95–103.

Herzlinger RE, Calkins D. How companies tackle health care costs: Part III. *Harvard Bus Rev* 1986;64:70–80.

Herzlinger RE, Schwartz J. How companies tackle health care costs: Part I. *Harvard Bus Rev* 1985;63:69–81.

Menzel NN. *Workers' Comp Management from A to Z—A "How To" Guide with Forms.* 3rd Ed. Beverly Farms, MA: OEM Press, 2006.

CHAPTER

16

Health and Productivity Enhancement

Ronald Loeppke and Pamela Hymel

"Good health is good business"—from the worksites of employees to the homes of consumers and from the exam rooms of providers to the board rooms of employers. The key stakeholders in the health care ecosystem are increasingly aligned around the need to lower total medical and health-related productivity costs. The solution that is emerging revolves around population health and productivity enhancement by reducing the burden of illness and the burden of health risks in society through closing quality gaps in the management of medical conditions as well as investing in prevention and improving health.

However, the current onslaught of patients suffering in the battlefields of daily life from chronic disease, illnesses, and injuries is utterly amazing. What is more disheartening is that many, if not most, of these medical conditions could have been avoided or significantly delayed—if only these patients would have known what to do to prevent their development and acted accordingly. Unfortunately, our current medical care system is focused on treatment and repair, with essentially no focus on prevention and promoting the health of people. That said, we are finally on the threshold of witnessing the transformation of our health care delivery system beyond being a reactive/illness-oriented *medical* care system to becoming a more proactive, wellness-oriented *health* system. As Joseph Malins expressed so eloquently in his 1895 poem "The Ambulance Down in the Valley," it is not good enough to just mobilize more ambulances "down in the valley" when we should also "put a fence 'round the edge of the cliff" as a means of prevention.

It has been said that we cannot alter the winds of change but we can adjust our sails and use the change to our advantage. The converging trends are clear and present, and there is definitely a change in our health care system gathering on the horizon. In fact, there is a new-value proposition in the health care marketplace that is emerging—one that those of you reading this book are uniquely positioned to leverage. That new value is the *business* value of health.

THE BUSINESS VALUE OF HEALTH

By investing in the health and human capital of their people, employers are looking to decrease their *total* health-related costs (medical/pharmacy costs as well as health-related productivity costs of absenteeism and presenteeism). The return on this investment transcends the traditional measures of medical costs, extending into the metrics of productivity improvement.

A once radical notion, the linkage between health and productivity is now being widely recognized and has made its way to mainstream media. For example, in March 2006, a Google search on health-related productivity produced 487,000 hits and had soared to 26,500,000 hits as of March 2007. Health and productivity is even discussed as front-line news in the likes of the *Wall Street Journal, Harvard Business Review,* the *Today* show, CNN, and CNBC.

In fact, in a 2006 CNBC interview, Paul Hemp, the senior editor of *Harvard Business Review,* remarked that "seeing the health in health care costs" was #2 on the publication's list of 20 breakthrough ideas. The essence of the story was that the crisis of runaway U.S. health care costs would be addressed more effectively and sustainably if the focus was on the health of populations rather than merely the cost of serving those populations. Put another way, there would be a reduction in the total health care cost (medical/pharmacy and health-related productivity costs) if the burden of illness and health risks on the population was reduced.

Why is this? Because today's reality is that health is a performance driver. Investing in health not only controls expenses but also protects, supports, and enhances human capital. Employee health and employee productivity are inextricably linked. By investing in their most important asset—their people, or "human capital"—progressive employers are showing the way to ultimately controlling health care costs by improving the health and well-being of their employees. Enlightened organizations are already profiting from leveraging this value proposition. Increasingly, business leaders are realizing that the health of the U.S. work force is directly related to the health of the American economy. In fact, the *global* economy would be well served by a bigger investment in preventive health measures. Primary (health risk screening), secondary (early diagnosis), and tertiary prevention (early intervention) are good for society, in general, as well as for business and industry.

Individuals do not leave the impacts of their personal health risks on the doorstep when they leave for work, just as they cannot leave the impacts of their workplace exposures when they return home. Therefore the occupational and environmental health professional is in a unique position to leverage coordinated and integrated health and productivity enhancement strategies that can deal with the whole person and the whole population.

SOUND REASONS FOR MEASURING HEALTH AND PRODUCTIVITY

Employers are concerned about the total impact of illness or wellness on their bottom line. The total health cost impact includes both direct medical costs and related productivity costs due to absenteeism and presenteeism. These productivity costs are typically two to three times the medical/pharmacy costs, as depicted in the health and productivity cost "iceberg" in Figure 16-1 [1–3]. Employers are recognizing that their employees' ability to function as high-performance "corporate athletes" is a key driver of enterprise performance and productivity. Health and productivity initiatives become investments to be leveraged so that these "corporate athletes" are empowered to deliver optimal performance [4].

As leaders in analyzing, understanding and managing the health and well-being of work force populations, occupational health

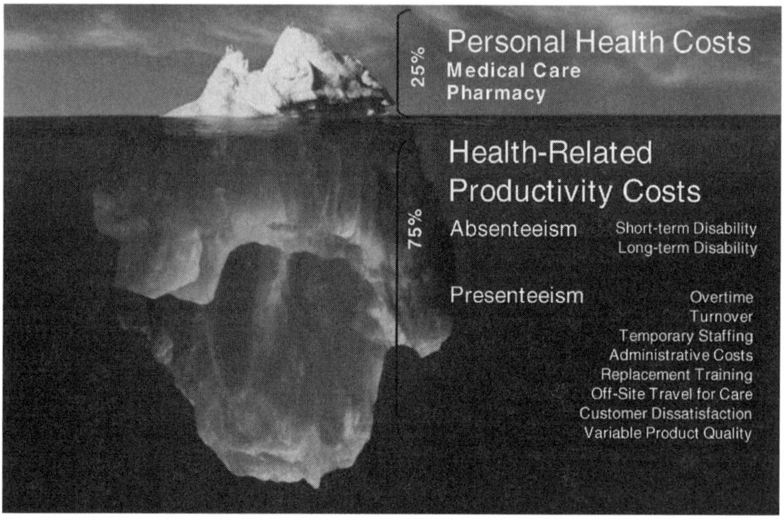

Figure 16-1. Full cost of employee poor health. (Adapted from Edington DW, Burton WN [3] and Loeppke R, Taitel M, Richling D, et al. [8].)

professionals have been on the front-lines of the health and productivity movement for many years [5]. The long-standing mission statement of the American College of Occupational and Environmental Medicine (ACOEM) reads, "Occupational medicine is the medical specialty devoted to prevention and management of occupational and environmental injury, illness and disability, and promotion of health and productivity of workers, their families, and communities."

Since the mid 1900s, occupational health physicians have received board certification through the American Board of Preventive Medicine and occupational medicine has been recognized as a medical specialty. However, in recent years the role of the occupational and environmental health professional has grown beyond preventing and treating work-related injuries and illnesses, including those covered by workers' compensation. The role has expanded to include managing the overall health-related issues of the work force, including personal health benefits and organizational health. This expanded role includes elements of health promotion, prevention, disease management, injury/medical care management, pharmaceutical management, disability/absence management, human resource/benefit management, and contract management with providers. This is reflected in the evolution of terms used to describe their role: from *industrial medicine* to *occupational medicine* to *occupational health* and *corporate health* [6].

Increasingly, the occupational health professional is being placed in the position of being the chief health officer for an employer (see Figure 16-2). He or she is responsible for managing the human capital related to the health and productivity assets of the work force in much the same way that a chief financial officer is responsible for managing the financial capital related to the economic assets of the company. However, managing this human capital in terms of health and productivity requires measuring it. Fulfilling this unique role requires accessing available health- and productivity-related information.

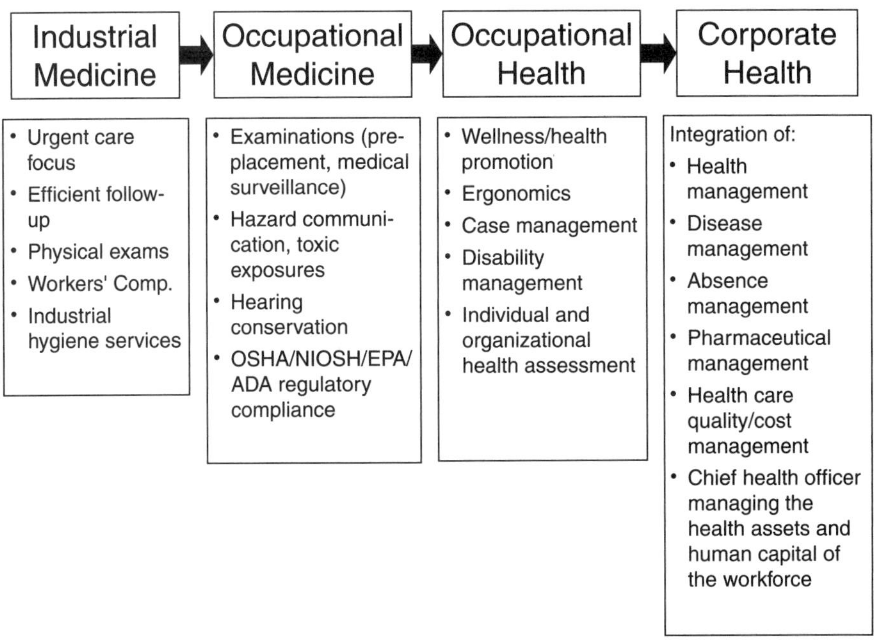

Figure 16-2. Transition to corporate health. (Adapted from Loeppke R [6].)

PRODUCTIVITY INFORMATION REQUIRED TO MAKE AN IMPACT ON AN ORGANIZATION

Occupational health and safety programs directly impact a company's workers' compensation experience. An organization must understand and analyze the number of recordable injuries/illnesses and of lost-time cases, severity of the cases, workers' compensation costs, and temporary disability days to effectively establish programs that impact these areas. Effective disability management programs are not possible without an understanding of employee time away from work due to workplace conditions. Assessment of safety in the workplace is contingent on measurements such as the number of recordable cases of injury and illness and the severity of those cases. One of the most direct ways to improve productivity in the workplace is to reduce the number of work-related injuries and illnesses.

Because of the improvement of workplace safety standards throughout the United States, there has been a significant increase in the productivity of workers with nonoccupational illnesses and injuries as well as those with work-related health problems. Increasingly, the occupational medicine professional is either leading or collaborating in a company's efforts to understand the nature of these nonoccupational injuries and illnesses and to craft programs that will most effectively reduce their occurrence. The ability to measure the costs of temporary or replacement workers is important in fully understanding the total corporate costs of disability and lost productivity.

Other areas impacted by the availability of productivity data include establishment of effective safety and industrial hygiene programs, medical surveillance programs, and integrated disability management programs (combining methods to decrease short-term disability [STD], long-term disability [LTD], and workers' compensation [WC]). Productivity data also facilitates evaluation of immunization programs, travel medicine programs, and general wellness programs offered by the occupational health and safety program.

It is recognized that many employers are struggling to identify the data that need to be captured, how to best record the data, and how to best analyze and interpret the information they have on hand. The occupational health professional is a valuable resource to whom the corporation can turn for data analysis. There are a number of good references that outline the data needed as well as provide a framework of the business rationale for linkage between health and productivity [1,7,8].

DEFINING ISSUES THAT IMPACT THE VALUE OF HEALTH FOR AN ORGANIZATION

Absenteeism

Individuals who are absent from the workplace cannot contribute. Absence can take many forms, and each has a somewhat different effect on the bottom line. For example, an unexpected death in the middle of a critical negotiation can change the direction of an entire company, while a planned pregnancy leave may have a net positive effect during a slack work period. The human capital method (e.g., job-specific human capital) and the friction method attempt to address these differences. Attempts have been made to quantify these business issues with metrics of lost-opportunity costs, training costs, replacement costs, and firm-specific human capital, but no system yet devised fully accounts for them. Absenteeism can be attributed to any of the following:

1. Workers' compensation (WC)
2. Short-term disability (STD)
3. Long-term disability (LTD)
4. Sick leave
5. Family Medical Leave Act (FMLA) absences
6. Absence for parent-teacher organization (PTO) attendance
7. Unpaid leave
8. Death (premature mortality costs)

Presenteeism

The term *presenteeism* describes an employee who is present at work but limited in some aspect of job performance by health problem(s) [4]. Presenteeism is a measure of presence at work and work capability. A decrease in the latter may range from slight deconditioning in a competitive athlete to mania in a stock trader. Presenteeism management includes:

1. Time on task: in the workplace, but not working (e.g., "sleeping it off")
2. Quality of work: incidence and magnitude of mistakes (e.g., injury rates, product waste, product defects)
3. Creativity
4. Executive functions (e.g., initiative)
5. Quantity of work: amount of work output
6. Capacity for peak performance
7. Interpersonal factors
    a. Social function (positive morale impact of optimally healthy)
    b. Addictions (manipulative behaviors)
    c. Personality disorders (criminal behaviors)
    d. Mood disorders (irritable with customers)
8. Motivational effect (Observing a sick employee's low output lowers the expectations of fully healthy workers for their own performance.)
9. Work culture (The collective health or impairment level creates a culture that has a positive or negative feedback on the health of the group.)

Other health issues that impact the productivity of the work force, and thus the profitability of the corporation, are described below:

1. Turnover/replacement costs are lower in a healthy work force and in a work force that has corporate loyalty because of the company's commitment to health.
2. A healthy work culture and good morale facilitate recruitment and allow for lower salary expenses.

3. Family health, that is poor health of a family member, can increase both absenteeism and presenteeism in an employee.

4. Poor employee and family health has a negative impact on experience-rated health and accident insurance rates.

APPLICATION OF HEALTH AND PRODUCTIVITY MEASUREMENTS

Conceptually, there are numerous applications of health and productivity measurements in occupational health management. In general, these can be grouped into two broad categories:

1. *Minimizing loss:* Applications intended to minimize loss of direct and indirect dollars or time associated with health and productivity
2. *Maximizing potential:* Applications intended to maximize the human potential and performance of the work force in an effort to produce greater up-side value for the organization through their optimal ability to contribute

Certainly there is a continuum that connects both categories, but the focus and underlying intent is important to note. In either of these broad categories, the following applications are identified:

1. Business case development
2. Quantification of health-related productivity costs
3. Data to direct priorities for investment in programs and services
4. Creation of a baseline from which to evaluate changes
5. Evaluation of overall efforts to impact health and productivity
6. Evaluation of a specific program or service
7. Part of performance metrics for a vendor
8. Part of a stress management system or other health surveillance

Overall, the productivity emphasis and data are very useful in the strategic and tactical level of planning and management for programs and services targeting the overall health of the work force. When integrated with the general array of metrics used by occupational health professionals, productivity data adds an important new dimension that is essential to managing overall investments and outcomes in work force health.

Measurement of Health-related Productivity

A distinction should be made between measurement of actual productivity and measurement of the health-related component of productivity. Many of the available instruments target the intersection between the impact of health on the ability to work and work output. There are certainly other factors that can impact productivity other than a person's health. When discussing measurement efforts with management and workers, care should be taken to acknowledge what is actually being assessed. Certainly there is great sensitivity on the part of the employee who knows that someone is measuring their productivity. In order to ensure high participation rates and accurate responses when self-reporting instruments are used, employees need to trust in the process and the maintenance of confidentiality, similar to that which applies to other personal health information.

A checklist for health and productivity management is available through ACOEM's toolkit [7], which facilitates a preliminary evaluation of a specific organization's level of health and productivity management. The components measured at a corporation or at an individual operating-unit level include:

1. General philosophy
2. Organizational structure
3. Goals, metrics, and other measurement tools
4. Interventions and outcomes
5. Integration and trends

As a prerequisite to applications of measurement, an organization or key leaders in the organization would have to recognize and appreciate the view that health is related to both health care costs preciate the view that health is related to both health care costs and productivity costs and that presenteeism is a very significant and valid construct. In the case of the second category, an organization would likely have adopted a perspective of human capital or health assets as essential to the success of the organization and would allow for the possibility that movement toward optimal productivity is potentially linked with profitability.

More specific information on the presenteeism within a corporation is obtained by using validated instruments. In selecting instruments for measurement of health-related productivity, it is important to consider the specifics of the application and the work force, since the focus of instruments varies. In fact, there is an increasing availability of resources to evaluate some of the instruments and survey tools designed to measure presenteeism and absenteeism [1,4,7].

The results of various measures can then be used to support decision making. When reviewing the burden of illness within an organization, it is helpful to focus on the aggregate information gathered from medical claims, disability data, health risk, and presenteeism/absenteeism measurement tools. This gives a broader view of the health of a company's population. Incorporating information from health assessments, which cover health risks and health-related productivity measures, can point to areas for preventive program development.

FINDINGS FROM AN ACOEM IBI MULTI-EMPLOYER RESEARCH STUDY ON TOTAL COST

A research study of employers' total health-related costs, funded by a grant from the National Pharmaceutical Council, is being conducted by the American College of Occupational and Environmental Medicine (ACOEM), the Integrated Benefits Institute (IBI), and the Harvard Medical School. The study was designed to describe the medical conditions in a work force and their impact on the particular employer's total health-related costs [8]. The study sample included 15,380 employees out of a total of 57,666 employees total among four employers.

Through the use of very sophisticated and complex analytic methodologies, the study was able to match the medical/pharmacy claims with the Harvard HPQ (World Health Organization's Health and Work Performance Questionnaire) survey tool presenteeism/absenteeism measures at the de-identified (i.e., individual identifying information removed), individual worker level, then aggregated that data across multiple employers' work forces to demonstrate the total mean health-related costs according to medical conditions. Figures 16-3 and 16-4, and Table 16-1 show the findings, which reveal that the top 10 medical conditions impacting the medical/pharmacy costs are not the same as the top 10 medical conditions impacting the total health-related costs, which include productivity losses [8].

VALUE-BASED HEALTH PROGRAM DESIGN

Corporations are looking for ways to foster employee engagement in their health and rewarding them for this engagement. It is important to look at the value in the overall design of the health pro-

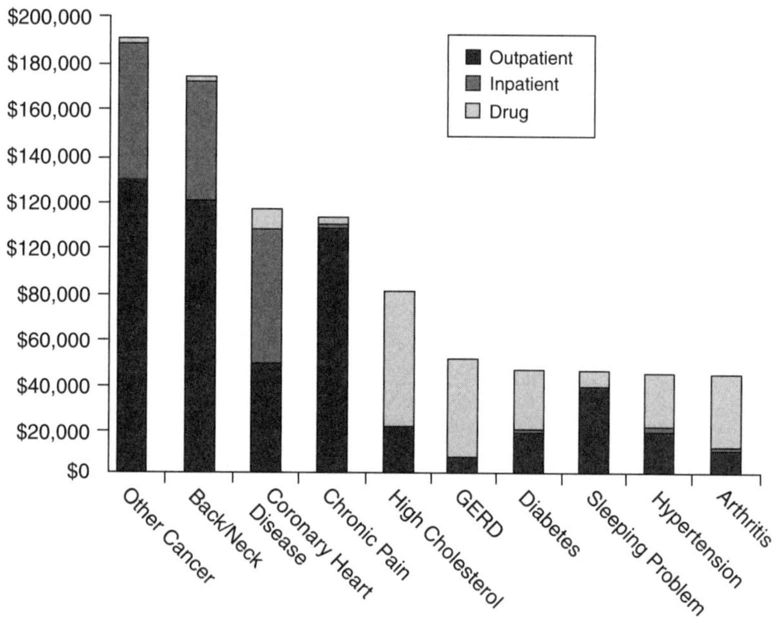

Figure 16-3. Top 10 medical conditions by medical and pharmacy cost per employee per year (4 employers; N = 15,380 employees). (From Loeppke R, Taitel M, Richling D, et al. [8].)

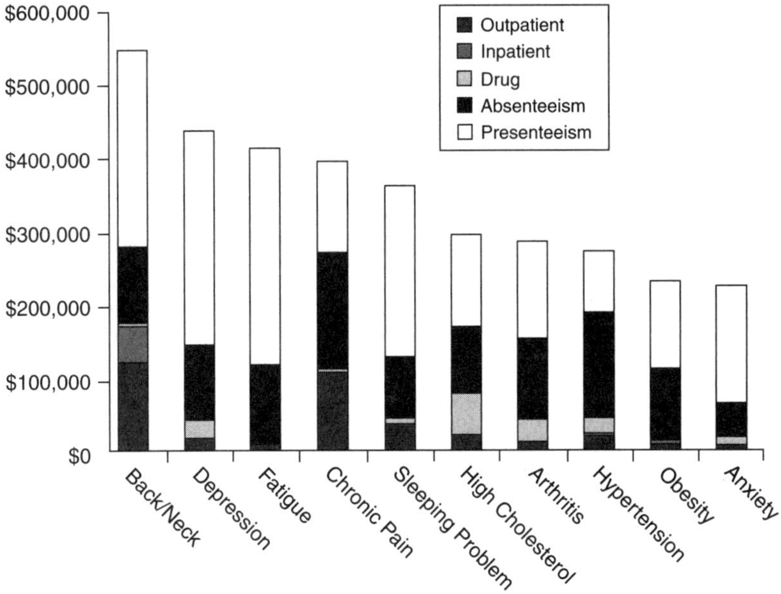

Figure 16-4. Top 10 medical conditions by total cost per employee per year: medical + pharmacy + presenteeism + absenteeism costs (4 employers; N = 15,380 employees). (From Loeppke R, Taitel M, Richling D, et al. [8].)

gram plan, not just at its cost. In 2004, a study by Lynch and colleagues reviewed the findings of a survey that evaluated how employers consider value when making policy and benefit design decisions about health [9]. A review of 32 value-focused activities was conducted, and it was found that employers, on average, used 5.2 activities. The primary activities used were providing access to flu shots, providing centers of excellence, and providing wellness programs.

Companies with the greatest focus on value more commonly chose to offer health promotion programs and incentives and to design their drug benefit to manage outcomes other than cost. For example, Pitney Bowes set about designing an innovative pharmacy-benefits program, which focused on ensuring that employees obtained the drugs and medical care they needed to attain optimal health. A data analysis showed that high-cost, poor-health populations included people with chronic diseases who were not taking

Table 16-1. Top 10 Medical Conditions by Cost Categories (4 Employers; N = 15,380 Employees)

	Medical	Pharmacy	Medical and Pharmacy	Productivity	Total Cost
1	Cancer	High cholesterol	Cancer	Fatigue	Back/neck
2	Back/neck	GERD	Back/neck	Depression	Depression
3	Chronic pain	Arthritis	Heart disease	Back/neck	Fatigue
4	Heart disease	Diabetes	Chronic pain	Sleeping problem	Chronic pain
5	Sleeping problem	Depression	High cholesterol	Chronic pain	Sleeping problem
6	High cholesterol	Hypertension	GERD	Arthritis	High cholesterol
7	Hypertension	Asthma	Diabetes	Hypertension	Arthritis
8	Diabetes	Allergy	Sleeping problem	Obesity	Hypertension
9	Headache	Anxiety	Hypertension	High cholesterol	Obesity
10	Depression	Heart disease	Arthritis	Anxiety	Anxiety

From Loeppke R, Taitel M, Richling D, et al. [8].

their medicine, those who did not have easy access to a doctor, and employees who never set foot in a physician's office. These were the employees most likely to be the source of higher corporate health-related costs [10].

David Hom and Jack Mahoney used the information to build a business case for changing the benefits design, eventually investing $4.5 million on top of the approximately $135 million the company was already spending. Two years later the company introduced its value-based benefits design. The company lowered the cost of drugs that employees needed for certain chronic conditions. Gone were controls that forced people to take only generic medications or to try generics before a brand-name drug. The goal was simply to lower the barriers that kept people from taking the drugs that, in the long run, would keep them healthy. The result has been a dramatic decrease in the cost of the corporation's medical benefits plan [10].

In a specific review of the impact of the program on diabetic patients, an article by Berger [11] showed that for the typical plan participant with diabetes, the average cost of a 30-day prescription decreased by 50 percent. Outcome assessments indicated that rates of compliance with prescribed diabetes medication increased significantly in response to the change in pharmacy benefits. Suboptimal compliance with insulin therapy decreased by two-thirds, and usage of blood-glucose meter test strips increased from 28 percent to 55 percent. An increase from 9 percent to 22 percent in compliance with single-pill oral diabetic regimens was seen [11]. While the company's total annual pharmacy costs per person increased slightly from about $26 per month to $35 per month, pharmacy costs for individuals with diabetes actually decreased by 7 percent [12].

ALIGNING PERFORMANCE INCENTIVES

The 2004 ACOEM value-focused activity survey found that despite publicity about pay-for-performance, value-focused provider strategies were generally not reported and were not among those activities being considered most often by employers [13]. However, this may have changed in the few years since completion of the survey. In a number of current efforts, pay-for-performance initiatives are being supported by numerous large employers.

The Institute of Medicine (IOM) report from the Committee on the Quality of Health Care in America addressed issues of quality care in the American health care system and made recommendations for a wide variety of key stakeholders to improve the effectiveness and efficiency of health care [14]. Employers are seeking an alternative health care delivery product focused on value-based contracting and integrated-population health-management programs with quality-driven cost-savings incentives. They are demanding effective care management and productivity management, using quality and total value as the drivers [1,7]. Poor quality includes unnecessary surgeries, duplication of services, and inadequate disease management as well as, for example, poor communication and coordination among physicians and other providers [15].

The Leapfrog Group (LeapFrogGroup.org), launched in 2000 to mobilize employer health-care purchasing power, was influenced by the 1999 IOM report on preventable medical mistakes [14]. This report recommended that large employers provide more market reinforcement for the quality and safety of health care. Their mission is to trigger giant leaps forward in the safety, quality, and affordability of health care by (1) supporting informed health care decisions by those who use and pay for health care and (2) promoting high-value health care through incentives and rewards. In addition, Bridges to Excellence, led by General Electric, Xerox, and other so-called pay-for-performance initiatives, is successfully breaking new ground in this arena [16].

With a focus on improved outcomes and information systems, three large corporations—Cisco, Intel, and Oracle—are supporting a pay-for-performance program for health IT (information technology) adoption among eight large medical groups in the Silicon Valley. The Silicon Valley Health Information Technology Initiative (SVHIT) focuses on the adoption of electronic medical records, e-prescribing, and a secure patient messaging system in order to enhance medical outcomes for physicians and patients. The pilot program is currently in its first year.

In an effort to increase data sharing, thus improving the quality of health care delivery and lower its costs, Secretary Leavitt, U.S. Department of Health and Human Services, launched the Value-Driven Health Care Initiative, a public–private effort. Secretary Leavitt's four cornerstones of value-driven health care are:

1. Interoperable health information technology
2. Standardized and transparent quality measures
3. Transparent price information
4. Incentives for high-quality and efficient health care

The first element of this initiative aims at national coordination by calling on all health care stakeholders to commit to public reporting on quality and costs, including recognition of consensus standards of care. The initiative also supports interoperable health information technology and incentives for value purchasing in health care.

National Business Group on Health, which represents over 250 employers, has urged its members to support transparency and the other cornerstones of Secretary Leavitt's initiative, including (1) a commitment to providing information about performance and quality of care delivered by providers who serve enrollee's in the company health plan, (2) a commitment to provide health care price and cost information to health plan enrollees, (3) a commitment to encouraging health plans and their providers to use standards of interoperability for health information technology products, and (4) a commitment to providing incentives to use these tools for achieving high-value health care for patients.

SUMMARY

There is a clarion call to action to engage in the battle to improve the health and productivity of our nation's work force. But make no mistake, our enemies in this battle are disease, disabling medical conditions, and serious, some dangerous, health risks. Our battlefields are the workplaces, the clinics, the hospitals, and the communities across the United States. Our weapons are quality-driven, cost-effective prevention-oriented health care initiatives that align incentives of the key stakeholders to improve the health and productivity of the U.S. population.

Fighting this battle requires inspired vision and focused leadership on the part of all health care professionals. In fact, together, we should assume a mantel of leadership and harness our intellectual capital as well as mobilize our collective consciousness in order to improve health and human performance. Based on the employer case studies and data described, one can define a compelling business case, delineate the tools to use, and determine an action agenda to implement. However, the battle we face in these times of health care cost crisis will be won not by better definition, delineation, and evaluation but by *action* on the front-lines. It is not what we know, but what we *do* that will make the difference.

So, let us, as health care professionals, meet the challenge. We are in the unique position to stimulate local market initiatives as well as influence national policies. We can also empower the cooperation of key stakeholders in a more rational, more accountable health care system—if we implement at a local level and influence at a national level. We have an unparalleled opportunity to accomplish this on behalf of workers, employers, patients, providers and the communities we serve. In fact, we can reposition our current health care system and transform it into a system *for health*—one built on the cornerstone of quality, on the foundation of health and productivity improvement, and on the pillars of prevention. As mentioned at the beginning of this chapter, "The Ambulance Down in the Valley," a poem written by Joseph Malins in 1895, is an eloquent statement of the value of prevention [17].

Instead of looking over our shoulder at the "good old days of medicine," let us look forward to the "good *new* days of medicine" on the horizon.

REFERENCES

1. Loeppke R, Hymel P. Good health is good business. *J Occup Environ Med* 2006; 48(5):533–537.

2. Brady W, Bass J, Moser R Jr, et al. Defining total corporate health and safety costs' significance and impact: review and recommendations. *Occup Environ Med* 1997; 39(3):224–231.

3. Edington DW, Burton WN. Health and productivity. In: R. McCunney (ed), *A Practical Approach to Occupational and Environmental Medicine,* 3rd Ed. Philadelphia: Lippincott, Williams & Wilkins, 2003. Chap. 11, pp 140–152.

4. Loeppke R, Hymel PA, Lofland JH, et al. Health-related workplace productivity measurement: general and migraine-specific recommendations from the ACOEM Expert Panel. *J Occup Environ Med* 2003;45(4):349–359.

5. Harris J, Loeppke R. *Integrated Health Management: The Key Role of Occupational Medicine in Managed Care, Disability Management, Productivity, Prevention and Integrated Delivery Systems.* Arlington Heights, IL: American College of Occupational and Environmental Medicine, 1998.

6. Loeppke R. Prevention and managed care: the next generation. *J Occup Environ Med* 1995;37(5):558–562.

7. American College of Occupational and Environmental Medicine. *ACOEM Health and Productivity Tool Kit.* Arlington Heights, IL: ACOEM, 2006.

8. Loeppke R, Taitel M, Richling D, et al. Health and productivity as a business strategy. *J Occup Environ Med* 2007;49(7):712–721.

9. Lynch WD, Riedel JE, Hymel PA, et al. Factors affecting the frequency of value-focused health activities and policies by employers. *J Occup Environ Med* 2004; 46(11):1103–1114.

10. *Workforce Management* January 29, 2007. Pp. 1, 12–17.

11. Berger J. Economic and clinical impact of innovative pharmacy benefit designs in management of diabetes pharmacotherapy. *Am J Managed Care* 2005;13(2):S55–S58.

12. Mahoney J. Reducing patient drug acquisition costs can lower diabetes health claims. *Am J Managed Care* 2005;11(5):Suppl.

13. Lynch WD, Riedel JE, Hymel PA, et al. Factors affecting the frequency of value-focused health activities and policies by employers. *J Occup Environ Med* 2004;46(11): 1103–1114.

14. Institute of Medicine. "Crossing the Quality Chasm: A New Health System for the 21st Century" Report, Washington, DC, 2001. Available at: www.nap.edu

15. Midwest Business Group on Health. *Reducing the Costs of Poor-Quality Health Care through Responsible Purchasing Leadership.* Chicago: MBGH, 2002.

16. Dudley RA. Pay-for-performance research: how to learn what clinicians and policy makers need to know. *JAMA* 2005;294(14):1821–1823.

17. Malins J. *The Ambulance Down in the Valley,* 1895. Available at: www.wealthandwant.com/docs/Malins_ambulance.html

CHAPTER

17

Health Promotion and Disease Prevention in the Workplace and Community

Beginning in the late 1960s and the early to mid-1970s, increasing interest in health promotion and disease prevention (HPDP) occurred. In part, this was because the social and political climate in the United States encouraged healthy lifestyles, with the result that jogging, playing tennis, concern about cholesterol, and similar aspects of HPDP programs became popular. Corporations supported and encouraged such activities by providing exercise and other lifestyle-enhancing equipment. A second aspect that favored HPDP efforts was the increasing costs of employee health care for employers. During the 1970s through the 1990s, the costs of employee health care benefits increased dramatically, with increases repeatedly exceeding the annual growth in the U.S. economy [1,2]. By 1990, corporate health benefits had increased to the point that they represented 45 percent of corporate after-tax profits [2,3]. During this period, the number of worksite health promotion and disease prevention (HPDP) programs also increased markedly [4–7]. Part of this increase reflected efforts by corporations to control the spiraling health care costs. Other factors that contributed to the increase included the Surgeon General's 1979 Healthy People Report, which emphasized the need for lifestyle changes to improve health [8]. The National Safety Council reported that work injuries alone cost Americans $132.1 billion, the equivalent of $970 per worker [9]. As noted in Chapter 15, earlier studies had indicated that the economic burden for occupational and nonoccupational illnesses and injuries in 119 million U.S. workers was $1.25 trillion, considering both direct and indirect costs [10].

Costs for workplace illnesses and injuries was estimated at $171 billion [11].

All of these factors combined to underscore the potential for worksite HPDP programs to improve the health and safety of workers and simultaneously reduce health care costs. The advantages of worksite efforts include the potential to focus on a single population group so that program outcomes could be evaluated and the support of management for the efforts. Similar situations may exist in other population groups and communities in which the effects of the interventions can be measured. Recent studies have considered methods to develop population-based public health promotion efforts [12] and implementation of evidence-based clinical and community strategies to improve health [13]. The challenges of implementing HPDP programs within communities may be complex, but the involvement of community leaders in the effort may be very advantageous. For example, challenging a city or community to lose several "tons" of weight collectively may pay important dividends in increasing awareness of healthy lifestyles and the benefits of such population-wide programs. In essence, the "lessons learned" from worksite HPDP programs can be translated to other populations, including communities [13].

WORKSITE HPDP PROGRAMS AS EXAMPLES

Approaches to implementing HPDP programs at worksites have varied greatly, and these approaches have included many new methods. As one investigator noted in 1993, "More significant and effective innovations in health promotion and disease prevention have taken place in worksites, from small to large, in the last 10 years than since the Industrial Revolution" [14]. Evaluation of many of these different approaches is now being accomplished. The increasing efforts to develop, implement, and evaluate HPDP programs in the workplace set the stage for OEHS managers to design effective programs that have the potential to greatly enhance the health, productivity, and enjoyment of life of large numbers of workers and their families. For purposes of this discussion, the term *HPDP programs* refers to those activities concerned with reducing health risks through changed human behavior. Specific

HPDP components include screening, health education, interventions, and evaluation. In a number of organizations, the term *wellness* is used to describe these programs, and wellness activities will be considered as equivalent to HPDP efforts. Although sometimes included as part of a company's overall HPDP program, employee assistance programs (EAP) will not be included in this discussion. These activities are designed to assist employees with personal problems such as alcohol and drug abuse, mental illness, financial difficulties, family conflicts, and similar concerns. While vitally important, EAP efforts are frequently structured and administered differently from other HPDP activities, and consideration of EAP programs is beyond the scope of this chapter.

Advantages And Benefits of Workplace HPDP

The workplace offers particularly attractive opportunities to accomplish effective HPDP activities. In such settings, there are opportunities for scheduled programs that permit effective utilization of the time of both workers and those conducting the programs. The population may be relatively stable and, particularly where stability does exist, there may be excellent opportunities for follow-up to ascertain the results of HPDP efforts. Frequently, corporations routinely provide personnel, facility, and equipment support, thus overcoming some of the financial constraints that can jeopardize community-based HPDP programs [15]. Since workers often play leadership roles in their families, improved health practices may be extended to other family members. Thus, efforts expended on worksite HPDP programs may pay significant health dividends beyond those that might be expected from investments in the worker participants. This consideration magnifies the importance of effective HPDP programs in the workplace.

A number of benefits have been reported from effective HPDP programs. These range from reduced health care costs to reduced absenteeism, improved morale, and increased productivity [7,8,16–19]. Table 17-1 lists some of the benefits that have been attributed to HPDP programs, although all are not necessarily well validated.

Table 17-1. Possible Benefits of HPDP Programs

I. Benefits to Employers
   A. Improved health and safety of workers
      1. May be extended to families of workers
   B. Reduced health care costs
      1. Direct, e.g., costs of therapy, insurance premiums, hospitalization, medications, and similar factors
      2. Indirect, including impact of health-related absences on productivity, cost of training replacements, litigation costs, and comparable items
   C. Improved worker morale, job-related attitudes, and goodwill toward the company
   D. Reduced absenteeism
   E. Reduced employee turnover and better retention
   F. Increased productivity
   G. Improved competitive advantage
   H. Enhanced corporate image, including evidence the company cares about the worker
II. Benefits to the Employee
   A. Reduced health-related out-of-pocket expenses
   B. Reduced use of sick leave
   C. Reduced loss of time (personal and work), due to transportation/waiting time for health care
   D. Increased co-worker and employer support due to demonstrated efforts to improve health
   E. Increased personal morale
   F. Improved family health
   G. Improved quality of life and sense of well-being
III. Potential Benefits to Society
   A. Reduced health care expenditures
   B. Increased productivity and international competitive advantages
   C. Expanded interest in HPDP activities

PROGRAM COMPONENTS

HPDP programs can vary from single intervention efforts to comprehensive prevention programs. The 1999 National Health Promotion Survey, conducted by the Office of Disease Prevention and Health Promotion, noted that 90 percent of workplaces with 50 or more employees offered one or more health promotion activities, compared to 81 percent in 1992 and 65 percent in 1985 [20]. (Table 17-2 lists some activities that have been used in worksite programs [1,4,5,7,8,18]. Selection of particular components depends on numerous factors, including applicability, available data

Table 17-2. Potential Components of Worksite Programs

1. Health Risk Assessment Questionnaire
2. Exercise and Fitness
3. Back Care/Injury Prevention
4. Smoking Cessation
5. Blood Pressure Screening and Control
6. Nutrition Guidance and Weight Control
7. Cholesterol Evaluation and Control
8. Cancer Screening
9. Stress Management and Reduction
10. Accident/Injury Prevention, particularly use of vehicle personal restraints
11. Glaucoma Screening
12. Specific Job-related Analyses and Preventive Intervention (e.g., Safety and Ergonomics Efforts)

regarding effectiveness, desires of corporate management, and worker interest.

The most frequent types of activities, as reported in one survey, are as follows [5]:

Activity	Percent of Respondent Worksites with Activity
Smoking cessation	35.6%
Health risk assessment	29.5%
Back care	28.6%
Stress management	26.6%
Exercise/fitness	22.1%
Off-the-job accident prevention	19.8%

Another survey of companies with 500 or more employees revealed a somewhat different picture [21]. Fifty (50) percent of respondents had some HPDP activity. Stress management was reported most frequently, and other activities (in decreasing frequency) were blood pressure screening, smoking cessation, nutrition education, and exercise/fitness enhancement. The 1999 survey reported blood pressure control, cholesterol evaluations, and cancer screening were the activities provided most frequently [20]. An often overlooked, but potentially important HPDP effort, is provision of annual influenza immunizations to workers and other population groups [22]. More than 85 percent of private worksites responding to a survey by the Office of Disease Prevention and Health Promotion (HPDP) opened activities to all employees, and approximately 30 percent also made these benefits available to dependents of workers and retirees [5].

STAFFING OPTIONS

Staffing of HPDP programs varies from place to place. Reflecting, in part, limited physician availability, a number of businesses have established HPDP programs that do not have any physician participation. For example, professionals such as occupational health nurses, health educators, and exercise physiologists may be well qualified to conduct such efforts; they have taken the lead in numerous locations [23]. Often, these programs involve a team of individuals. However, HPDP professionals tend to concentrate on activities involving larger populations that will support a full-time HPDP staff. In still other situations, and particularly in smaller organizations, interested members of the human resources (personnel) or other departments have initiated programs, on occasion without health professional input. Table 17-3 lists the professions and capabilities of some of those involved in HPDP programs, either individually or as part of a team [23,24].

Surveys of HPDP programs have provided information regarding the types of individuals involved in conducting HPDP activities. A study of the Fortune 500 companies revealed nurses were employed in 74 percent of the HPDP programs, physicians in 60 percent, health educators in 39 percent, and human resources/personnel officers in 39 percent [23]. Eighty-three (83) percent of the programs employed at least one health professional. Voluntary or nonprofit organizations, private (for-profit) providers and consultants, local hospitals, and insurance companies may also provide program assistance (see Table 17-3).

EXAMPLES OF HPDP PROGRAM EXPERIENCE

The following discussion reviews the experiences of several organizations to illustrate the scope of HPDP programs in the workplace and the type of results that can be achieved. Some of the concerns involved in attempting to evaluate the effectiveness of HPDP efforts involving workers are also delineated.

Table 17-3. Examples of Professionals Involved in Conducting HPDP Programs

Physicians, including occupational medicine, sports medicine, family practice, and other primary care physicians
Nurses, including occupational health nurses and nurse practitioners
Health educators
Exercise physiologists
Nutritionists
Ergonomists
Recreational therapists
Psychologists
Staff members of nonprofit health organizations
Consultants with varying backgrounds, e.g., MBA

The Beginnings

Procter and Gamble. Procter and Gamble implemented a work-site health promotion program called *Health Check* at its Cincinnati headquarters. Investigators compared total and lifestyle-related medical care costs for 3,993 employees who participated in the Health Check program with 4,341 employees who did not participate [1]. Age and gender adjustments revealed that participants in the third year of the program had significantly lower total and lifestyle-related health care costs, significantly lower inpatient costs, fewer hospital admissions, and fewer hospital days of care than nonparticipants. A greater reduction in lifestyle-related costs occurred than for total health care costs, although both reductions were significant. Of note is the finding that no significant differences were observed during the first two years of the study, emphasizing the importance of longer-term participation in HPDP programs as well as evaluation over longer time periods.

Johnson & Johnson. In 1979, Johnson & Johnson initiated "Live for Life" (LFL) programs, and these were progressively incorporated at various Johnson & Johnson locations during subsequent years. By 1986, LFL efforts were in place at 50 Johnson & Johnson locations employing 28,000 workers. Components of the program include a health screen, lifestyle seminar, and lifestyle improvement activities such as smoking cessation, weight control, stress management, nutrition education, fitness, and blood pressure control.

In an evaluation report [25], increase in inpatient costs of LFL participants over a five-year period were compared with those of a control, nonparticipant group. Mean annual increases were $43 and $42 for two LFL participant groups vs. $76 for non-LFL participants. The investigators noted it was not reasonable to project similar cost savings in future years, but the reported findings indicated that important benefits could result from LFL programs. Others have criticized the study because it primarily produced reductions in length of hospital stay, rather than admissions, and did not reduce outpatient costs [26]. It was also noted that the control groups were not comparable with respect to residence location and income [27]. Unfortunately, as the LFL investigators observed, diagnostic data were not available for the study. Such information could have aided in delineating specific areas wherein improvements occurred, such as respiratory problems related to smoking. In spite of these limitations, savings of over $980,000 were estimated for the five-year study period.

AT&T. At AT&T Communications, participants were given a health risk appraisal and offered health education modules on fitness, reduction of backache, weight control, stress reduction, smoking cessation, cholesterol reduction, cancer screening, nutrition, and interpersonal communication [28]. One comparison group was given the health risk appraisal, including a health promotion orientation, while a second control group was neither given the appraisal nor offered the modules. Evaluation one year after program initiation disclosed significant improvements in diastolic blood pressure, serum cholesterol, type A behavior, and body weight in participants at some of the locations. There were also increased exercise activity and decreased smoking among module participants. Participants were more committed to improving their health-related behaviors and felt more positive toward AT&T and their co-workers/supervisors.

Concerns and More Recent Answers

While such results are encouraging, various investigators have urged caution in utilizing the information. For example, one group reviewed the literature on workplace health promotion programs published through early 1986 [29]. The investigators found that, in general, "claims of HP programs' profitability are based on anecdotal evidence or analyses seriously flawed in terms of assumptions, data, or (study) methodology." It was also noted that the lack of sound studies should not necessarily be interpreted as a negative assessment of benefits of HPDP programs at the worksite. However, the deficiencies did indicate the need for sound, research-based information.

A subsequent review of HPDP programs, focusing on cardiovascular risk management programs, found that none of the 12 studies examined had adequate analyses of cost effectiveness [30]. However, the investigator noted that the majority of research on such programs indicated favorable clinical and cost outcomes and that more recent and more rigorously designed studies tended to support rather than refute the studies with design flaws. Further, the methodologic flaws may reflect the fact that programs have been initiated in many types of worksites with diverse employees, different scopes and types of interventions, and differing degrees of methodologic sophistication.

More recent reports of results of HPDP programs have been based on more extensive biostatistical analyses that enhance the validity of the results [1,2,4,8,31]. In their excellent discussion, McCauley and McCunney summarize data from multiple studies that have demonstrated the effectiveness of HPDP programs and risk analyses [8]. Pelletier published a series of articles that describe the increased quality of research design and data analysis in HPDP programs during the time periods 1991–1993, 1993–1995, and 1995–1998 [32–35].

Even at a worksite, problems exist in obtaining HPDP research data [16]. For example, employers may be reluctant to restrict one group from a program to serve as a control group, and appropriate matching of control and exposed groups may be difficult. Further, the costs of evaluation processes may be equal to or greater than the cost of the program itself [16]. Emphasizing the absence of evaluation data, one survey found that 47 percent of those with HPDP programs did not use needs assessment, evaluation, or cost analyses [23]. Other studies have suggested small companies, in particular, are likely to implement wellness programs in order to enhance morale or for other reasons not lending themselves to rigorous analysis [7,16,36]. As various investigators have observed, the decision to implement HPDP programs may require minimal resources, in relation to other personnel costs, and thus the decision does not justify extensive evaluations [36]. In such situations, those in charge of the resources necessary for HPDP programs may decide that they want to implement the activities even though no sound economic analyses are available to support the decision for their particular organization.

PROGRAM PLANNING REQUIREMENTS

Prerequisites will, of course, vary with different organizations, but a number of requirements are common to HPDP programs in the workplace [6,8,24]. These include:

1. Determine goals of management and employees.

2. Evaluate potential or available resources within the organization, the community, and among employees.

3. Ascertain the availability and validity of baseline data, e.g., health care costs, absenteeism, workers' compensation (WC) costs and multipliers, and similar items.

4. Decide on the program content, using input from management, employees or other targeted group, union representatives, human resources, potential HPDP staff members, community sources, and others.

5. Depending on the availability of baseline data, consider health risk assessment as an initial step.

6. Design the program, enlisting a sufficient number of HPDP personnel who have the appropriate expertise, and identify realistic goals, appropriate budget, reasonable implementation plan, and effective evaluation techniques.

7. Ensure that the HPDP director(s) obtain appropriate management support as well as approval.

8. Accomplish necessary communication with management, workers or other targeted group, HPDP staff, and others involved in the program.

9. Implement the program, utilizing the approved implementation plan.

10. Accomplish planned evaluation of program management and processes, outcome measurements, and other criteria, including user feedback, as established in the program plan.

11. Modify the program as indicated by evaluations and then re-evaluate.

PROBLEMS, CONSTRAINTS, AND OTHER CONSIDERATIONS IN IMPLEMENTING HPDP PROGRAMS

In carrying out the steps necessary to implement programs, directors can benefit from the experiences of predecessors who have defined a number of aspects that are of particular concern in establishing effective programs. The following discussion considers some of these factors.

Consider Modular Approaches in Planning

Recognize in planning that it may not be possible to implement all desired portions of a particular HPDP program at one time. In such instances, it may be beneficial to develop a modular approach. Under this concept, one or more facets are implemented initially and others are added as resources permit and program effectiveness is demonstrated to those sponsoring the program. Typically, the highest priority modules are implemented first. (See discussion of the Pareto approach in Chapters 3 and 15.)

Develop Realistic Expectations

In contrast to effective programs, poorly planned or designed programs can adversely affect opportunities to provide HPDP programs to populations. In many instances, unrealistic expectations have resulted in loss of credibility and cessation of activities. Thus, it should be recognized that 100-percent participation is not a reasonable expectation. Indeed, typical long-term participation rates at on-site locations have been reported to be as low as 20 to 40 percent and, for offsite activities, 10 to 20 percent [6,23]. Numerous factors may affect participation rates, including age, socioeconomic status, work schedules, costs, times the program is available as well as other aspects of accessibility, and similar factors [6,15,37]. Gender of the work force may also affect participation rates [38]. In establishing measures of success, it is reasonable to expect programs to produce measurable results in lowering cholesterol levels, decreasing the number of smokers, reducing elevated blood pressure levels, and increasing seat belt use, provided sufficient time is planned for accomplishing these results. It is *not* appropriate to expect programs to produce major reductions in health care costs within a brief period after implementation or to see rapid reductions in absenteeism rates [1].

Obtain Approval, Commitment, and Support from Management and Members of Targeted Groups

In many situations, programs have been greeted enthusiastically by workers or community members and management or community leaders. Over time, however, initial enthusiasm has waned, and HPDP activities have decreased in scope or have been eliminated. In some instances, this has occurred because initial "approval" by management was interpreted as active "support and commitment" by those directing the programs. Repeatedly, it has been observed that a successful HPDP effort must have both approval and dynamic commitment/support by management at all levels. One visible method of demonstrating support is for managers or community leaders to serve as appropriate role models for the HPDP activities. They, as well as HPDP staff and other medical personnel concerned with health support at the worksite, should reflect an appropriate image. A program's chances for success are markedly reduced if executives or health care providers caring for workers use cigarettes, are overweight, are known not to use seat belts, or question the benefits of HPDP activities.

Conversely, visible participation by a mayor or other community leader may significantly enhance participation rates. Even military groups may benefit from active involvement of senior officers, and such visibility may overcome reluctance of members to participate even though "ordered" or "required" to do so. For example, receiving annual influenza immunizations is a "mandatory" military requirement; however, each year participation rates at one Air Force base were approximately 70 percent or lower. Although the personnel agency was theoretically responsible for the program, the medical facility responsible for administering the immunizations was typically blamed for the low participation rate. I approached the commanding general at one base at the start of the immunization season to get his support, and he told me in no uncertain terms he was not interested, as the "shot" always made him sick! I replied that I had a "great deal" for him: If he would let me give him a shot that was "guaranteed" not to make him sick (since I know of no reaction to normal saline), I would let him give me a shot. His eyes lit up and he quickly agreed. At the next weekly meeting of senior officers, we exchanged "shots," with appropriate base press coverage and the general grinning as he "got even." After much laughter and applause after the event, the general announced that I had arranged for everyone present to get their "shot," and the group exited to line up in rows leading to tables covered with white sheets and upon which there were filled syringes ready for use by hospital personnel dressed in starched whites. The participation rate was over 96 percent that year, and the number of individuals who made excuses to avoid immunization was almost nil.

The need to have potential users actively involved in the design of the program is as essential as management support [24]. While workers may participate in programs developed and implemented without their input, the rates of participation are usually much higher if representatives have worked to develop programs and needs have been identified through worker surveys or similar needs assessments. The support of union leaders in unionized employee groups is a key variable. Additionally, experience indicates that programs may be more successful if participants directly support at least part of the cost of the program. Thus, we have repeatedly found participation rates and interest are markedly higher when participants pay some portion of the cost of the program on entry into the activity. By contrast, the ODPHP survey reported the majority of HPDP activities were paid for entirely by the company—that is, 76.9 percent of companies paid for group classes, 69.9 percent for health promotion counseling, and over 87 percent for screening/examinations [5]. Others have observed that emerging patterns seem to reflect practices whereby both employees and the company provide part of the time and finance costs for HPDP activities [23].

Ensure Compliance with the Americans with Disabilities Act (ADA)

The ADA establishes specific requirements for worksite HPDP programs. As noted in item 3 of the Program Planning Requirements, it is important to have human resources (or personnel) representatives involved in the planning process. In addition to providing other important inputs, these representatives can provide guidance on compliance with ADA requirements and obtain legal consultation as necessary. (See also Chapter 23.)

Allocate Available Resources Appropriately

Another constraint involves inappropriate allocation of resources. If available funds are put into facilities and staff without sufficient resources for promotion and marketing, excellent facilities may be underutilized. Conversely, so much effort may be expended on publicity and employee motivation that delivery of programs to meet expectations is compromised [24]. Either of these extremes, as well as lesser degrees of inappropriate resource allocation, can lead to the failure of what would otherwise have been excellent programs.

Develop Required Analyses of Costs and Benefits

Although some managers may decide to implement HPDP programs without requiring extensive analyses of cost and benefits, other executives will expect varying degrees of such analyses. Even when management does not want extensive program evaluation and analyses, the HPDP director may find it desirable at least to develop some indices to show appropriate "return" from the company's investment. Otherwise, the program may be disapproved either outright or when costs become more visible as the effort progresses. Both cost definition and evaluation techniques are important components and should be considered in developing plans for HPDP programs.

Accomplish Definition of Projected Program Costs

Regardless of the evaluation processes used, it is necessary that the HPDP program director provide reasonable estimates of the costs of HPDP programs. The projected budget should include not only personnel costs for those directly involved in the program but also any costs due to workers' leaving their normal activities in order to participate in HPDP programs. Some organizations may provide facilities, equipment, and staff to accomplish HPDP efforts but require use of these during non-work periods. Facilities and equipment costs should be included in program costs, even if provided by the company. Their use in the program represents lost opportunity for other uses, and thus, defining these costs is appropriate [39]. Accurate cost specification is, of course, particularly important if costs are to be compared to benefits.

Incorporate Evaluation Techniques in Plans

Evaluation approaches that include both short- and long-term evaluation are often attractive to those who have to provide funding for the HPDP programs. As noted earlier, blood pressure, smoking status, seat-belt use, and serum cholesterol levels are measurable. Changes in these parameters can be documented over a period of months. Documentation of changes in health care costs and morbidity may require periods of two to four years or longer [1]. Mortality changes, however, require extensive evaluation periods and generally should not be used for outcome criteria [39]. Evaluation processes may include Cost Benefit Analysis (CBA) and Cost Effectiveness Analysis (CEA) [40]. CBA considers the costs necessary to produce specific benefits, such as reduction in health insurance claims or reductions in cholesterol or smoking rates. CEA compares the costs of different programs to achieve the same end results, such as the cost per individual who stops smoking. If one program costs less than another to achieve a desired result, the less costly program is desirable if all other aspects are equal. As previously mentioned, significant investment is required to do valid CBA, CEA, or other evaluation analyses. If it is decided that such studies are warranted, the HPDP director should identify evaluation costs in the program proposal.

If economic analyses are planned, it will usually be necessary to obtain estimates of both direct and indirect health care costs. Chapter 15 provides extensive discussion of health care costs components and methods of estimating these costs. Previously cited references provide examples of more detailed methods of analyzing the economic impact of HPDP programs [1,2,4,31].

Use Effective Health Risk Assessment Instruments

On occasion, HPDP programs have suffered because long, detailed questionnaires were used to evaluate health risk. As in many other facets of medicine, simplicity in instrument design will increase participation and success rate. A complicated instrument may offer the promise of precisely defining risks for common and uncommon conditions. However, this goal may not be achieved because of the length, detail, and complexity of the questionnaire. By contrast, a simpler instrument of only two or three pages has proved to have excellent acceptance while providing desired risk information. Use of effective HRA techniques significantly increases the program's chances of success.

Emphasize Use of Validated Techniques

As discussed elsewhere in the literature, health-risk screening components should meet the criteria for such efforts. Table 17-4 is a summary of criteria that have been established by various authorities [41].

Table 17-4. Criteria for Implementing Prevention and Screening Activities

1. The activity is directed toward preventing a clearly defined condition that has a significant effect on the length or quality of life.
2. The activity is scientifically valid, appropriate to the health goals of the population of concern, and acceptable to that population.
3. The natural history of the condition is sufficiently understood to justify the activity as outweighing risks or adverse effects of either intervention or nonintervention.
4. The prevalence and consequences of the condition are sufficient to justify the costs and risks of intervention.
5. The activity is directed toward a condition that has an asymptomatic period during which intervention can decrease morbidity, early mortality, or both.
6. The activity is relatively easy to administer at a reasonable cost.
7. Appropriate methods of effectively treating detected conditions are available at acceptable costs. (If treatment is not available or it is not planned to provide available treatment, screening may not be appropriate. If screening is done in such circumstances, there may be adverse legal and other consequences.)

Source: Moser R Jr, Woolley, FR [41].

In addition, HPDP efforts ideally should incorporate Bayesian criteria, requiring that the prevalence of the condition in the population is appropriate for the selected screening and intervention efforts [41]. In some situations, adequate data to meet these requirements are not yet available, and the program director may choose to incorporate facets that have not yet been validated, with the goal of subsequently accomplishing such validation. If data are not available to support use of a particular HPDP modality, this limitation should be identified during program planning. As noted earlier, those funding a health and safety program may feel the overall benefits to the business are sufficient and that quantified validation is not necessary.

Even well-validated approaches must be appropriate for the setting in which they will be used. For example, an emphasis on aerobic exercise where there is no opportunity for exercise will indicate that the HPDP director is devising efforts that are not pertinent to the workplace. Unrealistic efforts may produce serious loss of credibility and lack of interest in the entire program.

Accomplish Appropriate and Effective Communication

As in numerous other aspects of health care, effective communication is critical for HPDP program success. Those involved in developing and implementing programs must be able to communicate frequently and effectively with colleagues, corporate executives, supervisors, and workers. Effective communication is essential throughout the life of the program. In addition to obtaining initial approval and support, directors must recognize that it is particularly important to keep management members advised of both successes and problems relating to the program. An unexpected, adverse "surprise" can result in immediate termination of an activity. Those concerned with HPDP work must also ensure their articulation of the program is readily understood by non-health-oriented people. While medical jargon may facilitate communication in a typical health care setting, it may be perceived as an ostentatious approach by those concerned with funding or participating in the program. Such impressions do not bode well for the success of HPDP activities.

Decide Whether Incentives Should Be Used

Incentives are sometimes used to encourage participation. The ODPHP survey found that approximately 20 percent of respondents offered prizes or other awards for enrolling [5]. Cash awards for desired behavior changes were provided by 13 percent, and over 27 percent recognized success with certificates, awards, or similar items. At the University of Utah, employees can obtain a reduction in their insurance premiums by completing a health risk assessment questionnaire. While recognition may encourage participation, an award program must be applied judiciously. Valuable awards may result in conflicts, disenchantment on the part of "losers," or other adverse effects. Conversely, recognition with items of minimal value, where qualification is based on an honor system of reporting efforts and progress, may engender group identification and promote participation. Somewhat surprisingly, awarding tee-shirts or baseball caps with appropriate logos has often generated more enthusiasm than a limited number of cash awards.

CONSIDER INCORPORATING HPDP PROGRAMS INTO COMPREHENSIVE HEALTH PRODUCTIVITY MANAGEMENT PROGRAMS

Recently there has been increased emphasis on integrating the HPDP program as a major component of a health productivity management (HPM) program rather than on a "stand alone" addition to a corporate health and safety program [42,43]. In addition, information regarding successful components of such programs should be more widely disseminated to stakeholders to enhance dissemination of "best practices" to permit broader application. In essence, as determined by a large-scale HPM Benchmarking Project sponsored by the Centers for Disease Control and Prevention and the National Association of Chronic Disease Directors, HPM promising practices include ". . . need-based, integrated medical, pharmacy management, health promotion, demand management, return-to-work, caregiver, employee assistance, and disease management programs . . ." [43].

This orientation complements the earlier discussions of cost containment and health and productivity enhancement discussed in the prior two chapters.

SUMMARY

The health and safety manager who designs and implements effective HPDP programs can play a vital role in reducing the corporation's or community's risk due to health costs and providing important benefits to involved individuals and their families.

REFERENCES

NOTE: More extensive references than usual have been provided for this chapter in an effort to assist the program director who may need such documentation to support a proposed program.

1. Goetzel RZ, Jacobson BH, Aldana SG, et al. Health care costs of worksite health promotion participants and non-participants. *J Occup Environ Med* 1998;40:341–346.
2. Yen LT, Edington DW, Witting P. Corporate medical claim cost distributions and factors associated with high-cost status. *J Occup Med* 1994;36:505–515.
3. Health Care Benefits Survey, *1991—Indemnity Plans: Cost, Design and Funding.* Princeton, NJ: A Foster Higgins & Company, 1992. Pp 2–3.
4. Edington DW, Yen LT, Witting P. The financial impact of changes in personal health practices. *J Occup Environ Med* 1997;39:1037–1046.
5. Christenson GM, Kiefhaber A. Highlights from the national survey of worksite health promotion activities. *Health Values* 1988;12(2):29–33.
6. Fielding JE. Health promotion and disease prevention at the worksite. *Ann Rev Public Health* 1984;5:237–265.
7. Chen MS Jr. Wellness in the workplace—beyond the point of no return. *AAOHN J* 1988;36:256–261.
8. McCauley MJ, McCunney RJ. Health promotion. In: McCunney RJ, *A Practical Approach to Occupational and Environmental Medicine,* 3rd Ed, New York: Lippincott, Williams & Wilkins, 2003.
9. American College of Occupational and Environmental Medicine, *ACOEM Health-related Productivity Roundtable,* March 24–25, 2003.
10. Brady W, Bass J, Moser, R Jr, et al. Defining total corporate health and safety costs—significance and impact. *J Occup Environ Med* 1997;39:224–231.
11. Leigh JP, Markowitz SB, Fahs M, et al. Occupational injury and illness in the United States. Estimates of costs, morbidity, and mortality. *Arch Intern Med* 1997;14:1557–1568.
12. Yancey AK, Fielding JE, Flores GR, et al. Creating a robust public health infrastructure for physical activity promotion. *Am J Prev Med* 2007;32(1):68–78.
13. Ockene JK, Edgerton EA, Teustsch SM, et al. Integrating evidence-based clinical and community strategies to improve health. *Am J Prev Med* 2007;32(3):244–252.
14. Pelletier KR. A review and analysis of the health and cost-effective outcome studies of comprehensive health promotion and disease prevention programs at the worksite: 1991–1993 update. *Am J Health Promotion* 1993;8(1):50–62.
15. Kruger J, Carlson SA, Kohl HW III. Fitness facilities for adults: differences in perceived access and usage. *Am J Prev Med* 2007; 32(6):500–505.
16. Fielding JE. The proof of the health promotion pudding is . . . . *J Occup Med* 1988;30:113–115.

17. Knight KK, Goetzel RZ, Fielding JE, et al. An evaluation of Duke University's LIVE FOR LIFE health promotion program on changes in worker absenteeism. *J Occup Med* 1994;36:533–536.

18. Goetzel R, Sepulveda M, Knight K, et al. Association of IBM's "A Plan for Life" health promotion program with changes in employees' health risk status. *J Occup Med* 1994;36:1005–1009.

19. Tsai S, Baun W, Bernacki E. Relationship of employee turnover to exercise adherence in a corporate fitness program. *J Occup Med* 1987;29:572–575.

20. Association for Worksite Health Promotion, William W. Mercer, Inc., U.S. Department of Health and Human Services—Offices of Disease Prevention and Health Promotion. 1999 National Work-site Health Promotion Survey: A Report of the Findings. 2000.

21. Roman PM, Blum TC. Formal intervention in employee health: comparisons of the nature and structure of employee assistance programs and health promotion programs. *Soc Sci Med* 1988;26(5):503–514.

22. Hymel P (moderator), Bunn W, Hudson TW, et al. Keeping employees healthy while cutting costs and improving productivity. American College of Occupational and Environmental Medicine 2007 Symposium, New Orleans, May 2007.

23. Hollander RB, Lengermann JJ. Corporate characteristics and worksite health promotion programs: survey findings from Fortune 500 companies. *Soc Sci Med* 1988;26(5):491–501.

24. O'Donnell MP. *Design of Workplace Health Promotion Programs. Health Promotion.* Royal Oak, MI: American Journal of Health Promotion, 1986, revised 2000.

25. Bly JL, Jones RC, Richardson JE. Impact of worksite health promotion on health care costs and utilization; evaluation of Johnson and Johnson's Live for Life program. *JAMA* 1986;256:3235–3240.

26. Frumkin H. Worksite health promotion and health care costs and utilization (letter). *JAMA* 1987;257:2756.

27. Kuller LH. Worksite health promotion and health care costs and utilization (letter). *JAMA* 1987;257:2757.

28. Spilman MA, Goetz A, Schultz J, et al. Effects of a corporate health promotion program. *J Occup Med* 1986;28:285–289.

29. Warner KE, Wickizer TM, Wolfe RA, et al. Economic implications of workplace health promotion programs: review of the literature. *J Occup Med* 1988;30:106–112.

30. Pelletier KR. Clinical and cost outcomes of multifactorial, cardiovascular risk management interventions in worksites: a comprehensive review and analysis. *J Occup Environ Med* 1997;39:1154–1169.

31. Goetzel RZ, Dunn RL, Ozminkowski RJ, et al. Difference between descriptive and multivariate estimates of the impact of Chevron Corporation's Health Quest Program on medical expenditures. *J Occup Environ Med* 1998;40:538–545.

32. Pelletier KR (ed). A review and analysis of the health and cost-effective outcome studies of comprehensive health promotion and disease prevention programs. *Am J Health Promotion* 1991;5(5):311.

33. Pelletier KR. A review and analysis of the health and cost-effective outcome studies of comprehensive health promotion and disease prevention programs at the worksite: 1991–1993 update. *Am J Health Promotion* 1993;8(1):50–62.

34. Pelletier KR. A review and analysis of the health and cost-effective outcome studies of comprehensive health promotion and disease prevention programs at the worksite: 1993–1995 update. *Am J Health Promotion* 1996;10(5):380–388.

35. Pelletier KR. A review and analysis of the health and cost-effective outcome studies of comprehensive health promotion and disease prevention programs at the worksite: 1995–1998 update. *Am J Health Promotion* 1999;13(6):333–345.

36. Walsh DC. Toward a sociology of worksite health promotion: a few reactions and reflections. *Soc Sci Med* 1988;26(5):569–575.

37. Conrad P. Who comes to work-site wellness programs? A preliminary review. *J Occup Med* 1987;29:317–320.

38. Spilman MA. Gender differences in worksite health promotion activities. *Soc Sci Med* 1988;26(5):525–535.

39. Dedmon RE. Serum lipid distributions in a corporate wellness program (including key elements for success in selling and implementing programs). Presented at the American Occupational Health Conference, Philadelphia, April 1987.

40. LaRosa JH, Haines CM, Kiefhaber A. Cost analysis and workplace health promotion programs. *OHN* 1985;33:234–236, 262.

41. Moser R Jr, Woolley FR. A prevention armamentarium for aeromedical practitioners. *Aviat Space Environ Med* 1987;58:590–599.

42. Schutt TMK, McGovern PM, Dowd B, et al. The future of health promotion/disease prevention programs: the incentives and barriers faced by stakeholders. *J Occup Environ Med* 2006;48:541–548.

43. Goetzel RZ, Shechter D, Ozminkowski RJ, et al. Promising practices in employer health and productivity management efforts: findings from a benchmarking study. *J Occup Environ Med* 2007;49:111–130.

ADDITIONAL READING

Aldana SG, Greenlaw RL, Diehl HA, et al. The effects of a worksite chronic disease prevention program. *J Occup Environ Med* 2005;47:558–564.

Brownson RC, Ballew P, Dieffenderfer B, et al. Evidence-based interventions to promote physical activity—what contributed to dissemination by state health departments. *Am J Prev Med* 2007;33(IS):S66–S78.

Finkelstein EA, Linnan LA, Tate DB, et al. A pilot study testing the effect of different levels of financial incentives on weight loss among overweight employees. *J Occup Environ Med* 2007;49:981–989.

Fuemmeler BF, Baffi C, Masse LC, et al. Employer and healthcare policy interventions aimed at adult obesity. *Am J Prev Med* 2007;32(1):44–51.

Halpern MT, Dirani R, Schmier JK. Impacts of a smoking cessation benefit among employed populations. *J Occup Environ Med* 2007;49(1):11–21.

Lynch WD, Chen C-Y, Bender J, et al. Documenting participation in an employer-sponsored disease management program: Selection, exclusion, attrition, and active engagement as possible metrics. *J Occup Environ Med* 2006;48:447–454.

CHAPTER

18

Health and Safety Consulting

Robert J. McCunney

Since the publication of the earlier editions of this text, consulting relationships between health care professionals and industry have come under increased scrutiny [1–3]. Links between academic institutions and the pharmaceutical industry, in particular, have raised questions about ethical challenges and conflicts of interest that may affect professionals [3]. In 2007, a national survey of physicians addressed consulting links between physicians and the pharmaceutical and medical device industries; results showed that most physicians (approximately 94 percent of 3,167) had some type of relationship with these industries, ranging from financial reimbursements for meetings to gifts to lecture fees [4]. Although occupational and environmental health and safety (OEHS) specialists were not surveyed in the study, the message for the discipline is none the less clear: In providing services to industry, the OEHS professional must be aware of potential ethical challenges and strive to demonstrate the highest principles of integrity. Although OEHS physicians do not have the attendant responsibilities and pressures of prescribing certain medications to the extent that primary care physicians have, other challenges to their professional integrity and ethics may surface. For example, industry-sponsored research, and professionals who participate in such work, face continued challenges regarding potential conflicts of interest. Nonetheless, as the science of occupational and environmental health expands to meet new challenges in the workplace such as nanotechnology, genetics, and biotechnology, considerable opportunities and challenges will be available for OEHS professionals who provide consulting services. In this chapter, the term *consultant* refers to professionals who deliver services within their domain of expertise to an organizational entity and in a non-employee status. The chapter primarily addresses the range of services OEHS professionals provide and gives practical guidance on both becoming and selecting a consultant.

Health and safety consulting services may involve short- and long-term projects but tend to comprise three major areas:

1. Provision of guidance and oversight on some aspect of the mission of the organization, such as product development, manufacturing, or use
2. Occupational health program development, operation, or evaluation
3. Project-oriented tasks

Consulting may take many forms, including full- and part-time, and may be conducted from a variety of perspectives, including those of academic institutions, consulting groups, hospitals, and other entities. In some cases, consulting issuing from a particular setting or environment, such as an academic center, offers special advantages. Regardless of the primary setting, the experiences that a part-time consultant may gain in problem solving and in contact with different people in various industries may improve his or her skills and effectiveness as an educator, researcher, or provider of services at the respective institution. The challenge is to gain consulting experience that aids professional development without compromising one's primary position.

On the other hand, professionals who devote full time to consulting come into contact with diverse settings, which often have similar challenges and solutions. This type of experience facilitates developing special areas of expertise that can become appealing qualities for future clients. A major challenge for most health and safety professionals is overcoming the hurdle to becoming a full-time solo consultant.

In contrast to business consulting, which has a rich body of supporting literature, surprisingly little has been published on consulting in occupational health, and nothing specific since the first edition of this text [5,6]. Nonetheless, fundamental principles of consulting, regardless of the discipline—whether engineering, business, or occupational health—prevail, including marketing, client affairs and communications, among others. Readers will find that other chapters of this book, including those related to financial planning, entrepreneurship, program evaluation, communication, legal aspects, personality factors, and presentations, contain interesting and helpful background for the health and safety professional contemplating the development of a consulting practice.

SCOPE OF OCCUPATIONAL HEALTH SERVICES IN CONSULTING

The activities of an occupational health and safety consultant can be placed into a number of different models, some of which have been proposed by Chapman-Walsh as a result of her work in the 1980s with corporate medical directors [6]. In Dr. Walsh's assessment, occupational health consultants operate as either "health care providers who have patients or as managers who have employees" [7]. Clearly, this model has evolved over the past 20 years as major organizations have reduced staff considerably and in turn relied on outside professionals for a variety of occupational health services. Nonetheless, descriptions of consulting by occupational health professionals, including industrial hygienists, nurses, epidemiologists, safety specialists, and toxicologists, as well as physicians, primarily discuss the provision of strategic advice and problem solving beyond routine professional services. For example, in a consultant capacity, an industrial hygienist may review all components of an exposure-monitoring program without having participated in the sampling or in analyzing the specimens, an occupational medicine physician may be asked to review diagnostic and treatment protocols, and an occupational health nurse may be asked to review the so-called standing orders at facilities without direct physician oversight. In some settings, however, a professional health and safety consultant may perform routine occupational health services as well as the higher level strategic and problem-solving functions, because the business entity does not have staff to provide the routine services. The services may need to be performed only on an infrequent basis, and thus it may be good business strategy in some settings to use a consultant for routine occupational health and safety tasks.

Based on Chapman-Walsh's work, a major role of the health and safety consultant is in "strategy formulation" [7,8]. Neither direct patient care in the form of medical treatment nor nursing care nor industrial hygiene sampling nor data management nor other such responsibilities are customarily part of the consulting role. Regardless of the role the consultant plays, occupational and environmental health and safety is a unique scientific discipline that can add strategic value to any organization—whether it is in the business, academic, or government sectors.

The driving force for health and safety consulting services is based on numerous factors such as the personal skills and interests of the consultant, the activities of the organization, including its primary purpose (e.g., manufacturing or service), the corresponding health risks of operations, and historical precedent, among others. In some cases, that stylish and amorphous concept the *culture* of the firm may greatly affect the type and scope of services that an organization may require or request. An influential CEO, for example, may stress the importance of prevention, an emphasis that will likely affect decisions of managers throughout the business.

The range of services usually within the domain of health and safety professionals is considerable. Although any categorization of consulting services in this dynamic field could be appropriately described as arbitrary, the following five areas cover the scope of many activities for which these professionals can contribute to an organization:

1. Plant and facility oversight, including office settings
2. Product stewardship
3. Health policy and program development
4. Academic and research efforts
5. Health and safety advice and problem solving

Plant Oversight

Plant oversight refers to activities that can involve oversight of clinical services, training and prevention programs, and other health and safety efforts. Corresponding duties may entail representing the organization through presentations to local community groups, unions, workers, and management. The consultant, despite the obvious inherent bias in being compensated by the respective organization, can engender a degree of credibility on contentious issues that an employee would have difficulty conveying. Although situations vary, most manufacturing facilities benefit from periodic "external" reviews of their policies, programs, and employee and community relation efforts, among other occupational health and safety services.

The type of health and safety consulting services appropriate for a company or organization depends primarily on its activities. Manufacturing operations, especially the petroleum, chemical, and pharmaceutical industries, often enlist consultants to provide regular oversight of medical surveillance programs, product safety initiatives, and workers' compensation activities. On the other hand, office settings may profit from the expertise of a consultant knowledgeable about ergonomics, indoor air quality, and employee relations, especially those related to human resource endeavors, such as the Americans with Disabilities Act (ADA).

The health and safety professional who is considering the establishment of a consulting practice might find smaller businesses, which tend *not* to have the in-house expertise of larger organizations, an appealing first step. Moreover, in working with a variety of small businesses, one can gain valuable experiences applicable to other settings. Certain responsibilities, such as safety, Occupational Safety and Health Administration (OSHA) regulations, medical services, and others apply to numerous employment settings. The range of experiences one gains in working with smaller businesses may also present the consultant with diverse challenges that can promote the individual's professional development and visibility in the business community, both critical factors in sustaining any consulting practice.

To determine the types of OEHS services appropriate for any organization, one should be familiar with the health risks associated with its operations and have access to key people within the business [8]. Decisions regarding health and safety services may rest within the human resources department or be assumed by another individual (e.g., plant or facility manager) who is responsible for requesting and overseeing consulting services not provided by the employer.

Product Stewardship

The term *product stewardship* broadly refers to the potential occupational and environmental health and safety implications of a "product," from its design and development to its production, use, and disposal. A considerable range of services may be necessary along this path, including toxicology, epidemiology, and industrial hygiene, as well as occupational medicine and nursing. In some cases, the occupational health professional is expected to have a broad understanding of the major functions of the related disciplines to coordinate the multidisciplinary challenges of some occupational health problems. For example, a toxicologist familiar with the expertise of epidemiologists, industrial hygienists, and occupational physicians, would ideally be more effective. Similarly, an industrial hygienist, through an awareness of certain exposure assessments in relation to specific symptoms, may request the interpretation and guidance of an occupational physician.

Consulting activities associated with product stewardship may also include the design and review of toxicology studies, medical surveillance of workers potentially exposed to hazardous products, regulatory interactions, and medicolegal services, among others.

The broad scope of services provided by health and safety professionals in product stewardship necessitates a full appreciation of the multidisciplinary aspect of the discipline. It is essential that the consultant recognize that the scientific expertise provided by one's own discipline is usually only one component of a total program.

Health Policy

Health and safety professionals, especially physicians and nurses, may provide policy assistance in the design of health care benefits, selection of clinical facilities for occupational health services, and advice in other areas of health care, such as preventive medicine screening programs and wellness initiatives. The roles of the consultant in policy may also include presentations to designated individuals regarding the state of affairs of occupational health in the organization and recommendations for policies based on the industry's risk, corresponding regulations, and corporate culture.

As health-care benefits plans change in response to market pressures and regulations, opportunities to develop and/or evaluate and operate managed care programs are likely to continue. The diverse tasks within such a broad responsibility depend on the availability of local expertise and resources as well as geographic factors.

At times, policy matters for which a consultant can provide support may arise in problem solving and project-oriented tasks. The challenge of policy formation can, however, be daunting because certain policies can significantly affect the operation of occupational health programs. Moreover, policies tend to seek the proverbial middle ground, which may not be acceptable to certain people. On the other hand, by assuming a major role in policy formulation, the consultant can offer some of the proverbial "cover" that a business manager may find helpful in making decisions, especially those involving significant change.

An area of increased focus in occupational health practice since publication of the last edition of this text is the link between health and productivity [9,10]. Astute occupational health professionals have recognized this area of increased importance to business and have designed, developed, and implemented various types of health-promotion initiatives. These programs have ranged from smoking cessation to fitness to weight control, among others, which can have substantial health benefits for employees as well as health care cost-reduction benefits for businesses. The well-trained consultant can assist organizations in setting priorities for health-promotion activities, often through the use of health risk appraisals to help establish the type of health problems that may exist in an employee group and the corresponding programs likely to be beneficial. After initiating health promotion, it is wise to monitor the effectiveness of the intervention by various measures, ranging from assessing the participation rate to evaluating the intervention's impact on measurable functions such as cholesterol, blood pressure, weight, and work attendance, among others. (See Chapters 15, 16, and 17 for further discussion.)

Academic and Research Initiatives

A health and safety consultant can facilitate interactions between the academic community and business and government entities. Through the initiatives and direction of a consultant, for example, students in various disciplines can gain practical training and experience in business operations and project-oriented tasks. Academic affiliations can benefit both parties: the organization that sponsors the activities and the student who gains "real world experience" in addressing occupational health and safety challenges. Moreover, consultants who facilitate the placement of students and trainees can enhance their own credibility through the academic affiliation. In addition, they can enhance their own capabilities if motivated students are properly supervised and contribute by performing needed services with educational value. Such student internship affiliations can also facilitate the development of applied research projects, such as assessing the workers' compensation experience of an organization, reviewing medical surveillance results, or preparing literature searches. Through affiliations with training programs in the health and safety disciplines, including medicine, nursing, ergonomics and safety, and industrial hygiene, the business can demonstrate a noble mission to competitors, regulators, and other interested parties, while displaying a commitment to promoting a healthy workplace.

Problem Solving/Advising

The problem solver/advisor role of health and safety consultants encompasses other activities that are difficult to categorize. Providing medical guidance about local health conditions to employees who are about to relocate to another country or participating with the local emergency planning commission are notable examples. Other examples are helping an employee select an appropriate physician, finding nursing care for an elderly parent, and responding to regulatory inquiries. The diverse opportunities depend to a large extent on challenges faced by the business as well as the interests, initiative, and professional credibility of the consultant.

One particular area of consulting—medicolegal support—deserves commentary. Health and safety professionals may be called upon to provide scientific expertise as expert witnesses in civil and criminal proceedings. These requests tend to be through invitation, however, and are generally not secured through self-promotional efforts. In fact, one's image in this arena may become tarnished if it becomes known that the professional is actively seeking such work or spends substantial time working in this area. On the other hand, assertive occupational health professionals usually recognize the potential value to the organization in contributing to the process in civil and criminal proceedings. The challenges of serving as an expert witness can be extraordinarily rewarding, in part, because of the intellectual stimulation of delving deeply into scientific matters. Those who provide medicolegal consulting services should recognize that time and scheduling demands, however, may be daunting and may necessitate a flexibility that can affect other responsibilities. (See Chapters 23 and 24 for further discussion.)

PRACTICAL CONSIDERATIONS

Working as a consultant requires attention to fundamental business principles, including planning, finance, marketing, and legal issues, along with program development, colleague recruitment, and training. In general, a consultant's practice depends on experience, local needs, and competition. Although it may be tempting to develop a long list of services to attract as many clients as possible, it is usually a better strategy to focus one's activities on a few areas and to develop the necessary skills and contacts accordingly.

In addition to the business matters described in previous chapters, the consultant needs to address the cost of establishing and maintaining an "office," even if the work is done from home. The type and cost of equipment needed to support the practice, including staff, computers, printers, copiers, and high-speed Internet access, among others, are important considerations.

As previously noted, a determination of the types of services to provide depends on the consultant's training, expertise, and interest, coupled with a marketing analysis of the specific geographic area to be served, with particular attention to the types of businesses and local competition in the area. A consultant needs to consider whether his/her services will be provided locally, regionally, nationally, or even internationally. The approach to determining the level of local competition, and the corresponding business need for services, can take many forms, including a comprehensive business analysis or an informal assessment based on discussions with personal contacts, especially influential business leaders and health and safety colleagues.

After deciding on the proper office setting and the services to provide, the aspiring consultant also needs to consider marketing and other efforts that can promote one's services. The best marketing is the provision of high-quality services to a client, who then recommends the consultant to another party. Similar to the selection of a physician, plumber, or a dentist, personal contacts play a major role in obtaining referrals. Prior to obtaining clients, however, one needs to consider an appropriate and effective means of promoting one's services. Notable examples include mailings, brochures, and presentations to local business groups, as well as publication of articles in trade and professional journals. Presentations at professional society meetings can also be helpful if one's colleagues are aware of the consulting practice and can initiate referrals. A well-designed web page may also be of assistance.

Legal concerns also need to be considered in health and safety consulting practice, including the type and extent of malpractice coverage. For example, one may need to determine whether the malpractice insurance for physicians, nurses, and other professionals covers the range of activities provided in a consulting practice. Physicians performing work as medical review officers in addressing drug testing results, and occupational health nurses who assess treatment guidelines at a local plant, may require supplemental insurance coverage such as errors and omissions (E & O) insurance. Since insurance policies vary, discussion with one's insurance provider is advisable to ensure that specific consultant activities that the individual plans to provide will be fully covered. (See Chapter 23.)

Other practical matters that any consultant needs to address are defining the work product and/or the service to be provided to the client, devising a corresponding timetable for their delivery, and calculating projected costs. (See discussion in Chapter 11.) A full discussion of one's availability and access can aid in establishing credibility and in avoiding misinterpretations that can damage one's reputation. It is essential that expectations of the consultant and the client be fully understood by both parties.

The consultant also has to consider whether to work full time or part time and whether to work solo or in a group. Whether the professional is developing a contract with another consulting firm or with individual organizations, some key factors should be kept in mind. Title, scope of activity, confidentiality of the work, and supplemented by the terms of agreement (including the fees and timetables) should be delineated. It is advisable that contracts undergo legal review to protect the interests of both parties, since unfair agreements rarely stand the test of time.

Fees for consulting vary considerably but are fundamentally either project or hourly based. The fees of a consultant depend not only on expertise, training, and reputation but also on the local competition and the organization requesting such services.

SELECTING A CONSULTANT

Whether a health and safety professional serves as a consultant or an employee, it is virtually inevitable that the individual will be called upon to help select a consultant for a particular activity or project. Ideally, the same considerations involved in recruiting a consultant will be of value to those considering becoming a consultant. In selecting a consultant, fundamental professional qualifications to consider are the person's reputation and experience, not only in the scientific discipline, but also in the designated industry and subject matter, and a review of publications, including the topics and the quality of journals in which they appeared. The latter can yield valuable information about the expertise and, perhaps more important, the perspective and biases of the consultant. Membership in professional societies, along with academic affiliations and service on government and scientific committees, are other factors that can be helpful in gauging the capabilities of a consultant.

In certain cases, especially for physicians and nurses, hospital affiliations may indicate an aspect of the consultant's quality. Moreover, through affiliations with hospital-based teaching centers, consultants can facilitate medical care and services for a client. Consideration of awards and honors also may aid in the decision-making process of selecting a consultant. Availability, personal traits, such as high ethical standards, punctuality, and the ability to deliver a project on time are other traits of a worthwhile consultant.

SUMMARY

Health and safety consultants are likely to continue to have considerable opportunities in the coming years. In fact, the complexity of running many businesses, especially manufacturing operations, includes diverse elements in health and safety, such as legal liability, regulatory accountability, and standards of care, among others. These challenges will likely result in further demand for high-quality health and safety consulting services. Businesses, however, need to be apprised of the capabilities of health and safety consultants, who, in turn, would be wise to enhance their skills in understanding business operations.

Stimulating consulting challenges are likely to appear in many areas of health and safety, including the academic/private sector (especially in conducting research) in education and in training. Although potential conflicts of interest may arise in any academic/industry relation, balancing the benefits and risk can overcome such impediments [11]. Notable concerns about ethical issues have focused generally on the biotechnology industry, genetics [12], and federally funded clinical trials. To avoid any potential conflict of interest, full disclosure of other professional commitments is advisable for anyone considering working as a health and safety consultant [13].

REFERENCES

1. Campbell EG, Claridge BR, Gokhale M, et al. Data withholding in academic genetics: evidence from a national survey. *JAMA* 2002;287:473–481.

2. Campbell EG, Weissman JS, Vogeli C, et al. Financial relations between institutional review board members and industry. *N Engl J Med* 2006;355:2321–2329.

3. Studdert DM, Mello M, Brennan T. Financial conflicts of interest with the pharmaceutical industry-self regulation in the shadow of federal prosecution *N Engl J Med* 2004;351:1891–1900.

4. Campbell EG, Russell LG, Mountfield J., et al. A national survey of physician-industry relationships. *N Engl J Med* 2007;356:1742–1750.

5. Ladou J. The occupational medicine consultant. *Am J Ind Med* 1991;19:257–266.

6. McCunney RJ. The academic occupational physician as consultant. A 10-year perspective. *J Occup Med* 1994;36:438–442.

7. Chapman-Walsh D. The vanguard and the rearguard: occupational medicine revisits its future. *J Occup Med* 1988;30:124–134.

8. Snyder TB, Himmelstein J, Pransky G, et al. Business analysis in occupational health and safety consultations. *J Occup Med* 1991;33:1040–1050.

9. McCunney RJ. Health and productivity: a role for occupational health. *J Occup Environ Med* 2001;43:30–35.

10. McCunney RJ, Burton W, Anstadt G, et al. The competitive advantage of a healthy work force: opportunities for occupational medicine (editorial). *J Occup Environ Med* 1997;39:611–613.

11. Blumenthal D. Academic-industry relations in the life sciences: extent, consequences and management. *JAMA* 1992;268:3344–3349.

12. McCunney RJ. Genetic testing: ethical implications in the workplace. *Occup Med: State Art Rev* 2002;17(4):665–672.

13. Palci J. Conflict of interest. PHS readies new rule. *Science* 1990 Dec 14;250(4987): 1515.

ADDITIONAL READING

American Industrial Hygiene Association. Critical steps in starting and sustaining an independent H&S consultancy. Available at: www.aiha.org/AIHA_Podcasts.htm; accessed November 27, 2007.

CHAPTER 19

Time Management

"Time is the scarcest resource, and unless it is managed, nothing else can be managed" [1]. This observation by Drucker is underscored by another manager, McCarthy, who has aptly noted that managing time is the most important management task [2]. He further observes that effective time management reflects effective self-management. If problems exist, their primary source should therefore be evident. In my experience, problems with time management, with communication, or with both have caused more health and safety program difficulties and failures than any other factors, including limitations regarding finances, number of people, facilities, or expertise. A manager who can use available time wisely can expect to achieve more effective and efficient results and also have more personal and professional satisfaction. This chapter considers some techniques to help use time wisely, and the next chapter reviews methods that can improve communication.

FUNDAMENTAL ASPECTS

Although time-management experts have proposed multiple approaches to managing time, there are four major components that the various approaches have in common. These are essential in using time most efficiently and effectively:

1. Know what you really want to do.
2. Know how you actually use your time.
3. Establish and use appropriate priorities to organize time.
4. Control time wasters.

The following discussion considers methods of accomplishing these crucial aspects of time management.

KNOW WHAT YOU REALLY WANT TO DO

In today's busy, sometimes hectic, environment, we frequently feel the pressure of having too much to do. Regardless of our capabilities, few of us are able to do everything we want, or are expected, to do. The temptation is to "press on" to get as much done as possible and hope for the best results. However, this very situation demands that we periodically decide what is of real importance to us—what do we really want to do. Covey stresses the importance of beginning with the end in mind, and this step is particularly important in time management [3]. What are our primary professional and personal goals? Do we want to develop extensive expertise in a particular health or safety area? Do we want to become an academician, or are we dedicated to a senior corporate management position? Is spending more time with family or significant others our major goal? Taking some planning time to decide on major, long-term goals will be of great value in establishing priorities for use of available time. Neglecting this step can result in a sense of dissatisfaction, even if we do manage our time well.

DETERMINE HOW YOU USE YOUR TIME

Most of us believe we are using our time effectively. In spite of a continuous stream of new demands, frequent interruptions, lack of support personnel, and similar factors, we are still "getting the job done." However, when we find that we are doing office work nights and weekends, arriving late at meetings that we forgot we were to attend, preparing our presentations as we walk down the hall to the conference room, and making errors because we did not have time to review our work, we may wonder how well we are getting the job done. Unfortunately, such difficulties may not only increase our personal stress level but also compromise our overall health and safety program.

In attempting to improve our use of our time, a useful step is to determine how we are currently using our time. Keeping track of how we actually spend our time for a week, as opposed to how we plan to spend it or think we spend it, can help define time needs and availability. The usual objection to doing this exercise is that the week was not "typical." Performing the process over several weeks may, however, demonstrate that no week is typical and, consequently, that attempting to keep to a rigid or "typical" schedule can never succeed.

Although keeping a log for a few weeks requires discipline and does take time, the resultant information can be invaluable in organizing time. The time log does not require elaborate mechanisms. Time can be logged on a simple appointment book, preferably kept by the telephone so that the nature of calls can be noted while in progress. On reviewing their time logs, most managers find that they spend much more time on routine or unimportant tasks than they thought. Finding that each day was predominantly occupied with nonessential, and often ineffective, work may reveal why it was never possible to find sufficient time for critical projects. Frequently, time logs produce surprises. One individual who complained that he was always behind because of frequent meetings was astonished to find that he spent only three hours per week in meetings, including travel time to and from the sessions. Informal "gripe" sessions, often to complain about the excess number of meetings, were actual time wasters for this person.

Typically, review of the logs will show some area of poorly organized time and one or more recognized time wasters. These might include interruptions for nonessential discussions, doing items rather than delegating, making mistakes due to rushed decisions, spending disproportionate time on low-priority items, or similar time wasters. Accurate time logs will usually clearly demonstrate the Pareto principle. As discussed in Chapters 3 and 15, this principle states that 20 percent of efforts produce 80 percent of results. Knowing that a few important efforts produce the bulk of intended results helps the manager to concentrate available time on the productive efforts.

USE PRIORITIES TO ORGANIZE TIME

With information from time logs in hand, the health and safety manager can develop approaches that facilitate use of available time more effectively. Some methods include the following:

Establish Priorities for Projects

In *The 7 Habits of Highly Effective People,* Stephen Covey observes that effective management is putting first things first [3]. He notes that a critical aspect of time management is to "organize and execute around priorities." Establishing priorities helps us to avoid the trap of relying on nonprioritized "to do" lists. We begin each day with our list and add to it as we get new tasks. A sense of relief occurs each time an item is crossed off, and we feel we are "making progress." If we are compulsive, we may have a list that reflects each item we have, or want, to do. At the end of the day, we may have a very real sense of satisfaction from having only two or three items left on the list. Of course, these often are the same items left over from the lists of prior days, such as "review literature to establish exposure levels," "provide commentary on proposed OSHA standards," "develop annual budget," or similar items of major significance.

The above describes the usual results of not establishing priorities. The day is spent "doing things," and the individual may feel exhausted at the end of the day. However, the things done are of low priority and may provide minimal benefit to the health and safety unit, worker or other person the health and safety unit supports, or the manager.

Various priority systems have been established to help prevent this situation. Covey groups items into "Important and Urgent" (e.g., crises), "Important but Not Urgent" (planning, finding new opportunities), "Not Important but Urgent" (some mail, phone calls, or meetings), "Not Important and Not Urgent" (trivia, busy work) [3]. Others suggest a 1,2,3 or A,B,C system. "Important," "1," or "A" priorities are the ones that contribute to your own or your unit's goals and missions. If not done, they could jeopardize a project or an entire health and safety program. Note that they may, indeed *should,* also include items of particular significance to the individual, such as spending more time with the family or significant other by not bringing work home. "Less important," "2," or "B" priorities may be efforts or projects that contribute to your program, but some delay will not pose major problems: obtaining funds to replace a still serviceable vision tester that is requiring repairs more frequently, for example. Over time, items in these categories may shift to a higher category. Conversely, with changes in goals or missions, the highest priority items could change to a lower level.

The "Unimportant," "3," or "C" items include trivia and some "nice to do" items. In contrast to higher priority items, these are often easy to do. Crossing them off a list can provide the sense of accomplishment mentioned above but will do little to benefit the manager or the health and safety organization.

Schedule Priorities

Prioritization is essential, but the process can produce serious problems for the manager. Many individuals do effective prioritization but find that listing the highest priority items each day demonstrates that there is not sufficient time available to accomplish the items. Since many particularly important efforts require significant amounts of time, none of them are done because a block of time never becomes available. At the end of a day, with its myriad other "requirements," including phone calls, interruptions, emergencies, and required meetings, the manager finds it necessary to list all that day's priority items on the next day's list. By definition, daily planning focuses on a day. Such an orientation can result in frustration for the manager as each day's activities still leaves major projects untouched. Activities that are accomplished are not necessarily the ones related to organizational or personal goals, missions, and objectives. All items on the schedule that were accomplished may be in the Important (A or 1) category, but few of them may have significant, long-term benefits. The Pareto principle applies in this situation, in that it is essential to be certain that time is provided for those few efforts that produce lasting benefits to the organization or to the manager.

Covey suggests that one method to avoid these difficulties is to organize efforts over a week, or longer, period rather than over a day [3]. Emphasis is placed on scheduling priorities rather than prioritizing what is on a daily schedule. By working a week or more in advance, there are better opportunities to help ensure that sufficient time will be available for the most critical, highest priority items. Without planning ahead, and blocking out specific time for important efforts, there will never be "enough time" to do the project. Getting to the item "as soon as I can find some free time" usually means the work is not done or done poorly in order to meet a rapidly approaching deadline.

In planning and organizing time, some find it helpful to group required tasks into categories. These categories might include managing (planning, evaluating opportunities, problem solving); special projects (preparing new exposure criteria, writing professional reports); outside requirements (professional meetings, speaking engagements, testimony); supervisory and personnel actions; routine administration; and personal. After assigning each task to the appropriate grouping, decisions on the priorities and approximate percent of time allocated for both the categories and the items in the categories, can be made. This effort clearly requires more than a day-by-day approach, and it facilitates planning time for major projects that may have a great impact on the organization.

The time log for a manager should help indicate the amount of time that will be needed each day of the week for the urgent interruptions and other unplanned impacts on the schedule. As subsequently discussed, it is sometimes possible to "plan" a grouping of "unplanned" events and thus allow for both high-priority work and the inevitable interruptions. Some managers include an "unplanned" category in their schedules to help provide time for interruptions and crises.

In scheduling time, a calendar or one of the many available "planners" (hard copy or computer assisted) may be of major benefit. Some corporations now provide each new junior manager with a planner and, in some cases, courses on use of the planner to increase time-management effectiveness. Regardless of what product is used, it should be readily available throughout the day. New requirements, appointments, thoughts about projects, and similar items can be entered at once. Doing so means that we do not have to clutter our memory banks with future dates, times, or needs. For those of us with a memory span measured in nanoseconds, this can be a distinct advantage. Attempting to remember an item until back at the office, or writing information on a slip of paper, which is subsequently included in items sent to the cleaners, sets the stage for errors, frustration, and embarrassment. Few things equal the feeling of receiving a call from the secretary of the CEO inquiring whether illness or some other problem has prevented attendance at the meeting that you had agreed to attend and which began 15 minutes ago. Recalling the agreement (made during an informal hallway conversation) at that time does little to relieve emotional stress. A planner or calendar can also be helpful in scheduling and organizing activities so that reasonable blocks of time will be available for a major project.

Of course, the calendar or planner has to be consulted daily if it is to be an effective time-management aid. Reviewing the document at the end of each day or the first thing each day, depending on personal preferences, is an essential habit. Many of us have experienced the embarrassment of coming to the office early to work on a project, becoming totally involved in the effort, and, on finally checking the daily schedule, finding that we have missed an important event. The results can be as disastrous as if we had relied on our memory and may produce a replay of the situation described in the preceding paragraph.

Another difficulty in scheduling time is attempting to keep too many calendars. It may seem beneficial to have a calendar at the office, one smaller one to carry, and one at home. An administrative assistant may have yet another one. However, problems will eventually occur with this "system." As Mayer has noted, the person "who has two calendars never knows his true appointment schedule" [4]. Obviously, there are some practical requirements that may necessitate use of more than one calendar, but it is essential that one "master" calendar serve as the reference for any others and be updated regularly, preferably daily, from any other calendar.

One final difficulty with the use of planners or calendars concerns fully scheduling each day. Just because 15-minute time blocks are available for each hour of the day, it is not necessary to fill each block! Crises will occur and unplanned priority items will have to be accomplished. If the health and safety manager facing such situations already has a completely "full" schedule, the new requirements will produce a sense of serious time pressures and frustration. Responses to the new events may be poorly done. As previously mentioned, some managers have found it beneficial to "plan for the unplanned" by blocking out some time each day or week for the unexpected. This will allow a more appropriate, and timely, response to unanticipated requirements, with lessened impacts on other high priority items.

Exercise Discipline and Avoid Procrastination

How well an individual manages an activity often reflects that person's willingness to work on priority tasks. Again, it is necessary to put first things first. A manager may not enjoy the tedious work of developing a budget for a health and safety unit, but recognition that the budget will affect all activities during the coming year should help provide the impetus to get the task accomplished. A high-priority task may take days to accomplish. Rather than starting on such a monumental task, a health and safety manager may fill the time with low-priority tasks, or even high-priority but less significant tasks, that can be accomplished more readily. Such forms of procrastination can result in either a rush effort that is poorly executed or failure to do the project. Both results can have a deleterious impact on the organization.

Breaking a project into manageable pieces may facilitate accomplishing major tasks that require weeks or months. Once broken down into necessary parts or steps, a PERT or milestone chart, as described in Chapter 12, may help the manager define what part of the project has to be done each week. The same approach will facilitate monitoring progress and thus assist in completing the effort on time, including time to review for and correct mistakes.

Review of time demands often (some would say "always") demonstrates that everything cannot be done. Prioritization helps the manager eliminate low-priority items or decide that, in accomplishing an essential but relatively insignificant task, "good enough is good enough." This disciplined approach helps ensure sufficient time is available for quality efforts on more important, higher priority tasks.

METHODS TO HELP CONTROL TIME WASTERS

General Principles

Drucker has commented that senior managers rarely have more than 25 percent of their time at their disposal [1]. The proportion may be somewhat higher for a more junior health and safety manager, but the basic concept—that a significant portion of the manager's time is beyond his or her control—is valid. Given this reality, it is essential that the manager use available time as effectively as possible.

In evaluating time use, the time log may clearly delineate some things that do not need to be done at all. For example, "update" meetings that have "always" been held every Monday, Wednesday, and Friday morning may no longer be useful and might be replaced by a meeting once a week or once every two weeks. Other efforts may be seen to be total time wasters, without any productive outcomes at all. In such situations, the manager may profit by asking, for each activity, "What would happen if I did not do this at all?" If nothing adverse would happen, it would make sense to stop spending time doing the activity [1].

Another area of possible time wasting is the occurrence of repeat crises. No matter how well organized the manager is, crises will still occur. However, each crisis should be a "new" event. Recurrence of the *same* crises, or crisis, usually reflects poor planning as well as poor use of time. For example, each year we are required to submit a grant renewal application. The document is several hundred pages in length, counting attachments. The due date is always the same (July 1), and failure to submit the application by the due date would result in an "automatic" loss of several hundred thousand grant dollars. The first year that I was responsible for the submission, the narrative (body) of the application was ready by the last week of June; however, I had not realized how time-consuming preparation of the attachments would be. As a result, we all had to work overtime, and the final result was not as high quality as desired. It would be nice to report that this particular crisis did not recur, but in spite of significant efforts to prevent a recurrence, we faced a similar, although not so extreme, situation the following year. Proper planning, prioritization, and time management would have prevented the recurrence, as demonstrated by the fact that we have not again experienced this particular crisis. "A recurrent crisis should always have been foreseen" [1].

Another general aspect of controlling time wasters is ascertaining what could be done by someone else. In part, this may result in delegation, as subsequently discussed. However, the evaluation may also result in reorganization or in modifying job descriptions or performance standards to assign responsibilities to more appropriate members.

Methods to Avoid Specific Time Wasters

Control Paperwork. Computers, electronic mail, local area networks, and similar "technology advances" were supposed to reduce paperwork. Instead, the health and safety professional finds that copy machines, fax and overnight mail, e-mail, increased regulatory and legal documentation requirements, and similar capabilities and needs have increased the paper blizzard. In addition, and in spite of increases in postal and overnight mail rates, our inboxes are filled with professional documents and advertisements for "must have" items if we are to stay abreast of the latest health and safety developments.

Typically, the health and safety manager is deluged with "action" correspondence each day. At the same time, the manager has to respond to other planned events and unplanned interruptions. Since incoming items require at least cursory review before taking action, it is often thought necessary to arrange the material in stacks so that it can be given appropriate attention at a later time. The end result of this process is often a desk covered with numerous stacks of "important" items. A call about the status of a "hot" piece of correspondence can result in wasting significant amounts of time, hurriedly looking through stack after stack. Often, more than one search is needed because the item was overlooked (sometimes because it was clipped to a miscellaneous communication) in the initial search. Estimating how long it takes to find an important piece of correspondence, especially if it is on your desk with other high-priority items, can indicate how well you are managing the paper deluge.

Such situations are typical of many of us. Obviously, the potential for wasting time is enormous, and doing so can cause serious problems for the director and the organization. Dealing with the influx of paper is not simple. Some of the following steps may help:

1. Have a large trash can by the side of the desk and use it. If an item, such as a health and safety meeting announcement, "sounds interesting," but you know you already have a commitment for that date, discard the brochure. If information items do not need follow-up or filing, use the wastebasket. A useful step in the screening process is to ask what the worst result could be if the item is discarded. If an item addressed to all health and safety managers in the country requests general information that will primarily benefit the requestor, but you know you will not have time to develop a reply "for a few weeks," add the query to the pile in the recycle basket. Decisions to discard items require discipline, but they can help avoid the tendency to keep material that "might be of interest" so that it can be studied "when there's time." The same decision-making process and self-discipline applies to items that have already accumulated on the desk.

2. Do something with the items you keep. The advice to take action on a piece of paper each time you handle it is sound. Instead of repeatedly shuffling through stacks of material to sort it into high-priority, priority, and less-important stacks, do something with the paper. At the least, send it to a colleague with the query "What do you think?" This is an excellent opportunity for the manager to practice delegation, with "For your information and action—please prepare response by . . ." notations. In other instances, it may be appropriate to return the item to the sender with the observation that the item is "definitely worth considering; please arrange a meeting with me and others concerned—prepare summary sheet for my review first." (This is especially helpful if the document is multi-paged and the notation to you is "What do you think?") In certain circumstances, you may want to file the

item, but this can result in voluminous files within a few months unless controlled. Can the item be readily obtained from the Office of Primary Responsibility (OPR), or whatever term is used in your company? If so, ask yourself if you really need to retain the item in your own files.

Note that the above-mentioned advice is to do something with a piece of paper *each time* you handle it. This is significantly different from the frequently quoted advice to "only handle a piece of paper once." In the real world, most of us have to handle an important item on several occasions. The essential action is to decide each time to do something with it [4]. If the item concerns an upcoming meeting, you may put it in a "tickler" or "reminder" file so that it will be readily available on the day of the meeting. In other instances, you may attach a request for background information or a literature review. If it requires a letter from you, you may schedule the action on your planner and then have the material filed until you come to the time to do the response.

Extending these efforts to do something each time you handle the paper may result in development of small "working files." I have found this approach particularly helpful. If an item is truly important, and if I have to do something with it personally, I assign it to a "working file" and simultaneously schedule a time on my calendar to work on it. If additional items pertaining to the effort are received, they are added to the original file. The working file may be of particular value in storing background or other items for an upcoming meeting or trip. Since several items of correspondence pertaining to the filed material may come to you during the days prior to the event, this procedure will help organize them and save the last-minute, often frantic, search for all pertinent items just before you depart for the meeting or trip.

The main point in managing paper is to do something—make a decision or take action—each time you pick up a piece of correspondence. Avoid the "I'll get to it later" syndrome that leads to stacks of material for "later." Appropriate paperwork discipline can save significant time and dramatically enhance effectiveness. (And remember that everything you contribute to the recycle basket can help improve the environment.)

Use the Telephone Effectively. Time logs of executives have repeatedly demonstrated that the telephone can be a major time waster. Playing "telephone tag" not only takes significant time but also raises frustration levels to the point that work efficiency is reduced. Attempting to agree on a time for call-back may help. In some instances, leaving information about the reason for the call may avoid repeated follow-ups. For example, leaving a message asking a colleague to just advise you whether the samples were under threshold limit values may save calls "returning his call returning your call returning his call returning your call."

One of the biggest telephone time wasters is the exchanging of extraneous information. In some instances, it may be appropriate to lead up to the discussion of a problem by engaging in social talk. However, necessary business can often be conducted without having to review the unusual weather at a caller's location or similar items. Another problem is not having background information available during a scheduled call on an agreed-upon subject. Having to leave the caller on hold, or call back, in order to obtain the information necessary to reach a decision can be frustrating as well as a time waster. Before the call is placed, taking a few moments to reflect on why you are making the call, what you hope to obtain from the conversation, and what information you need to make the call as effective and efficient as possible can pay significant dividends in saved time and reduced frustration.

It is often difficult for a visitor to understand why a routine phone call should interrupt a personal session. This is particularly true if the visitor had to travel any distance, even from the next building, for the meeting. In other instances, a phone call may be announced just as you are in the midst of an effort requiring intense concentration. If secretarial support is available, such problems can be prevented by having messages taken, unless the call is critical (or from the company president), and then scheduling a block of time to return calls. As discussed in the next paragraph, voice mail may be an option to reduce interruptions if secretarial support is not available. Although message taking and the use of voice mail may, through automated responses, result in some telephone tag, the results are usually less time wasted than if the caller were allowed to interrupt significant work.

Properly used, both e-mail and voice mail can help reduce wasted time due to the telephone. The direct personal touch is lost, but information can be provided succinctly through these mechanisms. However, it is essential that the user recognize the emphasis on *properly used.* Using voice mail to leave a message to "call me" without indicating what information is needed is only another form of telephone tag. E-mail can allow one to become very verbose, with resultant loss of both the originator's time and the desired impact on the recipient. In addition, it is disturbing to me to receive a call asking my thoughts about an e-mail message sent 30 minutes earlier. On too many occasions, I have had to explain to callers who were surprised that I had not immediately responded to their e-mail that the mere fact that the message was transmitted instantaneously did not mean that I would read it instantaneously. In such instances, when the message did reach its appropriate place in my priority system and I reviewed it, I found that I had to consciously work to review it objectively since my initial reaction was to dismiss it at once.

Manage Conversations. Unplanned, informal conversations can benefit organizations. Some organizations specifically arrange offices so there is more opportunity for interaction among workers. The management-by-walking (or wandering)-around (MBWA) approach, espoused in such books as *A Passion for Excellence: The Leadership Difference* [5] and discussed in Chapter 13, relies on personal conversations.

Unfortunately, some of these interchanges can become real time wasters. We all have had experience with the person who can extend any conversation indefinitely, either by repeating material or by conveying useless and irrelevant information. No one wants to avoid conversations and interactions with co-workers, but it must be recognized that some individuals can waste so much time that the health and safety professional is unable to accomplish necessary work. Undesired results may include taking work home and having to deal with increased stress levels.

In some instances, scheduling a specific amount of time for a visit may help, especially if the secretary calls to remind you of your next meeting. Setting or scheduling a pager to go off may also work. I had one senior staff member who repeatedly needed to see me "for just a minute" about a problem or concern. His need was always valid, but, once in the office, he routinely would discuss "one more thing," of minor significance and repeat the process for periods ranging up to 45 minutes, even though the decision on the basic problem was made in the first five minutes. I finally advised him that I wanted to be sure that I could give him undivided attention and would block out time for him. I scheduled the sessions 10 minutes before his quitting time. Subsequently, even though he knew I would still be at the office for some additional time, he made certain he ended every meeting within the 10-minute block of time so that his car could be one of the first out of the parking lot! This process can also work well in scheduling follow-up telephone calls, especially with individuals who are known for their verbosity.

Make Meetings Time Efficient. Few of us enjoy meetings, and if they are unproductive, our sense of frustration increases proportionately to the time spent in them. A few aspects to consider:

1. Meetings must begin and end on time. Delaying the start of a meeting "a few minutes" for late-comers wastes the time of those who arrived on time. As discussed in Chapter 21, the manager in charge of a meeting should begin and end the session on time to help ensure the most effective use of time by those attending.

2. The purpose of the meeting, and the responsibilities of those attending, should be known before the meeting. In addition, those expected to present at a meeting should have time to prepare. An agenda, distributed in advance of the meeting, can help achieve these goals.

3. Recognize that meetings do not have to fill the allotted time. I continue to be amazed by the surprised comments when a meeting that has accomplished its objectives is ended before its scheduled time. In such situations—when a meeting is adjourned rather than being continued, with ensuing extraneous discussions to fill the allocated time—we all appreciate the additional time provided for other activities. (See Chapter 21 for further discussion of meeting management.)

Organize Related Activities. It is surprising to see senior executives waste significant amounts of time by scheduling a meeting in one location, return to their office for a brief period, then chair another meeting in the same location as the original one. Certainly, there may be the need for a break between meetings, but often such schedules reflect someone else's convenience rather than the need to save time. Grouping activities, such as meetings, dictation, responding to telephone calls, and similar activities can help focus energies on similar efforts. As a result, "start-up" time is used only once for the group of activities, thus increasing efficiency and saving time.

Similarly, some people find they are more efficient at certain times of the day. A manager who is a "morning" person may find it most efficient to schedule time for the most complex and challenging tasks early in the day. Conversely, the time immediately after lunch may be one when no significant decisions should be made. Most managers are aware of their peak times. If uncertain, colleagues, secretaries, or others with whom the manager works can usually clarify the situation.

Delegate! The exclamation point in this heading emphasizes the fact that the failure to delegate remains a major management failure. Obviously, some things are the responsibility of, and can only be done by, the health and safety director. However, consciously evaluating what must be done will often reveal numerous activities that can be done by other members of the organization. Frequently, the rationale of managers who do not delegate effectively is that it takes more time to explain how to do the job than to accomplish the task themselves. This attitude, or belief, results in managers' spending time repeatedly on actions that should be done by others, with the result that the managers do not perform their own work activities as well as they should. As a result, both their careers and programs can suffer. Dorney aptly notes that delegation is the executive's greatest need—and greatest privilege [6].

Delegating is of such critical importance in saving time and improving management effectiveness that it is discussed in detail in Chapter 9.

Do Not Accept "Monkeys." Oncken and Wass emphasized the importance of not accepting other's problems, which are referred to as *monkeys* [7]. They observe that a subordinate with a problem (monkey) will greet a manager, often in the hallway, with the salutation "We've got a problem . . ." After explaining the problem, about which the manager ideally knows something but not enough to make an on-the-spot decision, the junior member waits expectantly. Frequently, the manager, wanting to appear helpful, responds that he is glad that the subordinate sought his advice, but since he is in a hurry for a meeting, he will have to think about it and get back later with a response. The junior member has adroitly had the monkey jump from his back to that of the manager, and, in effect, made the manager the junior member's subordinate. Some junior members are particularly adept at using this technique to "return" delegated items to the one who originally delegated the work. Recognizing the potential for this scenario will usually keep the monkey where it belongs. As a result, the health and safety manager will not spend time solving problems that should be the responsibility of someone else.

Learn to "Just Say No." As health and safety professionals, we are dedicated to helping others. If requested to do something, our natural tendency is to agree in order to help the person requesting the action. Remembering our overall priorities as well as our projected schedules for priority efforts will help us evaluate whether responding positively to the request is appropriate. Although difficult for us to do, just saying no may result in significantly less stress for us, especially if we can clearly justify, at least in our own mind, why such a response is absolutely necessary.

USE TIME TO SAVE TIME

Managers can benefit from self-study programs. This is particularly true in the case of time management. Many books provide valuable insights on methods to save time, and, fortunately, a number of them, such as *The One Minute Manager*, are succinctly written [8]. Managers will profit from perusing such publications.

REFERENCES

1. Drucker PF. *The Effective Executive.* New York: HarperBusiness, 1966, 1967. (HarperBusiness Edition, 1993.)

2. McCarthy MJ. Managing your own time: the most important management task. *J Nursing Admin* 1981(Nov–Dec):61–65.

3. Covey SR. *The 7 Habits of Highly Effective People—Restoring the Character Ethic.* New York: Simon & Schuster (Fireside), 1989.

4. Mayer JJ. *Time Management for Dummies.* Foster, CA: IDG Books Worldwide, 1995.

5. Peters TJ, Austin N. *A Passion for Excellence: The Leadership Difference.* New York: Warner Books, 1985.

6. Dorney RC. Making time to manage. *Harvard Bus Rev* 1988;66(1):38–40.

7. Oncken W Jr, Wass DL. Management time, who's got the monkey? In: *Harvard Business Review: Managing to Save Time,* No. 12201.

8. Blanchard K, Johnson S. *The One Minute Manager.* New York: Berkley Books, 1982.

ADDITIONAL READING

Ashkenas RN, Schaffer RH. Managers can avoid wasting time. *Harvard Bus Rev* 1982;60(3):98–104.

Mancini M. *Time Management.* New York: McGraw-Hill, 2003.

Matthews GH. Run your business or build an organization? In *Harvard Business Review: Managing to Save Time,* No. 12201 (reprints of articles copyright between 1968 and 1984).

CHAPTER 20

Effective Communication

The ability to communicate effectively remains *the* critical management skill. As mentioned in Chapter 19 on time management, poor time management and poor communication have been, in my experience, the two leading causes of program compromise or failure. Personnel managers have repeatedly ranked oral communication skills as number 1 and written communication skills as number 2 among all the factors most important in helping business school graduates obtain jobs [1]. These communication skills are "critical not only in obtaining a job, but also in performing effectively on the job" [1]. Health and safety managers engage in formal, informal, written, verbal, and nonverbal communication. How well the communication is accomplished often determines how successful an effort or program is. The manager's challenge is to ensure that communication is at the highest possible level of effectiveness.

This chapter covers some basic aspects of communication, highlighting those that have been found to be particularly important for health and safety managers. Chapters 21 and 22 contain additional information on effective communication.

BASIC REQUIREMENTS

Effective written and oral communication rests on meeting a number of basic requirements. The degree to which the requirements should be met increases with the importance of the communication. If the purpose of the communication is to convey a small amount of information that minimally impacts a program, poor communication technique may not be a significant problem. Conversely, if the communication is essential in obtaining health and safety program funding, a communication failure can have disastrous results. Some requirements for effective communication include the following:

Decide on the Purpose of the Communication

"What do I want my audience to do as a result of my communication?" is a useful question. If the communication is to provide information, the format and content may differ from those used to recommend a particular action. In this situation, the challenge for the presenter is to logically present sufficient information so that the desired action becomes an "obvious" choice by the decision maker.

Confusing the intent of the communication may compromise the process or make it fail altogether. For example, taking too much time to provide unnecessary, "nice to know" information in a document or presentation that is intended to culminate in a decision may instead result in loss of interest, insufficient time to support recommended actions, or similar difficulties. Deciding on the purpose of the communication may have an unexpected benefit: Communication may not even be necessary, saving time for both the presenter and audience!

For example, one altitude chamber training group at the U.S. Air Force School of Aerospace Medicine received a message from the office of the U.S. Air Force Surgeon General. The message had also been sent to all other Air Force physiologic training units (PTUs) and required amassing extensive information regarding physiologic training of pilots and navigators during the past three years. The data could be readily obtained from PTU databases, but the school did not routinely capture such data, since the training unit rarely evaluated pilots and navigators who had medical problems. The school's physiologic training focused on flight surgeons, flight nurses, and medical technicians. Consequently, members of the school spent a number of days trying to compile the information. Finally, when they brought the response for final signature, they complained that many of the questions "did not make sense" and asked for data that were not available. At that point, they were asked to provide the communication that had initiated the extensive effort. On review, it was noted that the school's unit was indeed listed along with the many other units to which the message had been sent. However, following the school's listing was the notation "INFO ONLY." The message had been provided only so that the school's educational unit would be aware of what the field units were being asked to provide! It required no action! Careful initial reading to decide on the purpose of the communication to the Surgeon General could have saved time and avoided substantial stress.

Know Your Audience

The communicator's approach may vary significantly, depending on the knowledge, experience, or background of the person or group receiving the communication. A presenter would not want to overuse professional jargon or provide extensive technical information to a group that is unfamiliar with the terms. A subtler problem may occur if one of the members listening to a presentation that requires action is actually opposed to the action. The presenter will be surprised, and perhaps unable to respond appropriately, when the objection is voiced. It is often helpful to discuss or coordinate a planned action with principals involved in the decision-making process before the presentation. If the content of objections, and the people who might raise objections, can be identified beforehand, the presenter will be better prepared to respond. This knowledge is beneficial in developing an approach or providing a rationale for action that recognizes the objections but still justifies the action.

Knowing the audience also helps avoid the embarrassment of having an expert point out a fallacy in the presentation. I was once scheduled to make a presentation on high-altitude hyperventilation to a group of flyers. The talk was designed to provide information on the pathophysiology of the condition and methods to prevent the onset of symptoms. I had prepared a number of overheads to show the shift of electrolytes that creates the symptoms fliers experience. Shortly before the presentation began, a flyer who had just joined the squadron happened to take a seat next to me. On learning of the topic of the talk, he casually noted that he had done his Master's thesis at the Air Force Academy on the subject! He mentioned a number of recent studies, including those dealing with subcellular processes, which were totally unfamiliar to me. As he continued to discuss his work and background, I shortened my talk by eliminating the planned pathophysiology portion. Instead, I limited my discussion to symptom recognition and corrective actions. Needless to say, this experience strongly impressed upon me the need to know the audience.

On occasion, health and safety managers have had similar experiences, finding that "nonmedical" managers have had graduate training in health and safety specialty areas. Learning about this before a presentation usually facilitates an enjoyable interactive exchange, while learning about this during the talk can be an unpleasant experience.

Organize Your Communication

One of the more frequent communication failures is poor organization of a document or presentation. Trying to sort through disconnected material, even though all desired information is included somewhere, frustrates the reader or audience. Further, poor organization does not support a recommended action, and a decision maker may lose interest or rapidly decide to disapprove the proposal. Conversely, a well-organized effort that flows logically toward a reasoned conclusion may be immediately accepted.

It is important to plan adequate time to organize effective communications. A frequent excuse for not organizing material better is that there was not enough time; however, the payoff in terms of enhanced effectiveness generally outweighs initial time expenditures. Frequently, organized efforts save time by eliminating the necessity for follow-up communications or for complete restructuring of the document or presentation.

Sometimes, the format and organization of a communication may be specified. For example, in some corporate settings, introductory portions of a talk or document are used to summarize the purpose of the communication and identify the recommended action. The remainder of the communication then supports the introduction. Other organizations may specify a different format and organization, such as use of the problem-solving format described in Chapter 14.

In some instances, organizations may require that written proposals must be limited to a one- or two-page "point paper." In many companies, no format is specified, and the health and safety manager can select thc most appropriate style for a particular communication. It is, of course, essential to determine before submitting the communication whether specific format or other requirements must be met.

Communicate Clearly and Concisely

An individual who communicates clearly and concisely usually has a better chance of the audience's registering important points than a person who does not. Just as organization takes time, significant time may be required to prepare a clear communication. However, the work can pay off in subsequent time savings by preventing misunderstandings or eliminating the need for clarifications and repetitions. Although some might separate presentation style from clarity, it should be recognized that how a presentation is made might affect clarity. Mumbling; reading word for word from a paper; speaking in a droning, monotone voice; showing unreadable slides; and similar problems can affect clarity as certainly as not clearly discussing a particular point. Similarly, using stilted, "plodding" language can be disastrous in oral and written communications, unless the purpose is to demonstrate the soporific effects of such use. These aspects of communication are considered further in Chapter 21, Effective Meetings and Presentations.

Conciseness is a virtue in most management settings. Even with the best time-management practices, there never seems to be enough time to do everything that is required. This is also true of communications. A communicator who wastes readers' time by preparing a document that is not clear and concise can expect that the effort will not be effective. Similar results can occur from overly long presentations filled with extraneous or poorly organized information.

Verbosity does not equate with effectiveness; there is usually an inverse relationship. Irrelevant discussions, ideas, or facts weaken the communication. The guidance that "when you are talking you are not learning anything" is applicable to the overly verbose speaker.

Meeting time limits requires both clarity and conciseness. Many presentations are restricted to a specified time period. If a speaker is obviously going to exceed the limit, the audience rapidly becomes more concerned with the time than with the talk. This practice is also unfair to any follow-on speakers. If the next speaker is the presenter's supervisor, serious consequences may result. Some CEOs are quick to note that a speaker who runs five minutes overtime, or begins five minutes late, costs the company a lot of money. The CEO can rapidly calculate the salary dollars paid per minute to each of the other senior managers attending the session and then provide the speaker the total cost of the extra five minutes. Such occurrences are rarely followed by accolades or approval of a proposal. (See further discussion in Chapter 21.)

Professional jargon can be a special problem for the health and safety professional. Programs and proposals have to be advocated by using language and data that those who are responsible for approving the programs can understand. The time is long past when health and safety programs were approved despite a disorganized, jargon-filled presentation because no one would risk disagreeing with or questioning the health and safety professional.

Use Proper Grammar

Written and oral grammar, punctuation, and construction errors continually plague professionals as well as other managers. For example, telling the CEO that a particular project is important "to you and *I*" or "will have a beneficial *a*ffect on workers may not result in disapproval of the project but can result in *personal* disapproval (i.e., disapproval of the individual who commits such grammatical errors). When a tenured associate professor signs a grant application as "Princip*le* Investigator," one has to wonder about the proposal's quality. Frequently, communicators dismiss their errors, claiming that the listeners "knew what I meant." This is not always correct. Further, regardless of whether the error was due to carelessness or ignorance, the result may still be an adverse response from the listener or reader.

If a manager's staff recognizes the need for use of proper grammar, they may work to "help" the manager. A personal "quirk" of mine is the unintentional misuse of *affect* and *effect*. As I was preparing to retire as director of a 900-employee organization, I mentioned to colleagues that I was pleased that I had not seen any misuse of *affect* and *effect* for the past several months. I noted my secretary smiling and my deputy grinning. When I asked the reason for their expressions, they advised that they had told all offices in the organization that they would not accept documents for my signature that had *affect* or *effect* in them—not exactly how I had intended to avoid improper usage of these words!

As in other aspects of a manager's efforts, setting standards for the proper use of grammar will assist in conveying the organization's professionalism.

Use One-way and Two-way Appropriately

In some organizations, communication is "top-down," or one-way, with no opportunity for feedback. This approach to communication can create significant difficulties due to misunderstandings or other factors. For example, I was assigned responsibility for providing new instructions to over 40 health and safety units assigned to our organization. I developed an instruction directive that I, and three colleagues in our office, thought was quite clear. Within a few days of sending the implementation message, I received multiple calls asking for clarification of my "clear" directives. I soon recognized that I had erred in not having some of the potential users of the new requirements involved in writing the implementation instructions. I had relied on one-way communication rather than getting feedback (two-way communication) to obtain input from the "users." Once such input was obtained, a revised message was distributed and implemented without difficulties.

Do Research to Support Your Position

It has been aptly noted that the secret of a good "extemporaneous" talk is extensive preparation and practice. Sufficient research helps ensure that the planned or extemporaneous communication accomplishes the desired result. Obviously, valid material can be presented to support a position. Conflicting information can also be noted and accompanied by the reason why one position is considered most appropriate. In fact, presenting both sides of a question often improves the presenter's credibility. It might also keep a member of the audience from negatively influencing the presentation. Thorough research helps alert the presenter to questions or objections that might come from the audience so that appropriate responses can be prepared ahead of time.

Revise and Edit

Once the communication has been prepared in initial form, it is important to revise and edit so it clearly accomplishes its desired purpose. Neglecting this process can jeopardize the entire effort. Some factors involved in editing include the following.

1. *Read what has been written.* Under pressure, authors may "read" what is intended rather than what is actually on the paper. Allowing some time to pass before proofreading may avert this problem. Both content and grammar should be critically reviewed. If possible, the author should read as a disinterested (not uninterested) party in order to judge the impact of the communication. A word processor's ability to check spelling can assist but is not a substitute for proofreading! If someone else has written a document for your signature, or if you are endorsing a document prepared by another, it is particularly important that you are comfortable with what has been written. For example, the Wing Commander at a U.S. Air Force base is Chief Executive Officer and, as such, is literally responsible for thousands of people and millions of dollars of resources. A newly arrived Wing Commander at one base received the Aerospace Medicine Report prepared by the senior flight surgeon on the base. The report was a multi-page document describing all facets of aeromedical activities, including flyer health and morale. On receiving the report, the Wing Commander asked the staff for options. The staff noted that the commander could merely provide a "noted-and-concur" endorsement or could provide more extensive comments on the report contents. The commander promptly directed his staff to prepare the noted-and-concur statement for his signature. Staff

members strongly recommended that either the commander or a staff representative read the report first. However, the commander, a "take charge" type of person as discussed in Chapter 8, emphatically stated that he did not have time to review such insignificant material and directed that the endorsement be prepared immediately for his signature. This was done, and the report was forwarded to higher headquarters.

Shortly after the report was sent, the Wing Commander received a call advising him that he would be "helped" in overcoming the morale problem that he had caused—he would be replaced the next day. He, in turn, would be placed in charge of supply operations at another, much smaller and more isolated base. On asking for clarification, the caller observed that the physician who had written the report noted that the new policies of the Wing Commander had had major deleterious effects on flyers' morale. Since the commander had "noted and concurred," it appeared that "assistance" was needed promptly. Attempting to explain that the report had been endorsed without reading did not change the results.

2. *Recognize that education or experience does not preclude making errors.* Some individuals may believe that they do not make communication errors or confusing statements because of their background or extensive experience in preparing presentations or documents. Often, this belief leads one who holds it not to read what is written and not to objectively critique talks. The result can be misunderstandings and rejections. Unfortunately, the communicator may attribute problems to deficiencies in the audience rather than recognize the true source of the problem.

3. *Avoid being "too busy" to revise and edit.* Health and safety managers experience continuous time pressures. Taking time to review and revise written material or practice oral presentations is difficult; however, the payoff in effectiveness, and often in time saved, makes this an essential practice.

Recognize the Spell-Check Hazard

In preparing documents, the spell-check capability of word processors is of great assistance in avoiding spelling and grammatical errors. However, reliance on spell-check can create serious problems for the user. The following poem, which appeared soon after the advent of computerized spell-check, stresses this hazard:

I have a spelling checker
It works with my PC
It plainly marks four my revue
Mistakes I can knot sea.
I've run this poem threw it,
I'm sure you're pleased too no.
It's perfect in every weigh
My checker tolled me so!

—*(Author unknown)*

For example, I decided to recheck a brochure I had prepared that would be distributed nationally and internationally. Only on this final check did I note that the copy read, "The Rocky Mountain Center for Occupational and Environmental Health provides highly effective and informative *curses.*" I would have been horrified if the document, with the error, had been printed and distributed. This example underscores the fact that computerized spell-checking is not fool-proof and therefore the user must still be vigilant and do careful proofreading.

Listen and Read Effectively

One of the most difficult aspects of communication for many of us is to listen and read effectively. Failure to accomplish these functions will directly compromise attempts to communicate effectively with others. It has been suggested that a manager may spend as much as 75 percent of work time in face-to-face communication and, of that time, approximately 50 percent is spent listening [1]. Effective reading and listening are especially valuable skills for the health and safety manager, and practice can enhance these capabilities. Too frequently, we begin formulating a response to another's comments, either verbal or written, as soon as we read or hear the first few words. We then become so engrossed in our efforts that we fail to "hear" what that individual is really saying. As noted by Rogers and Roethlisberger, a "major barrier to mutual interpersonal communication is our very natural tendency to judge, to evaluate, to approve (or disapprove) the statement of the other person or group" [2]. As a result, we do not really understand what the other side is stating. "Real communication occurs . . . when we listen with understanding" [2]. Allowing emotions to interfere with objective evaluations can result in extensive, but ineffective, activities as well as increased stress levels for all involved. A useful technique to assist understanding is to attempt to paraphrase the other person's statement. Such an effort will help highlight any gaps in understanding, assist in developing appropriate rebuttals, and also delineate areas of possible agreement or compromise. The paraphrasing may be difficult to do, especially in the midst of a heated discussion, but the result may prevent significant misunderstanding and enhance the work of all involved in the communication situation.

Closely aligned with the need to listen with understanding is the need to actually listen or read. Due to the press of time, I sometimes find that I have slipped into "listening" to someone while I am signing letters or scanning other documents. This is particularly likely to happen if I think I know what information the other individual plans to convey. It is only partway through the process that I realize such is not the case, and I have missed items that I have to ask to have repeated. Not only is such an approach rude, but it does not enhance one's image as an effective manager. I once had a supervisor who explained to me one day that he had found that in order to be effective, he had to devote full attention to a call, visitor, or letter and avoid trying to do two things at once. I thought it was astute and wise of him to recognize this. It was only some time later that I recognized he was describing a problem that I had. I came to this conclusion as I was talking on the phone, reviewing slides to be used in a presentation, and signing some documents. I suddenly realized that I was not certain what had been said, had signed documents without reading them, and had not critically evaluated the slides. It is easy to fall back into the habit of doing too much at the same time in order to "save time," and I, at least, find that I have to periodically reinforce my determination to avoid this trap. In today's work environment, where multitasking is considered an admirable trait, it may be difficult to avoid attempting multiple activities at the same time. However, it must be recognized that the quality of one's work efforts may be diminished if full attention is not directed to each individual activity.

A classic example of the adverse effects of multitasking is use of cell phones while driving. Attempting to concentrate on the conversation can lead to "inattention blindness," whether one is using hand-held or hands-free devices [3]. The outcome of such inattention is equivalent to that of a driver with a blood alcohol of 0.08 percent weight/volume [4]. Of course, the outcome of such multitasking may be much more severe that a "misunderstanding of the communication." In essence, "When humans attempt to perform two tasks at once, execution of the first task usually leads to postponement of the second one" [5].

Coordinate

In preparing proposals or correspondence, it is important to consider coordinating with *all* the other groups or departments that might be affected or may provide input. As previously mentioned, this action may illuminate problems or objections not previously recognized but which can be resolved. In addition, proper coordination can help prevent delays or embarrassment. For example, one health and safety unit had prepared a directive requiring that all individuals working in the warehouse and moving equipment attend a training session on back-injury prevention. The intent was to reduce the recent increase in those injuries. Shortly before the health and safety manager was to meet with a senior line executive to have the directive signed, he mentioned the effort, by chance, to a member of the personnel department. That individual offered to review the directive, and, on doing so, found that an error had been made in listing the job classifications of those who were required to attend. The error would have necessitated attendance by over 80 percent of all employees, the overwhelming majority of whom were at no increased risk of back injury. Further, the number of employees identified could not possibly have been accommodated in the scheduled prevention sessions.

Coordination, although fortuitous in this case, prevented a major error that would have distressed both the health and safety manager and the senior line executive. In spite of repeated personal experience with similar situations, I still find that failure to coordinate properly is, too frequently, a source of management problems for me.

Keep the Boss Advised

Closely related to coordinating with other groups or departments is the need to "keep the boss advised." No one, least of all one of our supervisors, likes unpleasant surprises. If your supervisor was asked why your unit is having such problems controlling the rapidly rising accident rate, and you had not previously advised the supervisor of the situation, it would be understandable if your superior was distressed. Most people are fair, and if you have advised the boss of the problem—and, especially, what you are doing about it—he or she will be prepared to respond appropriately to a question. A response like "Yes, the director told me about it and also what actions the unit is taking to correct the situation," is much more desirable from the perspective of the health and safety manager than having the boss respond, "What do you mean rising accident rate? I thought everything was under control!" Advising bosses of problems or bad news early prevents them from being blindsided by comments from others. The emphasis is on letting your boss know early if your program is having troubles and also providing specifics on what you are doing to correct the situation.

In addition to advising the boss about bad news in a timely manner, it is also important to update supervisors periodically on progress and good news. Many of us are, by nature, somewhat hesitant to discuss our achievements. In a business setting, the phrase "out of sight, out of mind" can be all too true. If supervisors are not aware of the contribution your health and safety unit is making to the organization, support for expanded activities may be more difficult to obtain. The unit may also be one of the first targeted for downsizing if senior management believes there will be little impact, since apparently it contributes so little. Deciding how much and how often to relate your successes to supervisors obviously requires a keen sense of balance and good judgment. However, scheduling periodic updates with them on a regular basis, every few weeks, may pay significant dividends. In addition to helping show "value added" activities, such updates will help ensure that evaluation processes, discussed in detail in Chapter 12, are proceeding as previously agreed.

Closely related to keeping the boss advised is the guidance "Always be ready to tell your story." An example is the "elevator story": A CEO was in the elevator with two senior managers, and he asked them how things were going. One of them replied, "You know our group—always busy, busy, busy!" The other chimed in, "We just completed a project that will save the company $140,000.00 and are nearing completion of another project that will reduce customer response time by over 40 percent." The question is then posed as to which of the two managers the CEO would be likely to pick when a promotion opportunity arises.

I am well acquainted with senior officers of a volunteer group. I heard them respond to questions about the purpose of the organization with "We do a lot to help the needy in the Salt Lake Valley." Since a questioner might be a donor, I suggested the following might be more effective: "Our organization is a 100-percent volunteer organization with no paid staff. Last year we provided complete sets of new clothes to over 6,000 children recommended by their school's social worker or principal. We read to and gave books to over 3,000 children and provided over 300 sets of clothes to women returning to the work force after being on welfare. (Depending on time, the speaker could provide other specifics.) My closing recommendation was "All of our overhead is covered by proceeds from our Thrift Shop, known as the Nordstrom's of Thrift Stores, so 100 percent of any funding donation goes directly to those needing our support." I was pleased that the officers adopted the approach I suggested.

A minor note of caution: It is possible to become known as too well prepared. I have been accused of having an on-off switch that turns on a verbose spiel when anyone asks about the Rocky Mountain Center for Occupational and Environmental Health!

Say "I Don't Know But Will Find Out" When Appropriate

As with advising the boss of bad news, and what the manager is doing to correct the situation, most health and safety managers will be asked a question or required to provide input that exceeds the manager's knowledge base at that time. In such a situation, use of "I don't know but will find out," or some similar phrase is much more appropriate than attempting to guess the answer or "bluff" the way through a response. If a senior manager recognizes that the health and safety director is not providing valid information, trust will be lost and possibly negatively impact on the organization.

Of course, the important aspect of using "I don't know" is that the manager find the answer and convey the information to the requestor. This action must be done on an "as soon as possible" basis.

Recognize When Communication Has Not Occurred

In Chapter 13 on leadership, I discussed the experience of one of the female graduates of our academic health and safety management course. She thought that she, as the safety and health and safety manager of the organization that hired her, had clearly communicated the goals and objectives of the organization to the professional staff, yet efforts seemed disorganized and not well coordinated. On questioning the staff, she was surprised to find that there was no agreement on the primary focus of the unit, on the unit's standards, or on the expected interactions among the members of the unit. Once extensive follow-up discussions clarified the expected activities of the unit, it began to function as a highly effective organization.

FORMAL AND INFORMAL COMMUNICATION

Every organization has formal and informal communication channels. How well they are recognized and utilized can affect the health and safety manager's overall success in developing and implementing programs.

Formal communication channels often follow the lines established in an organization chart. These communications frequently involve directives, policies, requirements, management information, and other "top down" items. Typically, these channels are also employed when material is sent "up the chain" for "further action."

As noted in Chapter 6, it is important for health and safety managers to recognize their position in the organizational structure in order to use formal communication channels effectively. In particular, it is imperative that the manager be sensitive to "line" and "staff" organizational and communication aspects.

Informal communication channels, including the "grapevine," can either be useful or disruptive. Typically, informal communications "travel" much faster than formal communications. Rumors of a cutback in an organization's activities can spread throughout an organization before a "Message to all Employees" can be prepared, even if started by the company's fastest typist immediately after the meeting where the cutback was decided.

It is important to be aware of what is transpiring in the informal communication channels. The health and safety manager can often get early indications of changing circumstances or problems through them. On the other hand, the program director will be alerted to rumors that have absolutely no basis in fact but which could potentially jeopardize health and safety efforts. Regardless of the situation, taking note of informal information can benefit both the manager and the health and safety program.

For example, a secretary in a CEO's office called the secretary to a health and safety director with the information that a department chief had just requested a topic be added to an upcoming meeting. The topic was "Medical Insurance Problems," and the meeting was one all senior managers would be attending in an hour. The secretary also provided the name of the patient who had precipitated the request. The call gave the director time to gather information about the situation and to review the medical insurance procedures. During the meeting, the department chief brought up the situation by describing how extensive the costs of a medical problem would be for the patient and commented about the need for better efforts to find an effective medical insurance program. Having been forewarned, the health and safety director was able to verify that the questioner had been provided misinformation, and that the insurance program covered all of the patient's bill, except for a small deductible. The director also briefly summarized the quality of the insurance program and commented on the fact that the employees rated the insurance as "highly satisfactory" during a recent survey. In this instance, informal communication allowed the health and safety director to defuse a potentially serious situation and also provide some favorable information about one aspect of the health and safety program.

The above example illustrates the significant roles the secretary or administrator for the CEO, or for other senior supervisors, can play in informal communications. These individuals should never be expected to violate a confidence, of course. However, establishing good rapport with them can greatly assist the health and safety manager by providing advance information on special needs or problem areas, requirements, difficulties, or similar matters. Often, these workers can review a preliminary proposal to help ensure that the effort is compatible with a senior manager's intent. Maintaining good working relationships with assistants to the senior managers can thus provide many benefits to the health and safety manager. Conversely, antagonizing them can result in unpleasant surprises, including the opportunity to attempt to establish rapport with new people in a different organization.

SUMMARY

Communication is a critical, multifaceted management skill. The more effective the health and safety manager is as a communicator, the more effective the health and safety program. Other chapters underscore this fact and amplify other aspects of effective communication, including meeting management, making presentations, communicating with the media, personality aspects of communicating with others, team building, and giving feedback.

REFERENCES

1. Hersey P, Blanchard KH, Johnson DE. *Management of Organizational Behavior—Utilizing Human Resources,* 7th Ed. Upper Saddle River, NJ: Prentice Hall, 1996.

2. Rogers CR, Roethlisberger FJ. Barriers and gateways to communication. *Harvard Bus Rev* 1952;30(4). Reprinted in *Harvard Business Review Business Classics* 1986:44–50.

3. Strayer DL, Drews FA, Johnston WA. Cell-phone induced failures of visual attention during simulated driving. *J Exp Psychol Appl* 2003;9(1):23–32.

4. Strayer DL, Drews FA, Crouch DJ. A comparison of the cell phone driver and the drunk driver. *Hum Factors* 2006;48(2):381–391.

5. Dux PE, Ivanoff J, Asplund CL, et al. Isolation of a central bottleneck of information processing with time-resolved fMRI. *Neuron* 2006;52(6):1109–1120.

ADDITIONAL READING

Argyris C. Good communication that blocks learning. *Harvard Bus Rev* 1994;72(4): 77–85.

Fielden J. What do you mean I can't write? *Harvard Bus Rev* 1964;42(3). Reprinted in *Harvard Business Review Business Classics* 1986:125–133.

Guffey ME. *Business Communication: Process and Product,* 5th Ed. Thomson South-Western, 2005. Available at: www.swlearning.com

Lohr S. Slow down multitasker, especially if you're reading this in traffic. *New York Times* March 25, 2007:1, 25.

CHAPTER

21

Effective Meetings and Presentations

Most health and safety managers spend significant time in meetings, including ones that they direct or chair. They also may have numerous opportunities to make presentations, ranging from formal ones before large audiences to informal sessions with only a few attenders. Both meetings and presentations may have significant impacts on health and safety programs, so facility in these areas can enhance the effectiveness of managers directing these programs.

MAKING MEETINGS WORK

Meetings have aptly been described as sessions where minutes are kept and hours are wasted. All of us have attended sessions that fit this description. On other occasions, however, meetings have been productive and provided a sense of accomplishment. It is hoped the following comments will increase probabilities of having ones that meet the latter description.

Decide on Purpose(s) of the Meeting

Although seemingly obvious, every meeting should have a clear-cut purpose. Having a meeting because one is "always" held every Monday morning can be a major time waster. Defining the purposes of the meeting facilitates allocating time for each purpose and for the overall meeting. It also helps avoid the consideration of so many major topics that none, except perhaps the initial one, receives appropriate consideration. As with other communication considerations, it is important to decide whether the meeting is to inform, make a decision, or accomplish another task. All involved should be clear on the purpose of the meeting. As Drucker notes, the effective executive states the purpose of the meeting at the outset, the contribution it is to make, and, at the end of the meeting, relates the final outcome to the purpose as originally stated [1].

Consider Use of an Agenda

An agenda may be of assistance in controlling discussion times and ensuring the purposes of the meeting are accomplished. Properly constructed and used, agendas have a number of advantages. If distributed prior to the meeting, an agenda helps those attending prepare for the session, especially if they are expected to present during the meeting. Specifying the time for each topic underscores the significance of the topic and facilitates keeping the discussions moving. However, agendas have some potential drawbacks. They may inhibit some meeting attendees and reduce spontaneity. If the times specified for discussion are exceeded significantly, many of the meeting attendees may begin to be concerned that the later topics will not be properly considered or that the meeting will extend indefinitely. Too many of us have been in meetings where "Welcome" as the first item on the agenda is listed as "1 minute." When the meeting chair drones on for 10 minutes just welcoming attendees, it is obvious that the meeting is going to be poorly conducted, possibly necessitating extension of the meeting or the scheduling of another meeting. The manager in charge needs to be able to focus discussions, summarize discussions, and ask for a group decision, all within a reasonable or specified time. Doing this requires skill but will markedly increase effectiveness of the meeting and also encourage attendance at future sessions.

Comply with Scheduled Start Time

It is essential that meetings begin and end on time. Few circumstances are as frustrating as arriving at a meeting on time and then watching time slip by while the group waits for late arrivals. The chair should recognize the significant productivity loss that such delays involve and begin the meeting at the scheduled time. Having 12 people wait "only" five minutes for a late arrival represents a loss of one person-hour. On more than one occasion, I have observed a senior manager chairing a meeting point out to a late arrival that the delay had cost the company "X" dollars! (As mentioned in Chapter 19, many senior managers seem particularly aware of the hourly "cost" of other managers sitting around the table.) The comment did not enhance the "comfort level" of the late comer, and it put the individual at a disadvantage in trying to participate in the session when it did start. Certainly, a health and safety manager would not want to be put in such a situation, particularly if one objective of the meeting is to gain approval for an increase in the budget for the health and safety program.

We often can do little about the late arrival of a chair, especially if he or she is the CEO. Nor is it politically wise to begin a meeting when the senior manager has not arrived, unless guidance to the contrary has been provided by that individual. However, in most other circumstances, it is advisable for a health and safety manager in charge of a meeting to begin the meeting promptly, as scheduled. Delaying "a few minutes" for late arrivals can begin a vicious circle that will result in more people being late for subsequent meetings, since they do not expect them to begin on time. Over time, meetings will start later and later, and eventually meetings will become totally ineffective, since little time will be available to accomplish the business of the meeting. I have found that consistently starting on time can help prevent wasting time waiting for late arrivals. Rather than take the time of prompt arrivals, I advise late-comers that they can catch up on actions by reading the minutes or talking with me after the meeting. Making decisions affecting late-comers before they arrive can also improve future attendance, especially if the decisions involve funding, staffing, or workload! One perennial late-comer was advised, after he once again arrived late at a meeting, with the excuse that he had been caught on a phone call with an East Coast client, that the group had just taken action to meet a financial shortfall by transferring $30,000 from his program! He was not late thereafter.

LEADING AND FACILITATING

The health and safety manager in charge of a meeting plays both leadership and facilitating roles. Although it is important to keep the meeting discussion focused and productive, it is also desirable to guide the group so that discussion is facilitated and decisions are *group* decisions. Group "buy-in" on decisions will generally be much greater than if the person in charge specifies both the question and the answer!

Meeting leadership responsibility requires tact and firmness. It is easy for meeting attenders to spend large amounts of time on relatively minor, non-threatening topics rather than focus on difficult subjects that are the real reason for the meeting. If not appropriately controlled, there may be extensive, prolonged discussion of whether to have an office party in the afternoon or early evening, with the result that minimal time is available to consider how to deal with a budget shortfall that could result in terminations.

It is also appropriate for the health and safety manager to remember the comment by Drucker, in his discussion of effective meetings [1]: "One can either direct a meeting and listen for the important things being said, or one can take part and talk; one cannot do both."

The leader will benefit by "reading" members' participation and actions. Is there productive interchange among all attending or do only a few speak? Have one or two members taken over the discussion and are attempting to force a decision? Is there a tendency to make decisions by default in that almost any decision is acceptable in order to conclude the meeting? Is there a real attempt to compromise and reach a group consensus? In evaluating group dynamics and interactions, it might be advantageous to consider the "types" of attenders. Recognizing the various types may aid in motivating them to be productive members of the group. A *few* examples are:

a. Initiator: Initiates discussion; may oppose new ideas, solutions, alternatives from others.

b. Energizer: Prods the group to action; maintains a high level of interest and enthusiasm; insists on covering the agenda.

c. "Expert": Expert on topic under discussion and often on any other topic; readily provides opinions regardless of real level of expertise.

d. Harmonizer: Friendly, warm, responsive; accepts ideas of others; attempts to clarify areas of disagreement and settle arguments; attempts to ensure all are treated fairly.

e. Listener: Says little but maintains a participatory attitude with body language; may be a true expert; summary comments may be particularly valuable.

f. Elaborator: Active in solving task at hand by clarifying remarks of others and giving own views; adds examples of how solutions might or might not work.

g. Dominator: Talks too much; attempts to monopolize conversation; autocratic; provides "the" solution and cannot understand if others do not accept it.

h. Withdrawer: Indifferent, bored, doodler; creates impression that topic is insignificant.

i. Sneak: Undermines the comments of others; sarcasm may be a trademark; works to undermine authority of chair.

j. "Agree-er": Agrees with everything the chair states *during* the meeting; derides action of group after the meeting; probably most dangerous to the organization.

Fortunately, the majority of those attending meetings chaired by the health and safety director will be action-oriented, constructive individuals. Recognizing the few times when obstructive individuals are present will aid significantly in dealing with them.

Establish and Maintain Appropriate Meeting Procedures

Many meetings that the health and safety manager directs or attends will be relatively informal, without any specified procedural rules. Even in this setting, however, it is not unusual to have motions presented and votes taken. Other meetings, especially in professional organizations or more formal settings, may operate under by-laws or other governing documents. Particular procedures, such as those of *Robert's Rules of Order* or Sturgis's *Standard Code of Parliamentary Procedure,* may be specified [2–4]. Many health and safety managers correctly consider parliamentary procedures too rigid or structured for the typical meetings they conduct. However, the underlying principle of parliamentary procedure—that all members are to be treated equally and fairly—is an appropriate one for any meeting, including those that are not required to comply with specified parliamentary procedures. Indeed, *Robert's* stipulation "No member can speak the second time to a question (motion) until every member choosing to speak has spoken" helps to preclude domination by a small group in favor or opposed to an action [2,3]. This is as appropriate in a small decision-making health and safety meeting as it is in a larger assembly.

On occasion, even in meetings without specified parliamentary procedures, those with parliamentary experience may attempt to apply such procedures inappropriately. A common "mistake" in such situations is to have a motion presented, then two or three speakers speak in favor of the motion, and one of those in favor immediately "calls the question," advising the chair that the motion must now be voted on without further debate. Such action can obviously work to the detriment of those who might oppose the motion. In actuality, assuming the chair recognized the individual calling for the question and it was seconded, the *call for the question,* not the main motion, must be voted upon without debate. Since approval will deny members the privilege of further debate, the call for the question must be approved by a two-thirds majority. Only if call for the question is approved is the main motion immediately voted upon. Of note is that failure of the person "calling the question" to be first recognized by the chair means that the request is out of order and should be ignored. The wise chair will recognize this inappropriate call for question and other parliamentary efforts to subvert appropriate consideration of an action. Having sufficient parliamentary capabilities to deal with such situations will enhance meeting effectiveness and, often, group harmony. A summary of some parliamentary procedures is provided in Appendix D.

Keep the Meeting Focused

During the meeting, it is often beneficial to summarize progress on the discussion item, possible approaches, and need to reach a conclusion. Some chairs use a problem-solving approach, as discussed in Chapters 9 and 14, periodically in discussing a complex item. Agreeing on the definition of the problem, reviewing background information and constraints, ascertaining progress the group has made, defining possible approaches, and selecting a course of action may help avoid prolonged "circular" discussions.

You May Have to "Force" a Decision

As noted in Chapter 14, it may be helpful to "force" a decision, particularly during brainstorming, retreats, or similar situations in which several attractive options are presented but only a few can be implemented. As discussed in Chapter 14, page 195, during brainstorming, "thinking," or similar sessions in early planning efforts, the group may develop a number of excellent ideas concerning health and safety efforts. Attempting to decide on "priority" goals or efforts may be difficult, especially since members may be hesitant to speak against another's position. One technique that may be helpful is to require each member to rank the top three (or more) items they consider most important or best. Members must use 1, 2, and 3 to indicate their priorities. Giving each of the three items an average of 2 is not permitted. By assigning appropriate "weights" to the rankings (e.g., 1 = 3, 2 = 2, and 3 = 1), it is then relatively easy to calculate which item has the highest overall score and continue the process of ranking the remaining items. Typically, some items may not be ranked at all, and this indicates those items' overall priority to the group. Some may be ranked 1 by one individual and 3 by another, but this is to be expected. Such rankings indicate that even though there is disagreement on the ranking, there is agreement that the item is one of the top 3. Through this process, a consensus can be developed on the most important items (recognizing the difference between consensus and unanimous agreement). This process has been helpful in bringing closure to prolonged discussions, which often occur during "retreat" or off-site meetings, and it facilitates adherence to time constraints. Since everyone had to provide rankings, there tends to be better buy-in on the "solutions" than if some other procedure was followed to select priority items.

Ensure Meeting Actions Are Implemented

If a meeting has been convened to decide one item, and a decision is reached, the follow-up is usually relatively straightforward. However, if several items are discussed, it is important that responsibilities for further action be clearly defined. Whether in minutes or in an informal summary by the chair, it is essential to identify *by name* the individual(s) who will accomplish the approved activities and time to accomplish the action. Frequently seen statements such as, "All members agreed that a new exposure history form would be developed" will usually result in no further action. Past experience suggests that the topic "need for development of a new exposure history form" will be found in the minutes of subsequent meetings of the group. A statement like "Members approved development of a new exposure history form. Dr. X will develop a draft of the new form and present it to the committee at its April meeting" is more likely to produce results and will facilitate tracking committee actions.

Honor Stop Times

Every effort should be made to stop the meeting at the specified time, even if some items have not been resolved. It may be necessary to have a second meeting for the unfinished business, but, in my opinion, this is preferable to carrying on the initial meeting "forever," especially if necessitated by poor meeting management. Conversely, as noted in Chapter 19, I have never experienced adverse comments when I concluded a meeting before its scheduled time when all planned business was accomplished.

Provide Appropriate Meeting Settings

Location, time, seating, and other aspects of a meeting's setting can impact its effectiveness. Locating a meeting of a small group in a large auditorium can impact efforts to build rapport or carry on cohesive discussions. Conversely, many of us have experienced the difficulties that occur when organizers try to fit many people into too small a room. Having some people standing, sitting on the floor, or squeezed into seats jammed against each other usually means the meeting's effectiveness will be minimal. The fact that such a crowd often "overwhelms" the room's temperature controls will further reduce the meeting's effectiveness.

Although a few managers favor the efficiency of stand-up meetings, where no seats are provided, most managers are more effective if attendees sit in comfortable chairs with convenient writing surfaces. Sufficient lighting also contributes to the success of meeting. Some attendees may make much of "power" seat locations, especially those next to the meeting chair. However, if these individuals are not in the direct field of view of the chair, those sitting across from the chair have more viability and "power." Overall, however, most senior managers focus more on the *content* of a meeting attendee's statement(s) than the seat position of the person making the comment.

Recognize Scheduling Aspects

Scheduling a meeting early in the morning may provide a more energetic meeting, especially if most attending are "morning" people, while holding a meeting right before lunch may help move the session along and avoid some extraneous comments. Holding a meeting right after lunch, especially if lights are dimmed for slides and the temperature is warm, will be very restful, although nonproductive, for many of us! Scheduling a meeting shortly before the close of business may markedly expedite progress.

MAKING PRESENTATIONS EFFECTIVE

Another special aspect of communications is making presentations. Most health and safety managers are responsible for presenting information to the organizations they support as well as to professional groups and to the public. Appropriate preparation will increase the effectiveness of these presentations. The following highlights a few pertinent elements in preparing a presentation.

Decide on Purpose(s) of the Presentation

This factor is important in any communication effort but is particularly significant in developing a presentation. Is the purpose to provide information, gain approval and funding for a new project, respond to a query, deal with a crisis, or to or accomplish something else? It is particularly important to carefully consider the presentation's *main* purpose, as it dictates the format and content of the presentation—that is, the presentation should be tailored and structured to fit the purpose of the presentation. Trying to accomplish several actions with a single presentation is not only difficult but may result in overall failure.

Ascertain Time Allocated

This is one of the most important aspects of preparing a presentation. In considering use of allotted time, plan for audience participation or questions, if appropriate for the session. In a corporate setting, running overtime typically results in adverse comments and reduced effectiveness of the session. In professional meetings, not adhering to the time requirements often results in being cut off before conclusions are presented or reducing the time for follow-up speakers. In public meetings, rapid loss of interest in the presentation usually ensues when the presentation is too long. In all settings, speaking longer than the scheduled time reflects rudeness to other presenters as well as to the audience. It also indicates poor planning and preparation.

Consider the Impact of Your Scheduled Time Slot

Your time slot on the schedule, as well as time allocated, may indicate the significance attributed to your comments. If you are to be a "keynote" speaker, your talk is obviously considered the, or one of the, most important and thus requires very thorough, "over and above" preparation. However, even if you are limited to a few minutes and scheduled right after lunch or just before the end of the day, you should make every effort to ensure that your words are effective and achieve the purpose of your session.

It is also important to determine whether your talk will be immediately after a break or other event in order to gauge whether there may be a delay in getting started. In such a situation, it is advisable to have sufficient flexibility built into your presentation so that you can modify your remarks in the event that the time allowed for your segment of the program is reduced.

It is also beneficial to learn who will be speaking just before and after your discussion. One of my most challenging presentations was the time I was slotted to speak after C. Everett Koop, a dynamic, forceful speaker who was then Surgeon General of the United States and consistently drew extensive media attention. Fortunately, organizers of the event scheduled a break right after Dr. Koop's presentation so I did not face the handicap of immediately following such a renowned and powerful speaker. Even if you are in the situation of following a noted speaker, it is advantageous to know this before the event rather than find out in the "final hour." This will help you prepare psychologically as well as organizationally.

Decide on Presentation Format

In some presentations, such as talks in corporate settings or at professional meetings, the format for your communication may be specified. If such is the case, it is obviously important to comply with these constraints. In other situations, there may be much more flexibility allowed. In these circumstances, selecting a format depends on the purpose of the event and the approach that you find most comfortable. With respect to speaker's notes, some prefer to dictate a talk and have a full script available. Others find cards with key points most effective. I personally prefer to speak from an outline of major points.

Decide on Use of Visual Aids

Appropriate use of visual aids can greatly heighten the effectiveness of a presentation, provided they are *properly prepared* and *appropriately used.* Hard copies can be converted to slides or overheads by commercial firms; however, most state-of-art word-processing products have capabilities to prepare slides with varying special effects. Microsoft PowerPoint and Corel Presentations 8 are two examples. (Depending on hardware capabilities, material can be printed directly as slides, colored overheads, or even as handouts, with three or six slides per page.) All slides should have a limited number of clear, "easy to read from the back of room" words. Everyone has their own preferences as to slide styles, borders, and colors. For example, I personally prefer dark letters on a light background. I have found boldface Times New

Roman font works well for me. I use 36-point size for the title at the top of the slide, 32-point size for the major headings and first-level subheadings, and 28 for the second or lower subheadings. (However, I try to avoid using lower than second-level subheadings whenever possible.) I use simple borders to avoid detracting from the content. High contrast between the letters and the background is particularly important if the room is dimly lit. Since I often cannot be certain of room lighting, I tend to make all my slides in a high-contrast format. If using computer-assisted presentations, it is usually possible to change the format or background of the slides if there is an opportunity to view the session in the presentation or "speakers' ready" room.

In general, I believe that a single slide should not include more than six to eight lines of large-print text in addition to the title. In my opinion, a speaker who apologizes for a "busy" slide—for example, one that contains a photocopy of a page of print or one that in some other way would challenge the eye of an eagle—simply lacked sufficient respect for the audience to prepare proper slides. An exception might be if the instructor/presenter has provided the audience with legibly printed handouts containing the slide material, and the slide is used only to refer the audience to the handout.

Although computer-generated slides often greatly enhance a presentation, in some cases they may be a distraction. Done properly, the color, graphics, video clips, and other special effects can add immensely to a presentation, whether using slides, overheads, or computer-projections during the session. Unfortunately, there are potential problems associated with their use. For example, I recently attended a presentation at which the presenter chose dark blue colors for the words and a black background. I suspect that the slides looked impressive on the computer screen, but they were impossible to read during the actual session.

Computer-generated slides can present other problems. It is essential to ensure that the projection equipment to be used at the session is compatible with the slide format. In the last few months, I have seen three major presentations seriously compromised because the computer-generated slides did not display, even though, in each instance, "It worked fine when we checked it out before the session began." The problem was that the testing had been done on a different computer, with different software. In another instance, the projections worked fine, but the presenter had so many slides with words and moving images expanding and coming into view from the left and then the right side of the screen, often accompanied by music or applause, that few in the audience were able to concentrate on the content of the talk. Developing computer-generated special-effects slides takes considerable time, particularly because there is always the temptation to add one more graphic component to the slide to achieve greater visual impact. However, more is often less. Relative to the potential for excessive use of slides and the use of excessively "busy" slides, it is of interest that one major software company reportedly banned any computer-generated slides from in-house presentations due to the enormous amount of time presenters were spending developing ever-more dramatic, complex slides in an attempt to outdo other presenters.

Slides should be used to *emphasize* points made orally and should visually *aid* the audience's comprehension of the information in the talk. Reading word from word from a slide is deadly: The audience can read much more rapidly. Constantly waving a laser or light pointer over a slide is another taboo, as it can have a dizzying effect on the audience and detract from the points made on the slide and in the oral presentation.

Both computer-generated and computer-projected technology are changing so rapidly that it is difficult to suggest any one system. As with slide style, it is perhaps most effective to evaluate and compare current systems if the number of presentations justifies procuring a system.

Even when using slides, overheads, or other visual aids, it is important to look at and speak to the audience, not to the visual aid. It is disconcerting to have comments suddenly become inaudible as the speaker turns from a microphone to look at a projected slide. A particular problem that may arise with use of overhead slides is the speaker's standing in a position that blocks the screen and interferes with the audience's view. If the speaker sees a portion of the audience suddenly leaning in one direction, this is invariably the problem.

In using visual aids, a good rule of thumb is that one slide will occupy approximately one minute's time in the presentation. Of course, there will always be exceptions and the time may seem shorter during any practice session, but this rule is a useful one for planning the actual talk. Someone who does believe that he or she can do much better and uses, for example, two full carousels, each containing 80 slides (160 slides for a 30-minute talk)—or a similar number of computer-generated slides—will, at the least, lose credibility with the audience. A senior government research person was invited to present the keynote address at a 2007 health and safety conference. Although he had been invited, and accepted, weeks before the conference, he rushed through a number of the slides, audibly noting that he had too many slides in his presentation. He commented that he knew the audience was already familiar with the material he was skipping in his talk. He obviously had not taken the time to develop a presentation tailored to the audience and the allotted time. Rather, he apparently used a "canned" talk, with the result that his session was not very informative or effective. Not surprisingly, he also ran over the scheduled time for his session. As previously noted, such a session conveys disrespect for the audience.

Decide on Use of Other Aids

Use of video or sound tapes may add to the effectiveness of a session, but it must be noted that the amount of time it takes to switch from one modality to another can break the rhythm of the presentation and create "dead space." Further, each additional medium that is used provides one more opportunity for something not to work properly.

In considering visual or other aids, it is essential to remember that the content should be the focus and whatever aids are used are to help convey the content, not highlight the aids. The story is told of an Air Force officer who had spent his entire career at the Pentagon after graduating from the Air Force Academy. The officer was known for the dramatic briefings he prepared, being one of the first to use dual screens, accompanying sound, and movie or video clips. Although some colleagues considered the content of many of the officer's session to be shallow, his impressive talks had contributed to his rapid promotion to lieutenant colonel. The officer was asked to present a briefing to General Curtis LeMay shortly after General LeMay became Chief of Staff of the U.S. Air Force, the highest position in the Air Force. Although the talk focused on a relatively minor subject, the presenter used all possible presentation modalities. The talk was acknowledged to be one of the most professional and theatrical that the audience had seen. At the conclusion, the lieutenant colonel asked General LeMay if there were any questions. The general responded that he had no questions but did have one comment. Expecting the typical congratulatory comments, the briefer queried, "And what is that, Sir?" General LeMay responded, "Neither one of us is going any higher in this organization!" The talk had all the "bells and whistles" but was weak on substance. Remember, spiffy visual and other aids without substance do not pass muster. Content counts.

Decide on Use of Handouts

Whether handouts are used depends on the purpose of the talk and any requirements specified by the organizers of the session. If handouts are used, keep in mind the need to keep it simple, i.e., the *KISS principle.* Handing out multiple-page documents and voluminous "background" items can result in total loss of the audience's attention as attendees page through the material. Perhaps providing a copy of the slides you are using will be of value, but there is the tendency for the audience to read ahead while you are presenting. Often, the best handouts are one-page summaries or point papers. If the main points of the talk cannot be *summarized* on a single page, the material may be too complex for a single session or simply not organized properly.

A Word About Introductions

If you are to introduce a guest speaker, it is advisable to check with the speaker regarding content and length of the introduction. Reading word for word from a long biography rapidly becomes boring and cuts into the speaker's time. I have repeatedly seen guest speakers cut off long-winded introductions, which reflects poorly on the person giving the introduction. Less verbiage is usually preferable in this situation, as in running effective meetings. My underlying philosophy has been "If you have to tell someone how important you are, you aren't!"

Consider Using Sources of Assistance

Professional and government organizations can be of immense benefit in preparing a presentation. These organizations may be able to provide "ready-made" slides, illustrations, or other previously prepared material. For example, the Association of Occupational and Environmental Clinics and the Aerospace Medical Association have lending libraries of PowerPoints slides that can be accessed by those preparing a health and safety presentation. The American College of Occupational and Environmental Medicine is planning to make slides that are used in some of its short courses available on its World Wide Web site in a format that will permit downloading. The National Institute for Occupational Safety and Health (NIOSH) and the NIOSH-supported Education and Research Centers, the American Society of Safety Engineers, the American Industrial Hygiene Association and other professional organizations typically have background and other information that can be used in preparing health and safety presentations. A listing of websites for selected sources of possible assistance is provided at thc cnd of this chapter.

Practice

Some OEHS health and safety professionals may be able to prepare a talk and give it without a "dry run," but most of us will benefit greatly from at least one practice session. A practice session often reveals problems with the presentation that need to be addressed—for example, reaching only the halfway point of the talk within the scheduled time, realizing the need for an additional slide to make a point, or discovering that the approach used to reach a conclusion has some serious flaws in logic.

I have found it advisable to project slides during the practice session in order to find errors that spell check did not catch and also to determine if the sequence of the talk really "flows." A practice session also provides the opportunity to see if your words are "alive" and your gestures appropriate. If you find that you are getting bored or sleepy listening to your words during a practice session, it is likely that you will feel likewise during the actual event. You want to convey excitement, or, at the least, generate real interest in the material you are presenting. Practicing, especially before a person who will honestly critique your efforts, can help achieve desired results. *A cardinal rule that bears special emphasis is never read your talk word for word.* Doing so is the "kiss of death" and will usually require significant sound amplification so that your remarks will be audible over the snores you induce. You will lose your audience's attention irrevocably. In the rare situations in which you are *required* to read from a report, it is imperative to keep your phrasing, voice levels, and eye contact with the audience engaging and dynamic.

In general, the amount of time spent on practicing the presentation should be commensurate with the significance of your session. For example, if the desired end result is approval or disapproval of your health and safety budget for the year or is a decision as to whether or not to hire your firm as a consultant for a corporate health and safety program, extensive preparation and practice may be warranted. In other instances, depending on the experience of the presenter, much less preparation and practice may be required. However, appearing unprepared or disorganized will result in poor response or evaluations. Recall the comment in Chapter 20 that the secret of a really good extemporaneous talk is lots and lots of preparation and *practice.*

Consider Factors Present at the Time of Presenting

The day of the event, it is often helpful to arrive early to evaluate room arrangement, how to control projectors (and where spare bulbs are located), how to control room lights (if an individual is not assigned to do so), and how to turn microphones on or off and control volume. If possible, it is also advisable to "run through" the slides or overhead-projection images. If the room itself is unavailable, a "speakers' ready room" may be provided for the run through. Even though I always checked my series of 35-mm slides and put them in a locked carousel marked "Final Presentation" before departing for the event, one or two will mysteriously be upside-down slides, will be out of order, or have typos when I rechecked them during the run through. PowerPoint and other presentation materials are not immune from similar problems.

It is also important to recall that intake of large amounts of food or fluids just before the session can create difficulties, particularly if the food is gas-producing or the fluids have diuretic or coordination-altering effects.

Finally, *do not apologize at the start of your talk!* Presumably, you have made your best effort to prepare the presentation, and apologizing to the audience at the start "sets them up" to be disappointed. Those listening will recognize if you have a cold or are hoarse from an allergy. Few participants have any real interest in your recent medical history, plane delays, or difficulty in sleeping in the hotel. In my opinion, focusing on personal difficulties like these and apologizing for the talk may lead some members of the audience to feel cheated or to "tune out" the rest of your comments. Your efforts and information are much too valuable to be compromised by apologizing for your session.

SELECTED SOURCES OF ASSISTANCE IN PREPARING PRESENTATIONS

(*Note:* These websites were current at the time of submission for publication in April 2007 but may have subsequently changed.)

Aerospace Medical Association
www.asma.org

American Association of Occupational Health Nurses, Inc.
www.aaohn.org

American College of Occupational and Environmental Medicine
www.acoem/.org

American Industrial Hygiene Association
www.aiha.org

American Society of Safety Engineers
www.asse.org/

Association of Occupational and Environmental Clinics
http://aoec.org/

Board of Certified Safety Professionals
www.bcsp.comorg/

National Institute for Occupational Safety and Health
www.cdc.gov/niosh/homepage.html

U.S. OSHA Salt Lake Technical Center
www.osha-slc.gov/SLTC/sltc.html
www.osha.gov/dts/sltc/index.html/ (See especially the link to Safety and Health Topics.)

Some academic institutions can provide sources of information through their own capabilities or through links. NIOSH (National Institute for Occupational Safety and Health) sponsors 16 academic Education and Research Centers (ERCs). Many of these have home pages that provide assistance—for example:

Rocky Mountain Center for Occupational and Environmental Health at the University of Utah
http://uuhsc.utah.edu/rmcoeh/

Other ERC home pages can be accessed from the NIOSH home page.

REFERENCES

1. Drucker PF. *The Effective Executive.* New York: HarperBusiness, 1966, 1967 (HarperBusiness Essentials ed. New York, 1993, 2002).

2. De Vries MA. *The New Robert's Rules of Order,* 2nd Ed. New York: Signet, 1998. (In addition to the rules, the book includes extensive discussions of planning and organizing meetings, including arranging for speakers, registration, facilities, promotion, and similar topics.)

3. Robert HM, Evans WJ, Honemann DH, Balch TJ. *Robert's Rules of Order in Brief.* Cambridge, MA: DaCapo Press, 2004.

4. Sturgis A. *The Standard Code of Parliamentary Procedures,* 4th Ed. Columbus, OH: McGraw-Hill, 2000.

ADDITIONAL READING

Allen J. *Event Planning: The Ultimate Guide to Successful Meetings, Corporate Events, Fundraising Galas, Conferences, Conventions, Incentives, and Other Special Events.* New York: Wiley, John & Sons, 2000.

Jay A. How to run a meeting. In *Harvard Business Review: Managing to Save Time,* No. 12201.

Robert SC, Robert HM III, Evans WJ, et al. (eds). *Robert's Rules of Order,* Newly Revised, 10th Ed. Reading, MA: Perseus Books, 2000. (10th version of the original (*The New Robert's Rules of Order*) with more detail than Reference 3 above, by the same authors.)

Running Meetings. Boston: Harvard Business School Press, 2006.

Zimmerman DP. *Robert's Rules in Plain English.* New York: HarperCollins, 2005.

CHAPTER

22

Interacting with the News Media

Ralph R. Cook

Regardless of their wishes, many health and safety managers have to deal with representatives of the media. How well this interaction is accomplished may impact, either negatively or positively, the health and safety unit and the organization it serves. The following information is intended to assist in responding to media needs.

BUILDING THE COMMUNICATION PYRAMID

Most health and safety professionals are trained in a particular style of communication. This is reflected in the way technical papers are written. In these papers, first there is a literature review that provides background to justify the research, ideally summarized in an articulated problem and a testable hypothesis. Next, the methods and materials are presented to give the reader a sense of how the data were collected and what technical biases might have been operational and how they were addressed. This is followed by a detailed discussion of the results and some alternative interpretations. Occasionally, but all too seldom, the authors critique their own work with a discussion of the strengths and weaknesses of the research and the respective implications of each on the findings. The bottom-line message, the conclusion, is saved for the final paragraph. This has been compared to building a pyramid. Each layer is placed upon the firm foundation of the previous one. Only when everything is in place, is the final point made. The process is one of nurturing the readers to the same conclusion as that derived by the authors.

In communicating with the news media, it is often advisable to invert the pyramid: Provide a summary of the findings and the conclusion in the first statement, the first paragraph—and then justify this statement with supporting data, interpretations, and references. In the process, it is wise to remember some key points.

KEY POINTS TO REMEMBER IN COMMUNICATING WITH THE NEWS MEDIA

Reporters Like to Deal with the Source

Journalists want to talk to the scientist who made the discovery, the manager who made the decision, the person responsible. They prefer not to deal with "official media contact persons," the public relations specialists. And they definitely do not like to talk to an organization through its lawyers—unless the story involves a high-profile trial (and usually not even then). If your organization is to be properly represented in the news, at some point you likely will be the best person to provide the interview.

Many companies and government agencies have recognized the preferences of the press and have provided their senior managers and scientists with formal media training. Some have even institutionalized the interface between their staff and the news media. For example, at my prior company, we had a "Visiting Scientist Program." Senior scientists in various chemical, environmental, and health disciplines proactively went out on planned media tours, providing interviews with the press and appearing on radio or television talk shows. In some situations, they concentrated on a single story. On other tours, they simply made themselves available to the news media for informal discussions and interviews on a wide variety of topics. Of course, the added benefit to this approach is that when there was a crisis that had to be handled reactively, our scientists were comfortable interfacing with journalists, and the journalists had the familiar name of a credible source in their address book.

Most Reporters Are Generalists

Reporters cover a wide variety of subjects. While they appreciate being educated so they can write or present a good story, they usually cannot afford to invest much time in learning the nuances of your discipline. This is especially the case with junior reporters. Offer to provide them in-depth background, but be prepared to keep your discussion succinct.

You also might consider writing a briefing piece and use this to augment the interview. Most reporters prefer to write their own stories, but if your write-up is well done and brief, it might be paraphrased or extensively quoted. This approach is especially useful if you are the one who is initiating the story—for example, when you are releasing the results of a new employee health study. In these situations, we often provided the news media not only a copy of the research abstract, but also a more in-depth executive summary. The latter was formatted in the style of an "inverted pyramid" and may include one or two key tables or graphs.

Your Audience Is Broad

In giving an interview, you will be talking to at least three different audiences simultaneously: your peers, the reporter, and the general public. The challenge is to satisfy all three with the same message.

As a result of your technical training, you may find this difficult to do. Health and safety professionals and other scientists tend to speak to their peers in technical jargon. To the reporter (and consequentially, the public), this can be viewed as obfuscation, or worse, it can produce some serious misunderstandings. For example, in epidemiology we speak of "positive" and "negative" studies. To the epidemiologist, a positive study is one in which there are statistically significant findings of adverse effects that likely have some biological significance. To most nonscientists, the term has exactly the opposite connotation. To them, such a study would be negative in the sense of providing bad or negative news.

In giving an interview, translate back to layman's language, but do it in a fashion that does not "speak down" to the reporter or the audience. I have often found it useful to state the message in a technically correct fashion and then immediately paraphrase in nonscientific terms. The sophisticated reporter will weave in some of the technical phrases to give the piece "authority," while the uninitiated will at least relay the simple message you are trying to deliver.

Most Reporters Are Proud of Their Profession and Practice It in an Ethical Fashion

With few exceptions, reporters are scrupulous in their attempts to present the facts, to report the truth. Having said that, you must recognize that their definition of fact or truth might differ from yours. To them, that someone made a statement and what they said is, by definition, fact. They may seek an analogous response from an alternative source, but rarely will there be any attempt to test the underlying validity of either statement. Actually, in many cases they will do just the opposite. They will purposefully try to find someone with an opposite opinion to "add balance" to a story. You must view a reporter not as an arbitrator of truth or a filter of fact, but as a conduit of opinion. Through them, you (and those with alternative viewpoints) speak to the public. Since it is the public who must discern fact from fiction, it is in your best interest to participate in the process.

Reporters Are Zealous Protectors of the Freedom of Speech

Reporters may call to double check some part of a story (you may even be that "reliable source" so often quoted) or read back direct quotes that will be attributed to you, but never expect them to submit a story for review and approval prior to publication.

The Selling of News Is Big Business

In the news media, there is significant competition for customers. Understandably, the press must be on the leading edge of an issue, for news is a perishable commodity. They cannot and do not wait for a consensus. As a result, they tend to report controversy. Indeed, some may even find it necessary to titillate, to imply controversy where there is little. For example, editors often write the headlines and the lead-in sentence or paragraph of a story as an enticement to the reader. These items may bear little resemblance to the overall content of the article. Do not be upset if this happens; remember that it likely was totally beyond the control of the reporter.

Understanding the process, the pressures under which the news media function can work for you. Proactively provide reports that are both timely and newsworthy. Be quotable. While your quote may not be used as a headline, it may direct the basic thrust of the article.

Reporters Generally Operate with Rather Severe Time and Space Constraints

Stories in the afternoon newspaper must be "put to bed" by late morning. If a reporter calls you at an inconvenient time or requests information that you must do research to find, ask about the deadline and offer to call back. And if the reporter accepts your offer, follow through. They will appreciate the courtesy, and it will probably be reflected in how they report your comments. Television can be much more spontaneous—now, reporting a story while it unfolds is common. But this communication vehicle must compress aggressively to fit everything into the format of "News at Eleven." While initially you might find the condensation of your life's work into a 30-second time byte somewhat frustrating, it can be an advantage. If you are being interviewed "live," you have the opportunity to deliver three or four key points in order to get your message across, sometimes irrespective of the questions that are asked. And if you are being prerecorded and are able to be brief and articulate, your comments will probably be used, and the public will likely remember them.

I personally am reluctant to participate in a prerecording session of great length. It provides too great an opportunity (or is it temptation?) to the program producers to edit my comments out of context. One memorable negative experience I had with the news media took place in such a session. An investigative reporter kept paraphrasing the same basic question. When asked why, she responded, "Because he has not provided the answer I want." If you must participate in a lengthy recorded interview on camera, try to establish some ground rules. Predetermine the topic, the participants, and the length of time for questions and answers. Arrange to videotape the session on your own in its entirety. If suitable arrangements cannot be negotiated, either don't participate (and then expect to hear something like, "The company refused to be interviewed") or agree to participate, but only for a "live" interview.

You Must Prepare for the Interview

Sometimes preparing for an interview may not be possible, but, as much as you can, think through the key messages that you hope to deliver and how you wish to deliver them. When asked a question, be innovative, weaving one of your key messages into the answer.

On the other hand, be truthful and don't try to stonewall. Reporters do not like "No comment." It makes them curious—or angry—especially if you agreed to provide an interview in the first place. If you cannot answer a question, say so and possibly offer to get back at a later date with an answer. But if you make this offer, do not fail to deliver.

You Should Personalize Your Comments

Call the reporter by his or her first name. Be attentive to the questions, an interested listener. Don't act as a representative of your organization, "We at XYZ organization . . . ;" instead, provide your own perspective, "It's been my experience . . . I think . . . My concerns . . . My family . . ."

Do not answer hypothetical questions, and do not speculate. Address the former by simply stating that you cannot answer the question. As for the latter, either indicate that the topic lies outside the area of your expertise, or that you have inadequate information with which to respond. Given knowledge of the deadlines under which the reporter is operating and your perception of the need to provide the information in order to obtain a balanced story, you may wish to try to identify someone more knowledgeable to provide an answer; but it is probably not wise to offer the reporter a name at the time of the interview. Instead talk to the alternative source and suggest they contact the reporter.

Impression Is as Important as Content, Especially on TV

After a short period of time, we all tend to forget facts and figures. What is remembered by the audience is the general message and whether the messenger was credible. Your appearance and demeanor are important. A comment about law and order from someone who looks like a Hell's Angel will be perceived as ludicrous, no matter how appropriate the content.

Seemingly insignificant aspects of your appearance can distract the viewer and detract from your underlying message. For example, television has limited resolution. Suits, dresses, shirts, etc., with fine patterns can lead to strange and disconcerting effects. It is not by accident that seasoned politicians usually appear on camera wearing a plain dark suit, a blue shirt, and a red tie. They are experts at mass communication. "Take a page from their book" and dress appropriately for the situation.

Image control is especially important with television. Try to train yourself to avoid involuntary movements like grimacing, lip biting, or the wringing of the hands. Saddam Hussein demonstrated his anxiety during the Gulf crisis by blinking his eyes. I have been told that experts were able to gauge the level of his distress (and the veracity of his comments) by the frequency of his blinking. The public also, however subliminally, will judge your comments by your body language. If standing, do not sway back and forth. If sitting, lean slightly forward with your hands comfortably in your lap. Use your hands to emphasize key points, but avoid expansive movements. Keep your facial expressions calm and friendly, and don't wear thick-framed glasses. They block facial communications. Maintain eye contact with the interviewer and not the camera. If you try to do the latter, you will always be looking at the wrong camera and will appear either disoriented or shifty. In a nutshell, it is possible that your competence may be determined by what you say and how you say it, but your credibility will be judged by your demeanor.

It is also wise to remember the vehicle that will be carrying your message. Provide the print journalist with metaphors, similes, and mental images. I can recall an interview I gave to the *Wall Street Journal.* In it, I discussed the importance to epidemiology of valid measures of exposure. During our lengthy conversation, I facetiously suggested to the reporter that ideally everyone at birth should have an exposure sensor implanted in their left buttock. As I recall, this quote appeared in the introductory paragraph.

Television, by way of contrast, is a visual medium. Consider providing the interview in a location with an interesting background—out in the field with a lot of interesting equipment or in a scholarly setting like a library with shelves of books. Otherwise, be prepared to have your voice dubbed into some potentially provocative video. I can recall one live interview that I gave for a midday news program. The topic was toxic chemicals and the potential hazards they posed to farm animals. In retrospect, it is obvious that my comments were to be merged with the video of dead farm animals from an entirely different episode. Fortunately, they got the wrong stock footage, one showing four cows contentedly chewing their cuds in an idyllic meadow.

You Must Maintain Some Control of the Interview

Controlling the interview is especially necessary in the case of "ambush" interviews, in which the reporter suddenly appears out of nowhere and "attacks" with a barrage of queries. The natural tendency is to match the speed of your responses to that of the questions. Do not do it. Make a conscious effort to be slow and deliberate. In this way, not only will you avoid mistakes, but the contrast in styles will work to your advantage. You will come across as calm and credible, and the reporter as frenetic and badgering.

In these types of interviews, the reporter may try to ask multiple questions, making inappropriate implications in some and talking over your responses in others. Do not allow the reporter to provide your answers for you—especially when such answers are contrary to your position. Without restating his opinion, simply disagree and give your own position. Present answers to one question before going on to another. Alternatively, if provided multiple questions, pick the one you wish to answer.

Depending on the situation, you may have to use different techniques to maintain control. For example, it can be particularly disconcerting to the investigative reporter if you provide one word or very short responses. On the other hand, if you are a member of a panel of differing opinions, it may be counterproductive to politely wait your turn. By the time your opponent has finished, you may be off the air and not have had the opportunity to voice any opinion. Furthermore, your silence may be viewed as acceptance of a position with which you disagree. In these situations, it might be most effective simply to start talking. This drowns out everything and forces the moderator to parcel out air time with balance. Former Secretary of Defense Weinberger was particularly effective with this approach. He used it with both the news media and Congress. Sometimes he would just say the same thing over and over.

You Must Remember That Anything You Say May Be "On the Record"

Different reporters use different techniques to obtain a story. Most reporters are relatively straightforward, but some can be nefarious. The editor of this book recalls one reporter who interviewed him on camera. When she apparently was finished, she indicated to the cameraman to stop the session. He turned off the lights and set the camera on the table. She then turned casually to the presenter and asked, "Tell me, what do you really think about this whole thing?" When he basically repeated what he had said "on camera," she turned to her sound man and indicated he also could turn off his equipment.

You Must Be Available for Follow-up

The reporter may feel at the time of the interview that he or she understands the topic only to later have confusion over some key point as the story develops. By being available, you increase the opportunity for clarity. You also develop a working relationship with the reporter that might provide opportunities for input later in some unrelated story.

You Must Measure the Results of Your Interview

This can be very difficult to do, but I have found it useful to get a copy of the article or a video of the television report. I review it critically from the perspective of both form and content to determine whether the report matches the interview. On occasion, I have asked friends and associates to review and comment. By far, my most useful, honest critic over the years has been my wife.

SUMMARY

Giving an interview can be an uncomfortable experience. Many professionals have made the avoidance of the news media a personal goal, but this is a mistake. Our society depends on a free press to mold public opinion, and public perception can be important to you, your profession, and your organization. If objective scientists refuse to provide perspective, the nonobjective advocates will control the message.

There are certain action steps you can take before, during, and after the interview that will make you more effective. If it appears that the job will give you multiple opportunities to work with the news media, get some formal training. In every case, develop key messages. Practice your delivery, both style and content. To give an effective interview, you must understand the process. Recognize that some techniques that are effective with the press will not work as well with television reporters. The opposite is also true. Radio and television tend to operate under more severe time constraints. Make this your advantage by utilizing the limited time available to present the messages you have prepared in advance. Dress properly. During the interview, be attentive, confident, and truthful. After the interview, evaluate the results. As with any other professional activity, learn from the experience. Much to your surprise you might find that you enjoy it.

Acknowledgment: I wish to thank Jennifer Drake for her helpful comments regarding both style and content.

Editor's Note: Dr. Cook, the author of this chapter, has interacted with the news media on numerous occasions, first as Director of Epidemiology for the Dow Chemical Company, later in a similar position for the Dow Corning Corporation, and most recently as a private consultant. He notes that while some of these "opportunities" were contentious, all of them were interesting.

ADDENDUM

Training in Interview Techniques and Other Comments

Dr. Cook has had extensive experience in interacting with the news media. Many of the rest of us have much more limited experience. As Dr. Cook has noted, it is useful to obtain training in media interactions if you will be working with the media on a frequent basis.

Various professional organizations offer training sessions for members that often include both instruction in communicating with the media and actual videotaping of practice sessions. For example, the American College of Occupational and Environmental Medicine provides basic and advanced seminars on media interactions at many of its annual scientific conferences. These sessions have been conducted by P. A. Clark, Director of Speech/Media Training Services of the American Medical Association, and the two sessions include videotaped interviews that typically are very illuminating to those taking the training. Her sessions underscore the importance of Dr. Cook's recommendations in preparing for an interview. The following are a few additional helpful points from Ms. Clark's sessions and in her 1998 book *Communicating With the Media, the Public and Our Peers* (American Medical Association):

1. Attempt to become familiar with the show on which you are appearing—particularly its host, format, and what is expected of you.

2. Rehearse possible questions and answers before the show. Recognize controversial areas that may be the focus of the questions even during a promised "friendly" interview. Can you provide succinct, "to the point" answers?

3. As noted by Dr. Cook, it is particularly important to explain the major point(s) you wish to make in 30 seconds.

4. Can you develop "bridge" responses? These are sentences that will enable you to respond to questions and then transition to the point you wish to make in spite of the flow of the show. For example, if you want to discuss the occupational health and safety programs in your community, you could respond to a question of "What is the major health problem in your area?" with "Low childhood immunization rates" (or whatever is correct) and then continue along the lines of "but we are fortunate that we have high-quality occupational health programs, or else injuries and disease at work could be the major problem." Then proceed to discuss the programs.

5. Have you developed a comfortable conversational style that will work under fire or confrontation? (Practice will help.)

6. Do you know who your real audience is, i.e., the make-up of the viewer audience, not the host? Such knowledge can help you

communicate appropriately to your viewing audience. Try to make your comments of real interest to the people at home watching the show, that is, make comments that will keep them watching the show rather than changing channels.

7. In addition to following Dr. Cook's recommendations regarding dress, posture, and facial expressions, my experience has demonstrated that it is important to avoid the swivel or rocking chair. Regardless of intentions, the tendency will be to go into motion (swivel or rock) as the interview progresses. This activity is not only distracting to viewers but may also suggest lack of interest in the session.

8. In considering the prior comments about dress, also consider whether you want to wear "status symbols" such as pens whose caps resemble a mountain peak or "the watch" used by celebrities. Most viewers can quickly identify such items and their approximate cost. Wearing them can detract from your message, especially if you are talking about dedicated health and safety professionals that need more financial support from the state for their programs.

9. If you are one of two or more being interviewed on television, remember the camera does not have to focus on the one speaking. Staring off into space, yawning, attempting to conceal a laugh, and similar actions while a colleague is speaking will almost guarantee that you, not your colleague, will be "on camera." Maintain a respectful, interested attitude and look at your colleague during his or her comments to avoid creating wrong impressions.

As a final comment, recall the difficulty in "really" being "off the record." This hazard is typified by Ms. Connie Chung's "Just between you and me" assurance to her subject, with the response then broadcast over national television! (Recall my experience depicted in Dr. Cook's section of "Now that the interview is over, what do you really think?")

Media Releases

Another aspect of media relations is assisting in preparing press releases. A few points to consider:

1. Prepare the release for the journalist, not the potential reader. "Why does the reader care about your information?"

2. Send your release only to journalists that may have an interest—do not release an "all points bulletin."

3. Because the author of a press release must be accessible, provide accurate and full contact information (preferably 24-7).

4. Avoid grammar errors in the press release.

5. Do not use all uppercase letters!

6. Avoid "best ever," "amazing," "first ever in the United States," and similar superlatives.

Some Additional Comments:

During a Public Officials' Conference, a state public information officer made the following points regarding interacting with the media [1]:

1. Remember the reporter is looking for the "5 Ws Plus 1." Who, What, Why, Where, When, How?

2. What makes news: Conflict, controversy, timeliness, impact, and proximity to the event.

3. Be educated before the interview so you can provide accurate, truthful information. Remember: "When the eyes say one thing and the tongue another, a practised man relies on the language of the first." (Ralph Waldo Emerson)

4. Have ready access to workers, management, and legal representatives.

5. Maintain confidentiality. Many reporters are facile at reading a document on a desk upside down. In a recent court hearing in Utah, a reporter took a photo of a document the defendant was attempting to pass to the judge. Although the judge refused to accept the document, the reporter enlarged the photo, resulting in the newspaper publication of visible information damaging to the defendant.

6. Keep your cool. (During a major accident or fatality situation, my caution is that you "Do not let them see you sweat!") Advise the media what has happened, that you do not yet know the full details, that many are working on the answers, and that you will get the information to them as soon as possible. Do not be afraid to make the media wait for *accurate* information. As in 3 above, make certain your eyes collaborate what you are saying. Above all, prevent media "nightmares": Avoid saying "No comment"; providing wrong information, especially a lie; and loosing your cool. Any of these responses can be damaging to the organization's image during a disaster and for prolonged periods thereafter.

7. Know the state law regarding media access, especially in the event of an emergency or disaster. In many situations, if the public is admitted, the media must be admitted. If the media infringes on response, interferes with work, or puts the public or the media themselves in danger, restrictions on media access may be necessary [1].

REFERENCE

1. Baird D, State of Utah Department of Human Services, Division of Substance Abuse and Mental Health. Media response to disasters. Comments at Utah Public Officials Conference, Department of Public Safety, Division of Homeland Security, Office of Emergency Services, Park City, Utah, August 2006.

ADDITIONAL READING

Mazur A. Media coverage and public opinion on scientific controversies. *J Communication* 1981:106–115.

Miller NE. *The Scientist's Responsibility for Public Information, A Guide to Effective Communication with the Media.* New York: Media Resource Service of the Scientists' Institute for Public Information, 1981.

Sandman PM. Medicine and mass communication. *Ann Intern Med* 1976:376–383.

Sandman PM. *Explaining Environmental Risk.* Special Publication of the United States Environmental Protection Agency Office of Toxic Substances, Washington, DC, 1986.

Stewart S. *Media Training 101: A Guide to Meeting the Press.* Indianapolis: John Wiley & Sons, 2003.

CHAPTER

23

Legal and Regulatory Aspects of Managing OEHS Programs*

Carolyn S. Langer

Occupational and environmental medicine is a branch of preventive medicine that "focuses on the health of workers, including the ability to perform work; the physical, chemical, and social environments of the workplace; and the health outcomes of environmental exposures" [1]. It is devoted to the prevention and management of occupational and environmental injury, illness and disability, and to the promotion of health and productivity of workers, their families, and communities [2].

From a medicolegal standpoint, occupational and environmental medicine is unique among the medical specialties. In no other specialty do regulatory and legislative mechanisms shape and drive the practice of medicine to the extent that they do in occupational health. Indeed an entire federal agency, the Occupational Safety and Health Administration (OSHA), has been established within the U.S. Department of Labor to safeguard the rights of this particular class of patients (the worker) and to specifically prevent or minimize the incidence of work-related disorders.

*From Langer CS. Occupational health law. In Sanbar SS, Gibofsky A, Firestone MH, LeBlang TR. *Legal Medicine,* 4th ed. St. Louis, Mosby, 1998.

Providers of occupational health services also face unique challenges arising out of their dual loyalty to patients and employers. Can the occupational health professional strive to uphold traditional notions of patient confidentiality, informed consent, and personal autonomy while simultaneously advancing the employer's goals of increased productivity; public goodwill; decreased workers' compensation costs; and promotion of worker, coworker, and customer health and safety? Although dual loyalties to employer and employee create the potential for conflict and impairment of medical judgment, in reality incidents of conflict occur far less frequently than might be anticipated [3]. Occupational health care providers (whether company employees, independent contractors, or private practitioners) can most effectively limit their own liability, promote the health and safety of their patients, and preserve the legal and moral rights of workers through familiarization with occupational health laws and regulations, and by adherence to principles of good medical and risk management.

THE OCCUPATIONAL SAFETY AND HEALTH ACT

In 1970, Congress passed the Occupational Safety and Health Act (OSH Act) "to assure so far as possible every working man and woman in the Nation safe and healthful working conditions and to preserve our human resources" [4]. This legislation created OSHA, the primary functions of which are to (1) encourage employers and employees to reduce workplace hazards, (2) to promulgate and enforce standards that lessen or prevent job-related injuries and illnesses, (3) to establish separate but interdependent responsibilities and rights for employers and employees with respect to achieving safe and healthful working conditions, (4) to maintain a reporting and recordkeeping system of occupational injuries and illnesses, (5) to establish research and training programs in occupational safety and health, and (6) to encourage development of state occupational safety and health programs.

Coverage

The OSH Act covers all employees and all employers (defined as any person engaged in a business affecting commerce who has employees) with the following exceptions:

1. Self-employed individuals
2. Farms on which only immediate family members of the employer work
3. Working conditions or workplaces regulated by other federal agencies under other federal statutes (e.g., Mine Safety and Health Act of 1969, Atomic Energy Act of 1954, Department of Transportation regulations)
4. Employees of state and local governments

Although OSHA provisions cover only the private sector, Section 19 of the OSH Act makes federal agency heads responsible for providing safe and healthful working conditions for their employees. The act also requires federal agencies to comply with standards consistent with those for private-sector employers. OSHA conducts federal workplace inspections in response to employee reports of hazards. Under a 1998 amendment to the Act, it covers the U.S. Postal Service the same as any private-sector employer. States with OSHA-approved state programs cover state and local government employees.

Standard Setting

OSHA standards encompass four major categories: general industry, construction, maritime, and agriculture. In the absence of a specific OSHA standard for a particular working condition or workplace, employers must adhere to Section 5(a)(1) of the OSH Act, or the general duty clause. The general duty clause directs each employer to furnish "to each of his employees employment and a place of employment that are free from recognized hazards that are causing or are likely to cause death or serious physical harm" [5]. Where OSHA has promulgated specific standards, the specific duty clause, Section 5(a)(2) of the OSH Act, mandates that employers "shall comply with occupational safety and health standards . . . promulgated under this chapter" [6].

Section 6(b) of the OSH Act authorizes the Secretary of Labor (hereafter referred to as *Secretary*) to promulgate, modify, or revoke any occupational safety and health standard. In adopting standards the Secretary must first publish a "Notice of Proposed

Rulemaking" in the *Federal Register* and allow for a period of public response and written comments (at least 30 days, although usually 60 days or more). OSHA may also schedule a public hearing to consider various points of view. After the close of the comment period and public hearing, the Secretary publishes the final standard in the *Federal Register.*

Enforcement

The OSH Act authorizes OSHA to conduct workplace inspections. As a result of a 1978 U.S. Supreme Court decision, however, OSHA compliance officers may no longer conduct warrantless inspections without the employer's consent [7]. Because OSHA and the state OSHA programs together employ only about 2,400 compliance safety and health officers to protect more than 115 million workers at over 7.2 million sites [8], the agency has established a priority system for inspections. From highest to lowest, these priorities are (1) imminent dangers, (2) fatalities or serious accidents resulting in three or more workers going to the hospital, (3) employee complaints, (4) programmed targeted inspections, and (5) follow-up inspections.

After an inspection, the area OSHA director may issue the employer a citation indicating the standards that have been violated and the length of time proposed for abatement of those violations. The area director also proposes penalties for these violations. The employer must post a copy of the citation in or near the cited work area for three days or until the violation is abated, whichever is longer. Table 23-1 summarizes the types of violations that OSHA may cite and the concomitant penalties that it may propose.

Employers may contest a citation, proposed penalty, or abatement period by filing a "Notice of Contest" within 15 working days of receipt of the citation and proposed penalty. The area OSHA director will forward the case to the Occupational Safety and Health Review Commission (OSHRC), an agency independent of the Department of Labor. An administrative law judge will rule on the case after a hearing. Any party may seek further review by the entire three-member OSHRC. Commission rulings, in turn, may be appealed to the U.S. Court of Appeals by any party.

Table 23-1. OSHA Violations and Penalties

Violation	Penalty*
Other than serious violation Directly related to job safety and health but unlikely to cause death or serious harm	**Discretionary penalty of up to $7,000 per violation**
Serious violation Substantial probability of death or serious harm. Employer knew or should have known of hazard	**Mandatory penalty of up to $7,000 per violation**
Willful violation Employer intentionally or knowingly committed a violation or was aware of a hazardous condition and failed to take steps to eliminate it	**Minimum penalty of $5,000 per violation, up to $70,000 per violation** (Willful violations resulting in a worker's death may lead to criminal conviction with fines up to $250,000 for an individual and $500,000 for a corporation, and/or imprisonment.
Repeat violation Violation of a previously cited violation	**Fine of up to $70,000 per violation**
Failure to correct prior violation	**Penalty of up to $7,000 for each day beyond the abatement date**

**NOTE*: OSHA also may issue citations and proposed penalties after conviction for falsification of records, reports, or applications; for violations of posting requirements; or for interference with a compliance officer's performance of duties.

KEEPING THE WORKER INFORMED

An estimated 80,000 chemicals are in commercial use today, and 2,000 new chemicals are introduced every year [9]. Yet the health effects of many of these substances remain unknown. It is estimated that more than 30 million workers may be exposed to hazardous chemicals in the workplace [10]. Following passage of the OSH Act, OSHA promulgated a series of regulations to provide workers with more information about the agents present in their workplaces and to enhance the detection, treatment, and prevention of occupational disorders. The following are three of these important provisions:

1. The Hazard Communication Standard [11]
2. The Recording and Reporting Occupational Injuries and Illnesses Standard [12]
3. The Access to Employee Exposure and Medical Records Standard [13]

Although employers are ultimately responsible for safeguarding the health and safety of their workers, these regulations have given employees, unions, health care providers, and governmental and nongovernmental agencies a more decisive role in the management of workplace health and safety.

The Hazard Communication Standard

OSHA promulgated the Hazard Communication Standard (HCS) to provide workers with the "right to know" the hazards of chemicals in their workplaces and to enable workers to take appropriate protective measures. Under the HCS, employers must (1) ensure the labeling of each container of hazardous chemicals in the workplace with appropriate identity and hazard warnings, (2) maintain and ensure employee access to Material Safety Data Sheets (MSDSs), and (3) provide employees with information and training on hazardous chemicals in their work areas. MSDSs list the chemical and physical properties of chemical substances, their health hazards, routes of exposure, emergency and first aid procedures, and protective measures for their handling and use. The HCS only applies to chemical agents. Furthermore, MSDSs are not subject to methodical review by regulatory agencies. Thus, their quality and adequacy of information vary widely, and they may be of limited use to the clinician in the treatment of exposed workers.

The HCS also contains important provisions for health care provider access to the identities of trade secret chemicals. When a medical emergency exists, the chemical manufacturer, distributor, or employer must immediately divulge the identity of trade-secret chemicals to the treating physician or nurse if this information is requested for purposes of emergency or first aid treatment. As soon as circumstances permit, the chemical manufacturer, importer, or employer may subsequently require the physician or nurse to sign a written statement of need and a confidentiality agreement.

In nonemergency situations, the chemical manufacturer, distributor, or employer must likewise disclose the identity of trade-secret chemicals when requested by a health professional (defined by the regulation as a physician, occupational health nurse, industrial hygienist, toxicologist, or epidemiologist). Before disclosure, however, the health professional must (1) submit a request in writing, (2) demonstrate an occupational health need for the information, (3) explain why disclosure of the chemical identity is essential (in lieu of other information, such as chemical properties, methods of exposure monitoring, and methods of diagnosing and treating harmful exposures to the chemical), (4) enter into a written confidentiality agreement, and (5) describe procedures to maintain confidentiality of the disclosed information. The HCS is a valuable instrument for providing health professionals with information about patient exposures.

Recording and Reporting Requirements

Soon after passage of the OSH Act, OSHA promulgated standards to fulfill the Act's mandate for the provision of recordkeeping and reporting by employers and for the development of information and a system of analysis of occupational accidents and illnesses. Under 29 C.F.R. 1904, employers in nonexempt industries with more than 10 workers must maintain a log and summaries and incident reports of all recordable occupational injuries and illnesses. Employers must record an injury or illness if it is work-related, is a new case, and results in (1) a fatality, (2) lost workdays, (3) restricted work or transfer to another job, (4) medical treatment beyond first aid, (5) loss of consciousness, or (6) significant injury or illness diagnosed by a physician or other licensed health care professional.

Employers must make entries for each recordable injury or illness on the OSHA 300 form (the OSHA 300 Log) and must also complete an OHSA 301 form (the 301 Incident Report form) or its equivalent for each recordable injury or illness on the OSHA 300 Log within seven calendar days after notification of a recordable injury or illness. At the end of the year, employers must also complete the OSHA 300-A form (Summary of Work-Related Injuries and Illnesses), summarizing all the recordable injuries and illnesses on the OSHA 300 Log. Businesses must retain the OSHA 300 Log, the 301 Incident Report forms, and the annual summary for five years following the end of the calendar year that these records cover.

In addition to these recordkeeping requirements, the standard further imposes a reporting requirement on working establishments. Employers must report all incidents to the nearest area OSHA office within eight hours when such accidents result in a fatality or the hospitalization of three or more employees. This requirement applies to all employers, regardless of the size of their work force.

This standard also grants OSHA, as well as other federal and state agencies, the authority to inspect and copy these logs of recordable injuries and illnesses (although some U.S. circuit courts have upheld the need to obtain a search warrant before such inspections). The regulations also ensure employee access to all such logs in their working establishment. These provisions have proved to be a useful source of epidemiologic data to employees, unions, researchers, the U.S. Bureau of Labor Statistics, and other agencies.

Access to Employee Exposure and Medical Records

Employers have no general duty to collect medical or exposure data on workers. Nonetheless, some specific OSHA standards (e.g., the lead standard) may require medical surveillance of workers exposed to specific agents. In other cases, employers may voluntarily institute biologic and environmental monitoring programs even when not mandated by OSHA. Regardless, to the extent that employers do compile medical and exposure records on workers exposed to toxic substances or harmful physical agents, they must ensure employee access to these data.

Employers shall make medical records available for examination and copying within 15 days of a request by an employee or his or her designated representative. The employer is obligated only to provide a worker with access to medical records relevant to that particular employee. Access to medical records of other employees requires the formal written consent of those other employees. Employers also must provide employees or their designated representatives access to employee exposure records. When an employer lacks exposure records on a particular worker, the employer must provide that worker with exposure data of other employees who have similar job duties and working conditions. Under these circumstances, access to co-workers' exposure records does not require written consent. OSHA has the authority to examine and copy any medical or exposure record without formal written consent.

When medical records do exist on an employee, the employer is required to preserve and maintain these records for the duration of employment plus 30 years. Although these record-retention requirements are the legal responsibility of the employer, it is recommended that independent contractors who provide occupational health or industrial hygiene services to companies clarify the ownership and disposition of medical or exposure records, specifying in advance the party to be charged with maintaining the medical or exposure records for the required duration.

The three standards discussed in this section—the HCS, the Recording and Reporting Occupational Injuries and Illnesses standard, and the Access to Employee Exposure and Medical Records standard—collectively keep employees informed of the nature and risks of hazardous materials in their workplaces and of resulting exposure or health effects. By communicating these risks to workers, OSHA seeks to (1) encourage employers to select safer materials and engineering controls, (2) enable workers to use better protective measures and handling procedures, (3) familiarize workers and health care providers with valuable emergency and first aid information, and (4) minimize health hazards through earlier detection, treatment, and prevention of occupational disorders.

DISCRIMINATION IN THE WORKPLACE

Americans "with disabilities are a discrete and insular minority who have been faced with restrictions and limitations . . . resulting from stereotypic assumptions not truly indicative of the individual ability . . . to participate in, and contribute to, society" [14]. The clinician functions as an important interface between persons with disabilities and the workplace in a variety of settings (e.g., primary care, pre-placement physicals, workers' compensation and disability evaluations). It is incumbent on health care providers to make a fair and accurate assessment of workers' functional capabilities in relation to job tasks so as to avoid reinforcing deep-rooted stereotypes about the disabled. Moreover, health care providers can play a key role in safeguarding the rights of workers and in educating employers through familiarization with recent legislative and judicial developments in the areas of employment and discrimination law.

THE AMERICANS WITH DISABILITIES ACT

Congress enacted the Americans with Disabilities Act (ADA) in 1990 to eliminate discrimination against the disabled. Although the ADA addresses five separate areas, this chapter focuses exclusively on Title I, employment discrimination [15].

Title I of the ADA prohibits discrimination against qualified individuals with disabilities in virtually all employment contexts (i.e., job application, hiring, discharge, promotion, compensation, job training, and other terms, conditions, and privileges of employment). The act defines *disability* as (1) a physical or mental impairment that substantially limits one or more of the major life activities, (2) a record of such an impairment, or (3) being regarded as having such an impairment. In addition, the act prohibits discrimination against those who are associated with an individual with a disability; e.g., an employee who has a family member with a disability. A *physical or mental impairment* is further defined as "any physiological disorder, or condition, cosmetic disfigurement, or anatomical loss affecting one or more . . . body systems" [16].

Under the ADA, qualification standards, tests, or selection criteria that employers administer to job applicants must be uniformly applied, job related, and consistent with business necessity. To be consistent with business necessity, a standard must concern an essential function of the job. An employer who denies an amputee a desk job on the basis of strength testing would be in violation of the ADA if, for example, lifting were not an essential job task. Moreover, even when the qualification standards are job related, the employer must first attempt to make reasonable accommodations before excluding disabled workers on the basis of strength testing or other pre-employment tests. Examples of reasonable accommodations may include modified work schedules, job restructuring, equipment modification, design of wheelchair-accessible workstations, and provision of readers or interpreters.

The ADA provides the following three major defenses to employers charged with discrimination:

1. Selection criteria are job related and consistent with business necessity, and a disabled individual cannot perform essential job tasks even with reasonable accommodation.
2. Reasonable accommodation would impose an undue hardship on the employer.
3. A disabled individual poses a direct threat to himself or herself or to the health and safety of others in the workplace.

The ADA has several important implications for health professionals—whether company employees or independent contractors—who are often called on to perform employment physicals of job applicants or return-to-work assessments of injured workers. First, clinicians and employers need to recognize that the ADA bars all pre-employment (pre-offer) physicals. Employers may continue to administer pre-offer nonmedical tests (e.g., language proficiency or strength and agility testing) that are job related, but they may require a medical examination only after extending an offer of employment. This offer may be conditional—dependent on the results of the post-offer, pre-placement physical examination—provided that all applicants for a particular position are subjected to such an examination regardless of disability.

Second, clinicians should recognize that one of their primary roles is to advise employers about workers' functional abilities and limitations in performing the essential functions of the job with or without reasonable accommodation and to determine whether these workers meet the employers' health and safety requirements. To make such determinations of functional abilities and limitations, health professionals should request that employers provide written job descriptions that accurately detail specific and essential job tasks. The employer has the responsibility to make all employment decisions and to determine the feasibility of reasonable accommodation. Nonetheless, health professionals may offer input on ways to achieve reasonable accommodation.

Third, the ADA imposes very strict limitations on the use of information from post-offer medical examinations and inquiries. Such information must be treated as confidential and must be maintained in separate medical files apart from personnel records. The employer must designate a specific person or persons to have access to the medical file. In some instances the release of confidential medical information may be allowable under the ADA, for example, to inform supervisors or first aid personnel about necessary restrictions or emergency treatment; however, a release form from the examinee is always advisable.*

Although the ADA discourages but does not explicitly prohibit clinicians from reporting actual diagnoses to employers, an action for breach of confidentiality may exist under certain state confidentiality laws and under the Health Insurance Portability and Accountability Act (HIPAA) if the clinician discloses medical information without a signed release from the examinee. In the absence of a release form, a clinician's best course of action is to inform the employer of functional abilities and necessary work restrictions (e.g., "no working at heights" and "no driving of company vehicles") rather than reporting actual diagnoses (e.g., "epilepsy").

Discrimination denies the disabled the many advantages of employment, including prestige, power, self-esteem, economic well-being, social outlets, and access to health insurance and other job benefits. The ADA fosters the protection of disabled workers' rights and their fuller integration into the workplace. Health professionals can play a critical role in furthering patient autonomy and educating employers while simultaneously promoting a safe and healthful workplace.

*Workers sometimes voluntarily disclose their diagnoses to supervisors or first responders on their shift to familiarize them with the signs and symptoms of their condition in the event of a medical emergency.

Gender Discrimination, Pregnancy, and Fetal Protection Policies

As with the disabled, pregnant workers have often represented a disenfranchised group within the workplace. Although the ADA is sweeping in its breadth, it does not shield these women from employment discrimination because "pregnancy" is not considered a physiologic disorder under the Act's definition of disability. Nonetheless, pregnant workers receive ample protection under both legislative and judicial avenues. The Civil Rights Act of 1964 (Title VII) prohibits discrimination on the basis of sex [17] and, as amended through the Pregnancy Discrimination Act of 1978, further prohibits discrimination against "women affected by pregnancy, childbirth, or related medical conditions . . . for all employment-related purposes" [18].

Despite the intent of these laws, a number of industries instituted fetal protection policies (FPPs) throughout the 1980s to exclude fertile or pregnant women from the workplace and to avert toxic exposures to the fetus. In some instances, companies went so far as to exclude all women, including postmenopausal women, from jobs or job tracks involving potential exposure to toxic substances unless these workers could provide documentation of surgical sterilization. These FPPs were unsound for several reasons.

1. They disregarded reproductive risks to male workers.
2. They assumed that all women in the workplace could or would become pregnant.
3. They essentially required female workers to proclaim their reproductive status to supervisors and coworkers (i.e., women remaining in the workplace were implicitly sterile).
4. They discouraged some women from applying for higher-paying jobs.
5. They encouraged other women to undergo unnecessary surgical sterilization solely to retain their jobs.
6. They overlooked the adverse health effects to the unemployed mother and child from foregone income and health benefits.

In 1991, the U.S. Supreme Court declared these FPPs to be unconstitutional in *International Union, UAW v. Johnson Controls, Inc.* [19]. The court held these policies to be discriminatory because they did not apply equally to the reproductive capacity of male employees. Furthermore, the court held that "decisions about the welfare of future children must be left to the parents . . . rather than to the employers who hire those parents" [20].

There are principally only two instances in which employers may discriminate on the basis of gender or pregnancy. Employers may deny employment when gender is a bona fide occupational qualification (BFOQ) reasonably necessary to the normal operation of that particular business or enterprise. Analogous to the ADA, the qualification standards must relate to the essence of the employer's business. For example, a movie producer would be justified in claiming that male gender is a BFOQ when hiring actors for male roles. Under the safety exception, employers may discriminate on the basis of gender in those instances in which gender or pregnancy interferes with the employee's ability to safely perform the job. For example, airlines are permitted to remove pregnant flight attendants from air travel at various stages of pregnancy to ensure the safety of passengers. In all other instances, women must be treated equally with men in the employment setting as long as they possess the necessary job-related skills and aptitudes.

In light of this legislative and judicial history, the following practices that screen out individual women or the entire class of women might be construed as discriminatory:

1. Implementing FPPs or other policies that exclude women from the workplace based on gender or reproductive status

2. Applying coercion by providing female but not male workers or job applicants with information on reproductive risks or by requiring only female workers to sign waivers absolving the employer of liability in the event of an adverse reproductive outcome

3. Administering special tests or medical examinations exclusively to women

4. Using physiologic parameters, such as muscle strength, as selection criteria when not a requirement for the job (For this reason, as with the ADA, job descriptions that accurately reflect job tasks are vital.)

5. Using gender as a proxy for physiologic parameters even when specific physiologic traits (e.g., anthropometric and muscle strength) are a requirement for the job and a high correlation exists between gender and ability to perform the job. (Employers must give each individual the opportunity to demonstrate that she meets the job parameters.)

Family and Medical Leave Act

The Family and Medical Leave Act (FMLA) [21] was enacted in 1993, and the final regulations, issued by the Department of Labor (DOL), went into effect in 1995. The FMLA enables employees to take up to 12 weeks of unpaid leave per 12-month period for their own medical care or for that of an immediate family member (child, spouse, or parent). To be covered by the FMLA, the employee must work for a public employer (regardless of the size of the work force), for a private employer with a worksite comprising 50 or more employees, or for an employer who has 50 or more employees within 75 miles of a worksite. In addition, to be eligible for FMLA leave, the employee must have worked for the employer for at least 12 months (although not necessarily consecutively) and must have worked at least 1,250 hours over the previous 12-month period. The employer may define in advance the method for calculating the 12-month period—for example, calendar year, rolling year, or a fixed 12-month period such as fiscal year or anniversary date of employment—as long as the same method is uniformly applied to all employees.

Employees may request FMLA for the following purposes: birth of a child or care for the newborn, placement with the employee of a child for adoption or foster care, a serious health condition that precludes the employee from performing his or her job functions, or care of an immediate family member with a serious health condition. The regulations further define *serious health condition* as inpatient care, incapacity of three or more days and treatment by a health care provider (HCP), pregnancy including prenatal care or incapacity due to pregnancy, chronic conditions requiring treatment, permanent or long-term conditions requiring ongoing supervision by a HCP, or any period of absence to undergo multiple treatments by a HCP for restorative surgery after an accident or injury or for a condition that, if left untreated, would result in an absence of more than three consecutive days.

Employers may require employees to submit a medical certification form, using either the DOL Certification of Health Care Provider form or their own forms. However, employers may not request additional information beyond that included in the DOL form and must safeguard all medical information as confidential. Furthermore, a health care provider representing the employer may, with permission of the employee, contact the employee's HCP in order to clarify or confirm the certification. Employers may seek a second opinion at their own expense, provided they do not use an HCP with whom they regularly contract for services. If a second opinion differs from the employee's own HCP, the employer may require a third opinion from an HCP mutually approved by the employer and employee. This third opinion is, again, at the employer's expense and is final and binding. Importantly, employers cannot require an employee to perform light or modified duty in lieu of FMLA leave. Furthermore, upon return from FMLA leave, the employee is entitled to reinstatement in the same or an equivalent job. Finally, employers must retain FMLA records for at least three years.

WORKERS' COMPENSATION

Clinicians who evaluate and treat workers with job-related injuries or illnesses must acquire a broad understanding of the workers' compensation (WC) system and an appreciation of their role in the legal disposition of WC claims. WC systems exist in all 50 states and in three federal jurisdictions. WC is a no-fault system that evolved in the earlier part of this century to promote expeditious resolution of work-related claims. Injured workers relinquish their rights to bring a tort action in a civil court in exchange for a rapid, fixed, and automatic payment. WC pays for medical and rehabilitation expenses and typically up to two-thirds of wage replacement. The *quid pro quo* for the employer is a limited and predictable award.

Unlike tort actions, WC claims do not require a showing of employer negligence. Therefore, even injuries or illnesses resulting from the employee's own negligence are compensable. Regardless of the cause of the injury, the worker carries the burden of proving, by a preponderance of the evidence, that a causal relationship exists between an occupational exposure and the resulting injury or illness—that is, the injury or illness must "arise out of or in the course of employment." Employees are also generally entitled to compensation for work-related aggravation of preexisting disorders. To qualify for WC payments, the worker must prove damages, typically by demonstrating a disability or loss in earning capacity. Even when the occupational exposure results in an injury or illness that produces no disability, the employee may still be eligible for an award if the WC laws in that jurisdiction explicitly provide for such coverage, such as payment for scarring, disfigurement, or damage or loss of function of specific organs or body systems.

Clinicians who evaluate and treat injured workers must strive to be objective when documenting physical findings and impairments. If health professionals are called upon to proffer an opinion about causality, they must be prepared to support their conclusions in a deposition or courtroom in the event the claim leads to litigation. It is also incumbent on clinicians to familiarize themselves with alternative work and transitional duty programs to minimize the length of disability.

Confidentiality and WC Claims

Patients often present treating clinicians with signed authorization forms requesting release of medical records to the patient's employer, the employer's attorney, or the employer's insurance company in support of a WC claim. By signing these release forms, patients do not, however, waive all rights of confidentiality. It is critical that health professionals disclose only information related to the disorder that forms the basis of the WC claim. Several physicians have been sued for releasing confidential information about HIV status that was unrelated to the WC claim, such as ear and sinus problems or head injury with back pain [22,23]. These suits were filed under various claims, such as negligence, breach of confidentiality, breach of contract, and invasion of privacy.

TORT LIABILITY

Many WC statutes contain "exclusive remedy provisions" that hold that WC shall provide the exclusive remedy for injuries and illnesses arising out of or in the course of employment. The goal of these provisions is to foreclose litigation and to limit the employer's liability to WC. During the last few decades, however, workers have periodically sought to circumvent the exclusive remedy provisions of WC laws and pursue tort suits. As in WC cases, plaintiffs in tort suits may recover medical expenses and lost earnings from employers, but because tort actions are more likely to take into account future promotions and job advancements, they may yield a higher award for wage replacement. Moreover, tort suits offer the additional advantage of recovering certain types of damages often unavailable under WC, such as pain and suffering, punitive damages, and loss of consortium.

Despite the greater financial incentive for an employee to bring a tort suit, the courts have been reluctant to carve out exceptions to the WC exclusive-remedy provisions. The most common doctrines under which employees file tort suits are as follows [24]:

1. Third-party and product liability suits
2. Intentional harm committed by an employer
3. Injury by a co-employee (Injuries by co-employee health care providers are discussed in the following section on medical malpractice.)
4. Dual capacity doctrine (The employer assumes a second role or capacity sufficiently distinct from its role as employer such that workers injured by the employer while acting in this second capacity may recover outside of the WC system.)

Theoretically, both WC and tort liability should provide employers with the incentive to promote a safe and healthful workplace. However, given the employer's ability to obtain WC and liability insurance and to pass on some of these costs to consumers, the extent to which these goals are accomplished remains unclear.

CRIMINAL LIABILITY

Under the OSH Act, OSHA may recommend criminal prosecution by the U.S. Department of Justice (DOJ)—typically only in cases involving a worker fatality or fraudulent recordkeeping. Notwithstanding, of the more than 150 cases referred by OSHA to the DOJ since 1971, only a handful have resulted in imprisonment. Gradually, there has been an expansion, although slow, of criminal liability for workplace health and safety violations at the state and local levels. Several factors may account for increased criminal prosecution in the occupational health and safety arena, including precedent and active prosecution under environmental regulations, a heightened duty of care owed by employers, increased public awareness, and inadequate staffing and enforcement by OSHA.

Often relegated to the occupational health and safety professional, workplace health and safety has been underscored once again as the responsibility of employers, managers, and supervisors. The rationale behind criminal liability for company officials is that they have decision-making authority and are in the best position to ensure compliance with OSHA regulations and to safeguard workers from occupational hazards. As a result of criminal liability initiatives, prosecutors have attempted to hold these company officials individually liable under a number of legal doctrines (depending on the state),* such as aggravated battery, second-degree assault, reckless conduct, reckless endangerment, conspiracy, criminal negligence, manslaughter, involuntary manslaughter, murder, falsification of business records, and falsification of exposure and medical surveillance data. Despite increased attempts to pursue criminal prosecution, jail sentences are relatively short.

The Clean Air Act and other federal legislation likewise delineate criminal penalties—including fines, imprisonment, or both—as

*Some business and government leaders favor federal preemption of state criminal prosecution, citing the possibility that employers would be subject to disparate criminal laws depending on the state(s) in which they are located or do business. Others support state criminal prosecution, noting that laws for other types of criminal activity vary among the states.

enforcement mechanisms. For example, under the Clean Air Act, a person may be subject to criminal penalties for knowingly making false material statements; altering, concealing or failing to maintain required documents; tampering with or failing to install required monitoring devices; and releasing into the ambient air certain hazardous or extremely hazardous substances that place another person in imminent danger of death or serious bodily injury. Occupational and environmental health and safety professionals and corporate management should be cognizant of their potential for criminal liability exposure under applicable laws and regulations.

Box 23-1. Workplace Criminal Prosecution

In one of the most prominent workplace criminal prosecution cases, the two owners of a thermometer company were found guilty of felony charges of second-degree assault and falsification of business records after concealing the operation of a basement mercury-reclamation facility. It is alleged that operation of this facility led to permanent neurologic impairment and brain damage of a worker from exposure to mercury fumes. After a jury conviction, the owners received a sentence of weekends in the city jail for six months and a fine of $10,000. They were also sentenced to a concurrent four-month weekend sentence for misdemeanor charges of reckless endangerment and conspiracy. Many viewed the punishment as lenient, reinforcing their notion that courts still do not take workplace safety violations seriously enough.

LEGAL LIABILITY OF THE OCCUPATIONAL HEALTH CARE PROVIDER: MEDICAL MALPRACTICE

What is the legal liability of a health care provider who commits medical malpractice while under contract to provide occupational health services to a company's employees? Management hires physicians, nurses, and other health care professionals—company salaried and independently contracted—for the benefit of the company (i.e., to ensure workers' fitness for duty).

Historically, health care providers salaried by a company were considered co-employees of other workers whom they evaluated and/or treated. Consequently, their negligent acts and the resulting injuries were regarded as arising out of, and in the course of, employment. Injured workers were therefore limited by the exclusive remedy provisions of WC laws to recovery under the WC system. However, if the workers could establish that the health care providers were not under the control of the company but rather were functioning as independent contractors, they could avail themselves of a remedy in torts.

The distinctions between co-employees and independent contractors have been problematic for the courts. Health care providers often have greater latitude than other employees in exercising their judgment. At what point do their actions transcend the control of their employers and exceed the scope of employment? Some companies authorize and encourage occupational health care providers to furnish primary care services to employees and even to their families. At what point does the occupational health professional establish a health care provider–patient relationship, no longer acting solely for the benefit of the employer but also for the benefit of the employee?

The courts have developed two tests to determine whether an occupational health care provider is the co-employee of an injured worker (in which case the negligent act was historically covered by the employer's WC policy) or an independent contractor (and therefore subject to a malpractice claim). Under the *control test*, health care providers are more likely to be presumed company employees when management exerts greater control over their function, operation, and judgment. For example, clinicians act under the control of the company when they must follow predetermined guidelines and protocols in conducting physical examinations (e.g., which forms to use or which lab tests to obtain).

Under the *indicia test*, the court analyzes various indices of control that normally signify employee status. For example, health care providers are more likely to be categorized as company employees rather than independent contractors if they receive a salary, health insurance, and other company benefits; fall under the company's WC and pension programs; work out of company offices; have regularly scheduled work hours; and report up through the company's chain of command [25].

Despite these two tests, the impunity of occupational health care providers continues to erode. Although the control and indicia tests provide useful guidelines concerning the potential liability of the occupational health care provider, their application results in some uncertainty because many professionals only partially meet these criteria. For example, some clinicians may have a part-time private practice and request workers to follow up after duty hours in their private offices. Other clinicians may work out of their own offices but see exclusively company employees. Furthermore, courts appreciate that workers often rely on the results of employment physicals to their detriment and that health professionals have a high degree of skill and training and are in a better position to warn patients of harm and to insure against losses. Thus, although courts were traditionally reluctant (even in the independent contractor setting) to hold that a doctor-patient relationship existed between a prospective or actual employee and the clinician conducting the examination at the employer's request and for the employer's benefit, more courts are willing to recognize that the "examination creates a relationship between the examining physician and the examinee, at least to the extent of the tests conducted" [26].

Therefore, the wisest approach for occupational health care providers, whether company employees or independent contractors, is to conduct examinations and treatment with due care, to arrange for qualified clinicians to interpret diagnostic tests (such as radiographs), and to disclose the results from any tests or examinations to the patient. Health care providers should act under the assumption that a health care provider–patient relationship exists to the extent of the examination and should ensure follow-up appropriate to the circumstances. For example, a clinician who notes an incidental and non-work-related finding on a preplacement or periodic examination, such as poorly controlled hypertension or a suspicious lesion on the skin, should advise the patient to follow up with his or her primary care doctor or specialist and should also document this conversation in the medical records or even send patients a letter advising them of the need for follow-up. (See Box 23-2 for more detailed risk management principles.) Health care providers would be prudent to delineate these responsibilities in their contracts with employers. Furthermore, occupational health care providers employed by companies should not rely solely on employers' WC policies but should also obtain malpractice insurance.

LIABILITY FOR CONSULTING AND ADMINISTRATIVE SERVICES

In recent years, employees have also sought to hold consultants, inspectors, and physician-administrators liable for an array of alleged injuries. For example, some job applicants found medically unfit on preplacement medical examinations have attempted to sue not only the prospective employer, but also the occupational medicine physician performing the evaluation under a number of state or federal claims including medical malpractice and discrimination.* In other cases, injured employees have sued contractors or private consultants hired by the employer for their role in allegedly creating a hazardous or unsafe workplace or for negligence in conducting workplace inspections and audits. For example, some employees with symptoms attributed to poor indoor air quality have sued builders, HVAC contractors, subcontractors, and engineering and maintenance companies claiming negligence in construction or maintenance of the workplace facility. These claims are often difficult to pursue because the injured worker not only has difficulty in establishing the contractor's duty but also because so many other causative or intervening factors can contribute to poor air

*In at least one case, a claim that the examining physician's actions constituted discrimination based on the prospective employee's disability was dismissed in the pre-trial stages in part because the court recognized that hiring decisions rest solely with the employer.

quality. Claims against OSHA and state agencies generally do not prevail due to the agency's recognized "discretionary" function. For example, in one case* an employee of a shoe company unsuccessfully sued OSHA after a compliance officer failed to identify an unguarded rotating drive shaft that allegedly resulted in a scalping injury and severe neurologic damage.

Box 23-2. Risk Management Principles for Occupational Health Care Providers

1. Request job descriptions that reflect essential job functions.
2. Coordinate with treating clinicians:
   - Communicate with clinicians treating workers *and/or*
   - Obtain old medical records as necessary.
3. Adhere to company policies.
4. Evaluate necessity and purpose of components of physical examinations and medical surveillance programs.
5. Do not order unnecessary tests.
6. Ensure all tests are interpreted by qualified individuals.
7. Inform patients of results of all tests.
8. Ensure adequate/reasonable follow-up (appropriate to circumstances).
9. Keep good records—document, document, document.

There are a number of steps that occupational health and safety professionals can take to limit their liability when functioning as consultants. First, if providing clinical services, they should carry medical malpractice insurance. Second, when the malpractice insurance excludes coverage for consulting duties, it may be advis-

***Irving v. U.S.*, 17 OSHC 1705 (1996), *reversed,* 162 F.3d 154 (1st Cir. 1998).

able to secure error and omissions insurance. Third, consultants might be able to contractually seek indemnification by clients for potential injuries to workers alleged to arise out of the negligent performance of consulting duties. Finally, consultants should take care not to contractually expand their duties by making certain representations or warranties. For example, some consultants guarantee or represent that they will conduct wall-to-wall inspections and will uncover all potential deficiencies in mechanical systems, work practices, safety measures, and regulatory compliance. Whether through written disclaimers or informal discussion, consultants should educate employers and employees to ensure that they understand the limitations of a workplace inspection, audit, or other consulting services.

SUMMARY

Occupational health care providers are in a unique position to ensure worker health and safety. Because they interact with injured workers at every phase of employment—preplacement, periodic, and post-injury—they have a profound impact on the disposition of job applicants and employees in the workplace. Occupational health professionals place a strong emphasis on prevention. Familiarization with judicial, legislative, and regulatory mandates will enable them to preserve the health and uphold the rights of workers, to educate employers, and to minimize their own liability.

REFERENCES

1. From the ACGME Special Requirements for Residency Education in Occupational Medicine. Available at: acgme.org/acWebsite/downloads/RRC_progreg/380pro7012007.pdf; accessed September 2007.

2. From the ACOEM Policies and Position Statements. Available at: www.acoem.org/guidelines.aspx?id=992; accessed September 2007.

3. J. A. Gold, The Physician and the Corporation. 3 *Bioethics Bull* 1 (fall 1989).

4. 29 U.S.C. § 651(b).

5. 29 U.S.C. § 654(a)(1).

6. 29 U.S.C. § 654(a)(2).

7. *Marshall v. Barlow's, Inc.*, 436 U.S. 307 (1978).

8. Available at: www.osha.gov/Publications/all_about_OSHA.pdf

9. Available at: www.ncsl.org/programs/pubs/601dises.htm

10. Available at: www.osha.gov/pls/oshaweb/owadisp.show_document?p_table=TESTIMONIES&p_id=349

11. 29 C.F.R. 1910.1200.

12. 29 C.F.R. 1904.

13. 29 C.F.R. 1910.1020.

14. 42 U.S.C. § 12101(a)(7) (Suppl IV 1992).

15. Americans with Disabilities Act of 1990: Title I, Employment; Title II, Public Service/Public Transportation; Title III, Public Accommodation & Services Operated by Private Entities; Title IV, Telecommunications; Title V, Miscellaneous Provisions.

16. Note that under the ADA current drug abusers receive no protection because illegal drug use is not considered a disability under the act. However, alcoholics and fully rehabilitated drug abusers may be protected under the ADA (unless they fail to meet productivity and other performance standards that cannot be corrected by reasonable accommodation).

17. Pub. L. 88-352, 78 Stat. 241, July 2, 1964; codified as 42 U.S.C. § 2000e-2(a).

18. P.L. 95-555, 92 Stat. 2076 (amended section 701 of the Civil Rights Act of 1964); codified as 42 U.S.C. § 2000e(k).

19. *International Union, UAW v. Johnson Controls, Inc.,* 111 S. Ct. 1196 (1991).

20. *Id.* at 1207.

21. P.L. 103-3, February 5, 1993.

22. *Doe v. Roe,* 190 A.D. 2d 463 (1993).

23. *Urbaniak v. Newton,* 277 Cal. Rptr. 354 (Cal. App. 1 Dist. 1991).

24. Modified from S. L. Birnbaum & B. Wrubel, *Workers' Compensation and the Employer's Immunity Shield: Recent Exceptions to Exclusivity,* 50 J of Products Liability 119 (1982).

25. See, for example, *Garcia v. Iserson,* 33 N.Y.2d 421, 309 N.E. 2d 420 (1970), holding that worker's exclusive remedy fell under WC law. The plaintiff could not maintain a malpractice action against a company physician because the injuries arose out of and in the course of employment, since the company employed the physician at a weekly salary and took the usual payroll deductions, required the physician to work on company premises during certain scheduled hours, and included the physician in the company's medical plan and WC policy. See also *Golini v. Nachtifall,* 38 N.Y.2d 745, 343 N.E.2d 762 (1975), holding that WC provided the exclusive remedy to the plaintiff injured by a physician who received company salary and benefits and worked in company facilities.

26. *Green v. Walker,* 910 F.2d 291 (5th Cir. 1990), holding that physician who was under contract to perform annual employment physicals was liable for malpractice in failing to diagnose and report cancer to examinee.

CHAPTER

24

Preparing for and Providing Health and Safety Legal Testimony

Donald C. Sinclair II

Occupational and environmental medicine physicians are frequently embroiled in personal injury litigation related to occupational exposures to physical or chemical environmental factors. Typically, the plaintiff alleges that the defendant negligently exposed the plaintiff to a physical or chemical agent that has proximately caused an adverse health consequence for which the plaintiff seeks monetary damages. Because the relationship between exposure to an occupational and environmental factor and an adverse health outcome is not only disputed but also beyond the purview of common knowledge, determination of causation is uniquely within the province of the occupational and environmental medicine physician—expert opinion testimony is required to resolve the medicolegal dispute.

Challenges to the admissibility of "scientific" testimony are ubiquitous in litigation involving exposure to occupational and environmental factors. Ordinarily, such contests are predicated on the deposition testimony of the respective parties' expert witnesses. While not calculated to provide an exhaustive treatise on the admissibility of scientific evidence in legal proceedings, this chapter is designed to provide occupational and environmental medicine physicians and other health and safety professionals preparing testimony with a fundamental understanding of the principles governing the admissibility of scientific testimony in civil proceedings.

The chapter is predicated on the Federal Rules of Evidence and the legal decisions construing the Federal Rules, since these rules have been adopted in a majority of states and are therefore generalizable. Although the described rules of law are not specifically applicable to workers' compensation claims in a majority of states, the methodologic prerequisites for credible (valid and reliable) testimony in administrative proceedings are relevant.

Even if *qualified* by virtue of education, training, experience, and specialized knowledge in the relevant field, the expert's opinion must be based on facts or data that others in the relevant field of endeavor would ordinarily rely on in forming opinions and inferences on the subject matter. Scientific evidence is subject to an exclusionary rule unless it is demonstrably *reliable* and *relevant*. An expert's opinion must not be cumulative, confusing, or wasteful of the court's time and monetary resources ("judicial economy"). An expert's opinions are admissible if the testimony is based on reliable facts or data and is the product of reliable principles and methods, provided the witness has applied the principles and methods reliably to the facts of the case. Each of these requirements must necessarily be established by the witness' testimony.

Whether expert testimony is admissible mandates a critical examination of the nonexhaustive factors articulated in the landmark decision *Daubert v. Merrell Dow Pharmaceuticals, Inc.* [1] and its progeny, including whether the technique can be tested, whether the technique has been subject to peer review and publication, the existence of standards controlling the technique and its operation, the known or potential rate of error, the existence of standards controlling the technique's operation, whether the technique is generally accepted in the scientific community, and the nonjudicial uses to which the method has been subjected. A rigorous, logical method is recommended for the determination of work relatedness of occupational injury and illness claims in compensations environments.

In March 1993, the United States Supreme Court rendered its decision in *Daubert* [2], creating a new standard for the admissibility of scientific evidence and overruling the common-law standard established in *Frye v. U.S.* [3]. The *Daubert* decision arose largely in response to the rapid expansion of novel scientific theories that fostered burgeoning controversy in the legal community over the admissibility of "junk science" in the courtroom.

"'Junk Science' [is] the mirror-image of real science with much of the same form but none of the same substance. . . . On the legal side, junk science is marked by what might be called liability science, a speculative theory that expects lawyers, judges, and juries to search for causes at the far fringes of science and beyond" [4].

SCIENTIFIC EVIDENCE PRE-*DAUBERT*: THE *FRYE* STANDARD

The admissibility of scientific evidence in judicial proceedings had classically been a "source of confusion for the judiciary and a source of contention for adversaries before the courts" [5]. In *Frye* [6], the Court of Appeals for the District of Columbia attempted to resolve these issues by articulating a guideline for the admission of expert testimony [7]. The court excluded the testimony after an inquiry into whether the proffered expert testimony was predicated on "generally accepted" scientific principles [8]. Justice Van Orsdel authored the opinion for the court [9]:

> Just when a scientific principle or discovery crosses the line between the experimental and demonstrable stages is difficult to define. Somewhere in this twilight zone the evidential forces of the principle must be recognized, and while courts will go a long way in admitting expert testimony deduced from a well-recognized scientific principle or discovery, the thing from which the deduction is made must be sufficiently established to have gained general acceptance in the particular field in which it belongs.

For an opinion that remained the bastion of evidentiary doctrine for nearly 70 years, the decision is remarkably concise and does not reference any precedential legal authority. This standard, dubbed the *general acceptance test,* was nearly universally adopted to determine the admissibility of expert testimony. Unfortunately, the decision essentially relied upon an opinion-poll methodology to determine the reliability of scientific testimony. (Remember that popular scientific opinion once held that the Earth was flat.)

A 2002 article in the *Pennsylvania Bar Association Quarterly* [10] stated:

> Although *Frye* became the prevailing test for the admissibility of scientific evidence, commentators [observed] that it was adopted largely without any analysis or discussion of its rationale. The test has received considerable criticism, and has been described as "archaic" and "antiquated on the day of its pronouncement." Justice Blackmun observed in *Daubert* that, "[l]ike the question of *Frye's* merit, the dispute over its survival has divided courts and commentators."

Despite the foibles of the decision, particularly its inability to adapt to rapid expansions in science and technology [11], the *Frye* test reigned as the classic criterion regarding the admissibility of scientific evidence until the adoption of the Federal Rules of Evidence.

THE FEDERAL RULES OF EVIDENCE—THE CONUNDRUM RESOLVED?

With the promulgation of the Federal Rules of Evidence in 1975, courts began to re-examine the "general acceptance test." Whether the "*Frye* test" had survived engendered fractious debate. Although several lower tribunals attempted to address this issue, it was not until the Supreme Court's decision in *Daubert* that the matter was ultimately resolved in favor of the Federal Rules of Evidence.

The admissibility of scientific evidence is a threshold determination for the court: "Preliminary questions concerning the qualification of a person to be a witness . . . or the admissibility of evidence shall be determined by the court. . . ." [12]. The party proffering an expert witness' scientific testimony bears the burden of establishing its admissibility by a preponderance of proof [13]. Ordinarily, the court will convene a hearing outside the presence of the jury and the proponent of the testimony will attempt to establish the admissibility of the scientific evidence. The opponent of the evidence is then afforded an opportunity to cross-examine the proponent's witnesses and may present the testimony of counter-experts to establish that the evidence is unreliable. A hearing is not always required, however. Depositions of "experts" may be sufficient for a determination; hence, an expert witness' deposition may afford the sole opportunity to establish the admissibility of the testimony [14].

As a prerequisite to testifying on scientific subjects, a person must be qualified as an expert witness. Rule 702 of the Federal Rules of Evidence governs the admissibility of expert testimony [15]:

> If scientific, technical or other specialized knowledge beyond that possessed by a layperson will assist the trier-of-fact to understand the evidence or to determine a fact in issue, a witness qualified as an expert by knowledge, skill, training or education may testify thereto in the form of an opinion or otherwise.

Rule 702 "contemplates a broad conception of expert qualifications" [16]. Although the threshold for qualification as an expert may be nominal, an impressive curriculum vitae alone does not render an expert's testimony admissible.

The jeopardy inherent in expert testimony is the mirage created when an apparently qualified expert witness proffers pseudo-scientific testimony cloaked in the guise of scientific veracity. Jurors are inclined to assign special significance to "scientific testimony," and therein lies the peril: "Under the regime of *Daubert,* . . . a . . . judge asked to admit scientific evidence must determine whether the evidence is genuinely scientific, as distinct from being unscientific speculations offered by a genuine scientist" [17]. Hence, the inquiry is not whether the witness is a true scientist, but rather, whether the testimony is truly scientific. Cognizant that expert testimony is often the proverbial "wolf in sheep's clothing," Rule 702 permits testimony from a "qualified" expert only if (1) the testimony is sufficiently based upon reliable facts or data, (2) the testimony is the product of reliable principles and methods, and (3) the witness has applied the principles and methods reliably to the facts of the case [18].

Rule 703 of the Federal Rules of Evidence establishes parameters for the bases of admissibility of expert opinion testimony [19]:

> The facts or data . . . upon which an expert bases an opinion or inference may be those perceived by or made known to the expert at or before the hearing. If of a type reasonably relied upon by experts in the particular field in forming opinions or inference upon the subject, the facts or data need not be admissible in evidence in order for the opinion or inference to be admitted. Facts or data that are otherwise inadmissible shall not be disclosed to the jury by the proponent of the opinion or inference unless the court determines that their probative value in assisting the jury to evaluate the expert's opinion substantially outweighs their prejudicial impact.

Upon promulgation of the Federal Rules of Evidence, experts could permissibly rely not only on facts and data admitted in evidence but also on facts or data that would not be independently admissible, provided that the facts or data were information reasonably relied upon by experts in the relevant field of endeavor to reach opinions or inferences on the subject matter. A disconcerting, but pervasive, practice arose among litigants—the "bootstrapping" of otherwise inadmissible facts, data, or information merely because an expert ostensibly relied upon it in forming an opinion. In response to this abusive practice, Rule 703 was revised to clarify that otherwise inadmissible facts, data, or other information may be admitted only if the evidence is sufficiently reliable.

Even if expert opinion testimony passes muster under Rules 702 and 703 of the Federal Rules of Evidence, expert opinion testimony remains subject to the requirements of the rules defining relevance. Rule 402 of the Federal Rules of Evidence simplistically states that "relevant" evidence is admissible, and "irrelevant" evidence is not. "Relevant" evidence is evidence that tends to prove that a fact in issue is more probable than not—a seemingly liberal standard for admissibility. Yet, even relevant evidence may be inadmissible! [20] Rule 403 of the Federal Rules of Evidence qualifies the rule on relevancy [21]:

> Although relevant, evidence may be excluded if its probative value is substantially outweighed by the danger of unfair prejudice, confusion of the issues, or misleading the jury, or by considerations of undue delay, waste of time, or needless presentation of cumulative evidence.

Rule 403 implicates Rule 601 of the Federal Rules of Evidence, which provides presiding judges with broad discretion in the presentation of evidence and order of proof [22].

DAUBERT V. MERRELL PHARMACEUTICALS, INC.: RESOLUTION AT LAST?

Daubert [23] was an action prosecuted by two minor children born with limb-reduction birth defects and their parents against Merrell Dow. The plaintiffs alleged that the birth defects resulted from the mother's prenatal ingestion of Bendectin. The plaintiffs relied upon the testimony of eight "experts" who claimed that Bendectin could cause human birth defects. Their conclusions were founded on animal studies, chemical structure analysis, and the reanalysis of previously conducted epidemiologic studies [24]. The District Court for the Southern District of California held that Daubert's evidence failed to meet the general acceptance standard for the admission of expert testimony [25]. The United State Court of Appeals for the Ninth Circuit affirmed the District Court's opinion, citing the now arcane *Frye* test. The appellate panel found that since the technique was not "generally accepted as a reliable technique among the scientific community," the plaintiff's expert testimony was inadmissible [26].

The Supreme Court granted *certiorari* [27] to resolve the issue of the proper standard for the admission of expert scientific testimony. The Supreme Court reversed the Ninth Circuits' decision, declaring that Federal Rule of Evidence 702 governs the admissibility of scientific testimony [28]. Since the "general acceptance test" is incompatible with Rule 702, the court concluded that the *Frye* test was no longer a proper evidentiary standard [29]. The court adopted revised guidelines for determining the admissibility of scientific evidence.

The Court held that expert testimony must be relevant and rest on a reliable scientific foundation [30]. In order for the trial judge to determine whether the threshold requisites for the admissibility of scientific evidence have been met, the court must focus on (1) whether testimony is based on "scientific knowledge" and (2) whether the testimony "assists the trier of fact to understand or determine a fact in issue" [31]. To qualify as scientific knowledge, the expert opinions expressed must "be determined by the scientific method" [32]. The court enunciated four nonexclusive factors for the trial judge to consider in determining whether evidence is scientific knowledge: (1) whether the technique can be tested, (2) whether the technique has been subject to peer review and publication, (3) the existence of standards controlling the technique and its operation, and (4) and the acceptance of the technique within the scientific community [33].

Even if expert testimony is reliable and relevant, the testimony may still be inadmissible on other grounds. The court must determine "whether the admission would create unfair prejudice to one of the litigants, confuse the jury in regard to the issues, or completely mislead the jury in its assessment of the facts of the case" [34], under the Federal Rules of Evidence [35].

DAUBERT II: "A BRAVE NEW WORLD"

Nearly two years after the Supreme Court rendered its hallmark decision in *Daubert,* the Ninth Circuit was "charged with the task of ensuring that an expert's testimony both rests on a reliable foundation and is relevant to the task at hand" on remand [36]. Once again addressing the admissibility of plaintiff's proffer of scientific evidence, the Ninth Circuit observed, "[F]ederal judges ruling on the admissibility of expert scientific testimony face a far more complex and daunting task in a post-*Daubert* world than before" [37]. Undertaking this task, the court discussed means to determine whether expert scientific testimony is admissible under Rule 702 [38]. While acknowledging the "illustrative" list of factors to be considered in such an analysis under *Daubert,* the Ninth Circuit concluded that the list was not and that other factors should be considered as well [39]:

> One very significant factor to be considered is whether the experts are proposing to testify about matters growing naturally and directly out of research they have conducted independent of the litigation, or whether they have developed their opinions expressly for purposes of testifying . . .

"That the testimony proffered by an expert is based directly on legitimate, preexisting research unrelated to the litigation provides the most persuasive basis for concluding that the opinions he expresses were 'derived by the scientific method'" [40]. When the "proffered expert testimony is not based on independent research, the party proffering it must come forward with other objective, verifiable evidence that the testimony is based on 'scientifically valid principals'" through proof that the research and analysis supporting the proffered conclusions "have been subjected to normal scientific scrutiny through peer review and publication" [41]. Peer review and publication of the underlying scientific theory lends validity to the evidence because it indicates that the theory is "taken seriously by other scientists and . . . that it meets at least the minimal criteria of good science" [42]. Although peer review and publication of the theory supporting scientific testimony does not guarantee the correctness of an expert's conclusions, "[t]he test under *Daubert* is not the correctness of the expert's conclusion, but the soundness of his methodology." Peer review and publication increases the likelihood of detecting flaws in methodology [43].

Daubert provides an evidentiary paradigm meaningful in law and medicine—disciplines that have become inextricably intertwined in the modern courtroom. At once, *Daubert* demands of both physicians and lawyers a systematic approach to legal and medical evidentiary conundrums—a "scientific method" for the resolution of medicolegal disputes.

EVIDENCE-BASED MEDICINE IN THE MEDICOLEGAL ARENA

Against this palette of legal principles, how does the occupational and environmental physician or other health and safety professional prepare for deposition or trial testimony? How does one ensure that his or her testimony is admissible in evidence? With what scientific literature should one come armed? What methodologies for the determination of causation pass muster under *Daubert* and its progeny? What is a "reasonable" degree of medical certainty? The answers to these and other relevant questions lie in the practice of evidence-based medicine.

The term *evidence* is one that has special meaning to occupational and environmental medicine physicians and attorneys at law. Evidence is "information or data that tends to furnish proof." Evidence is of two types: "direct" and "circumstantial." *Direct evidence* is evidence that tends to prove without requirement of inference. *Circumstantial evidence* is evidence from which proof may be rationally inferred [44]. *Relevant evidence* is evidence having a tendency to make any fact of consequence to the determination of truth or falsity more probable or less probable than it would be without the evidence [45]. Evidence stands in contrast with belief: Belief is a conviction that certain things are true; examples include religion, faith, trust, doctrine, opinion, expectation, and dogma [46]. In demonstrating causation, the occupational and environmental medicine physician must rely upon evidence rather than conviction, no matter how deeply held—*Daubert* and its progeny demand this.

Evidence of Causation

Despite the convenient notion of evidence-based medicine, reliable evidence of causation is frequently absent in medicolegal proceedings. The classic definition of *evidence-based medicine* is Sackett's: "The conscientious, explicit, and judicious use of current best evidence in making decisions about the care of individual patients" [47]. Nearly a decade has passed since Sackett originally proposed this definition of evidence-based medicine. As both the medical and legal sciences have evolved exponentially in the last decade, consider the following proposed definition: The conscientious and methodical application of state-of-the-art knowledge of the medical, epidemiologic, technical, psychological, and clinical sciences to all decisions relating to the diagnosis, etiology, and treatment of injury and illness among individual patients—a multidisciplinary approach that may require judicious consultation with colleagues in other disciplines. A methodologic, multidisciplinary approach to determinations of causality in the medicolegal environment is not novel.

In 1979, the National Institute for Occupational Safety and Health (NIOSH) proposed an elegantly logical methodology for the determination of work relatedness: *The Guide to the Work-Relatedness of Disease* [48]. The Guide "presents one method for assembling and evaluating *evidence* that may be relevant in determining the work-relatedness of a disease in an *individual* [49]. The Guide "discusses various factors associated with establishing the relationship between disease and occupation. Prepared as an aid to ". . . physicians . . . concerned with workers' compensation for occupational disease, the publication describes a method for collecting, organizing, and appraising medical, occupational, and other *evidence* with the aim of determining the probable work-relatedness of a given disease" [50].

> In order for the compensation system to treat both injury and disease in a uniform manner, disease must be related to the workplace as effectively as injury. For the decision-maker, however, establishing the causality of a disease is often a difficult task, especially if it becomes necessary to decide if an employee's disease resulted from, or was aggravated by, employment-related factors. In contrast to a traumatic injury . . . a cause-effect relationship between disease and an agent or conditions in the workplace may not be clear. Occupational disease may be slow to develop. Symptoms of disease may be confused with changes that are due to the aging process, or with the effects of [other relevant factors]. [I]nformation on past work exposures is often unavailable, inadequate, or incomplete. Not all individuals react in the same way to similar exposures to disease-producing agents. Off-the-job exposures may contribute or be a primary cause of illnesses and accidents. These are but some of the factors which must be considered in the decision making process. [W]hen appropriate evidence is presented in a logical and orderly sequence . . . then the decision-making process is facilitated and an equitable decision is likely to result [51].

The Guide remains remarkably in accord with the principles articulated in *Amorgianos,* (see page 391), particularly as it pertains to evaluation of exposure dose (i.e., the frequency, intensity, duration, and temporal pattern of exposure). Generally, a disease is occupational if all of the following are present [52]:

1. The medical findings of disease are compatible with the effects of a disease-producing agent or agents to which the worker has been exposed.
2. There exists in the worker's occupational environment exposure to an agent or agents sufficient to have caused the disease.
3. The weight of evidence supports that the disease is of occupational rather than nonoccupational origin.

The Guide describes a step-by-step method for the determination of work relatedness [53]:

1. Consideration of evidence of disease.
2. Consideration of epidemiologic data.
3. Consideration of evidence of exposure.
4. Consideration of validity of testimony.
5. Consideration of other relevant factors.
6. Evaluation and conclusion.

As the Guide aptly observes, "[w]ith occupational diseases, what appears to be 'obvious' is often subject to controversy [particularly in a compensation or litigation environment], and it is important to assemble complete information wherever possible in order to assure an equitable decision" [54]. Nearly 30 years after its publication, however, is the Guide relevant or relic? The contemporaneousness of the Guide was recently affirmed by the American College of Occupational and Environmental Medicine [55].

The Guide is recommended as an algorithm for preparation of the occupational and environmental physician's expert report. The Federal Rules of Civil Procedure govern the orderly conduct of pre-trial litigation. The court rules require the disclosure of expert witnesses expected to testify at trial. The required disclosure must include a written report prepared and signed by the expert(s). The report must contain a complete statement of all opinions to be expressed; the bases and reasons for the opinions; the data or information considered by the expert in forming opinions; any exhibits to be used as summary of or support for the opinions; the qualifications of the expert, including list of the witness' publications in the preceding 10 years; the compensation to be paid for the study and testimony; and a list of other cases in which the expert has testified by deposition or at trial within the preceding four years [56].

The rules of civil procedure also permit the court to order a party who has put his mental or physical condition in controversy by alleging a personal injury or illness, to submit to a physical examination by the adverse party's expert or experts (an *independent medical examination* (*IME*)). The notice of examination is required to specify the time, place, manner, conditions, and scope of examination and the identity of the physicians or other examiners conducting the examination. The examinee may request a detailed written report containing the examiner's finding, the results of testing, and the examiner's diagnoses and conclusions. Failure or refusal to provide the required report upon request may result in exclusion of the examiner's testimony at trial. If the examinee requests a copy of the report or obtains the deposition of the examiner, he waives any privilege that would otherwise preclude disclosure of the information in the lawsuit or other legal proceeding relating to the alleged personal injury or illness [57].

Since the primary role of the occupational and environmental medicine physician is to determine causation, addressing each element of the Guide in the witness' report, together with appropriate appendices (including the witness' *curriculum vitae,* bibliography, fee schedule, illustrative exhibits, and a roster of legal proceedings in which the expert has testified) satisfies the requirements of the court rules.

The Guide also has utility in preparing for deposition and trial testimony. *Daubert* mandates a scientific methodology for the determination of causation. The Guide provides just such a method, together with a rational and orderly exposition of the evidence.

Depositions commence with a written notice of deposition. Reasonable advance notice of the deposition must be provided to every party to the lawsuit. The notice must state the time and place for the conduct of the deposition and the name and address of the deponent. The notice is typically accompanied by a subpoena or request for production of documents enumerating by paragraph various categories of documents that the deponent must produce at or before the deposition [58]. The deponent should review the notice with counsel for the party by whom he or she has been retained well in advance of the deposition, since considerable preparation is often required. The general categories of documents requested usually coincide with the NIOSH Guide factors; hence the Guide is apropos for the organization of both one's thoughts and files.

Evidence of Disease

Consideration of evidence of disease requires the physician-expert witness to provide a diagnosis for the contested injury or illness or, alternatively, to refute the evidence of disease. Based upon review of the medical records, a physical examination, and objective testing, the expert witness should describe the adverse health outcome in accordance with generally accepted diagnostic criteria—a "case definition." An ideal case definition may include the following features:

1. Symptoms consistent with the adverse health outcome of interest.

2. Clinical signs consistent with the adverse health outcome of interest.

3. Manifestation of signs and symptoms in an anatomic pattern (and conversely, an absence of nonphysiologic signs and symptoms).

4. Confirmatory objective testing, if applicable.

Consideration of evidence of disease is confined to the *diagnosis* of a disputed disease or disorder [59]. Diagnosis is concerned with the description of an adverse health outcome and its diagnostic features; causality is not relevant. Conversely, *etiology* is concerned with the determination of the cause of a disease or disorder [60]. So-called differential diagnosis is often confused with an etiologic determination; such confounding has been rejected by courts of law. Consideration of the validity of testimony is discussed in more detail in a subsequent section of this chapter.

Epidemiology in the Courtroom: *Amorgianos v. Nat'l RR. Pass. Corp.*

In *Amorgianos v. Nat'l RR. Passenger Corp.* [61], the court had occasion to consider whether reliable epidemiologic evidence is a condition precedent to recovery of damages in occupational injury and illness litigation. A painter prosecuted an action against his employer to recover damages for central and peripheral nervous system symptoms allegedly suffered due to exposure to deleterious concentrations of fumes accumulating in a confined area while spraying paint. The employer sought to exclude the testimony of the plaintiff's expert witness, an industrial hygienist, arguing that the expert's opinions were not founded upon reliable epidemiologic evidence.

The District Court addressed the admissibility of the plaintiff's expert testimony in accordance with the standards established in *Daubert* and its progeny. The court reiterated that the *Daubert* factors are illustrative, not exhaustive [62–64]:

> [A]dditional factors may be appropriate in a given case, and a district court enjoys the same "broad latitude" in deciding what are the "reasonable measures of reliability in a particular" case as it does in reaching its ultimate determination of reliability. [The] standard of reliability is, however, "exacting," and does not, as some pre-*Daubert* courts held, turn simply on the quality of the proffered expert's credentials.

Observing that "[t]he fundamental principles of epidemiology, which are now becoming well known to the courts, provide additional guidance," the court articulated relevant epidemiologic precepts that courts should consider in assessing whether the plaintiff's experts had reliably extrapolated their causal hypothesis from the evidence [65]:

> There are several different forms that epidemiological studies take in the areas of occupational medicine. . . . These various research designs differ in the evidentiary weight they lend to a hypothesis that exposure to a given substance [or physical factor] causes a given condition. [Citation omitted] An uncontrolled case study, or cases-series report, is not actually a formal epidemiologic investigation but simply the identification of an unusual occurrence or disease. [Citation omitted] In an occupational cross-sectional study, a survey is conducted to determine and compare the prevalence of disease or health status between groups of workers classified with respect to exposure status. [Citation omitted]
>
> Because of their inherent limitations, these two study designs "usually represent preliminary or pilot investigations used to screen for possible workplace hazards or to generate hypotheses for testing in more complex designs." [Citation omitted] In contrast, cohort studies, which test for the incidence of a health condition in a randomly selected group of exposed workers and a randomly selected group of unexposed workers overtime [citation omitted], and case-control studies, which compare the exposure histories of a group of workers who exhibit a particular health condition with a group of workers who do not exhibit the health condition but who are comparable in characteristics other than exposure [citation omitted] are "the most informative investigations used to test specific etiological hypotheses and to confirm and quantify degrees of health risk related to causal exposures. . . ." [Citation omitted]

Modern jurists clearly have sufficient understanding of epidemiology to recognize not only the necessity of epidemiologic evidence, but also to critically evaluate the quality of the epidemiologic evidence.

The sophistication of the court's analysis was not confined solely to an evaluation of the evidentiary weight of the epidemiologic studies relied upon by the expert witness. The court further observed, "Even when an appropriately designed study yields evidence of a statistical association between a given substance [or physical factor] and a given health outcome, epidemiologists generally do not accept such an association by itself as proof of a causal relationship between the exposure and the outcome. [Citations omitted] Epidemiologists generally look to several additional criteria to determine whether a statistical association is indeed causal" [66]. The court recognized the time-honored guidelines for inferring causation, commonly referred to as the *Bradford Hill criteria,* after the author of the most recognized statement of the principles [67]:

1. *Strength of the association:* What is the magnitude of the association between the suspected risk factor and the observed outcome?

2. *Consistency of the association:* Have different investigators consistently observed the association among different groups under different conditions of exposure?

3. *Specificity of the association:* Is the specific exposure factor uniquely associated with the specific health outcome of interest?

4. *Temporality of the association:* Does exposure to the putative risk factor always precede, never follow, the onset of the adverse health outcome?

5. *Biologic plausibility of the association:* Whether the observed association is logical, given contemporary knowledge of the physiologic response of the anatomic region to external stimuli?

6. *Coherence of the association:* Is the association consistent (i.e., does not conflict) with contemporary knowledge concerning the natural history of the health outcome?

7. *Experimental evidence for the association:* Has observation of the association led to some preventive action that has, in fact, prevented or diminished the prevalence of the disease?

8. *Dose-response relationship:* Does the occurrence or severity of the adverse health outcome increase or decrease in relative proportion to the magnitude of the exposure?

The testifying expert should, therefore, identify a finite body of quality epidemiologic literature in support of his opinions. The epidemiologic evidence should either establish a reasonable inference of causality or, alternatively, refute a causal association, predicated upon insufficient credible evidence. It is improvident to produce reams of epidemiologic literature relating to the subject matter of the lawsuit in an attempt to overwhelm the adverse attorney. Rather, select the seminal relevant literature bolstering your opinions, know it well, and rely upon it to defend your opinions.

Evidence of Individual Exposure: "The Dose Is Still the Poison"

In *Amorgianos* the court recognized that epidemiologic evidence establishing that the risk factor of interest is *generally* capable of causing the plaintiff's adverse health outcome still fails to prove that an individual's adverse health consequence was *specifically* caused by the exposures of interest. The outcome remains critically dependent on the dose to which a human being is exposed and the duration of the exposure: "[E]xpert testimony . . . meets *Daubert's* 'fit' requirement only if the testimony includes an opinion that (1) exposure to the particular substance at issue, (2) in the dose to which the plaintiff was exposed, (3) for the duration in which plaintiff was exposed, (4) can cause the particular condition(s) of which the plaintiff complains" [68].

The mere presence of a putative risk factor is insufficient evidence that the risk factor was the cause-in-fact of the disputed adverse health consequence. Whether a risk factor has contributed to causing a disease or disorder depends on the frequency, intensity, duration, and temporal pattern of the exposure. Exposures should be independently measured and quantified according to valid and reliable scientific techniques. Consensus standards and guidelines and scientifically well-founded threshold limit values (permissible exposure limits) may provide evidentiary substantiation for expert opinion. In the absence of such metrics, an individual's quantified exposure should be compared to those reported in the epidemiologic literature to determine if they are comparable. Consideration of exposure frequently requires the occupational and environmental medicine physician to rely upon exposure data acquired by colleagues in other disciplines.

Consideration of Other Relevant Factors: The Workplace and Beyond

As the NIOSH Guide recognizes, avocational exposures, personal health attributes, and lifestyle may contribute, modulate, or constitute a primary cause of injury or illness. The natural progression (natural history) of pre-existing disease or prior traumatic injury may act singly or in combination with other contributory factors to cause adverse health outcomes. Hence, the weight of evidence must support that the disease is of occupational rather than nonoccupational origin [69].

This element of the Guide requires the expert medical witness to critically evaluate the epidemiologic literature examining exposures foreign to the workplace to determine whether there is credible evidence of a causal association unrelated to the occupational environment. Examples of such factors include body *habitus* (e.g., obesity), personal health attributes (e.g., diabetes mellitus), lifestyle choices (e.g., nicotine abuse, lack of exercise), pre-existing injury or illness, and the magnitude of nonoccupational exposures to environmental or physical factors. Is it more probable than not that exposures unrelated to the workplace caused or substantially contributed to the alleged "occupational" injury or illness? If so, then work relatedness is not proved to a reasonable degree of medical certainty.

Consideration of Validity of Testimony

In most cases, professionals in several disciplines necessarily provide testimony relevant to the determination of causation. These professionals must consider all pertinent data in order to present an accurate and meaningful evaluation of the available data. The professional qualifications and competency of those testifying should be evaluated. Critical evaluation of the scientific bases of the testimony, the weight accorded to various data, and *the rationale for conclusions derived* is required [70].

In *General Electric Co. v. Joiner,* the U.S. Supreme Court expanded the scope of the court's inquiry into an expert's *conclusions* where an opinion bears no *nexus* (connection) to the expert's methodology [71]. The District Court had found the testimony of the plaintiff's experts was inadmissible, since it failed to rise above "subjective belief or unsupported speculation" [72]. The Court of Appeals for the Eleventh Circuit reversed, holding that a district court should limit its role to determining the "legal reliability of proffered expert testimony, leaving the jury to decide the correctness of competing expert opinions" [73]. General Electric appealed the Eleventh Circuit's reversal to the Supreme Court.

Joiner argued that the District Court had erred by substituting its conclusions from the studies for those of the experts themselves, citing *Daubert's* language that the "[f]ocus, of course, must be solely on principles and methodology, and *not on the conclusions* that they generate" [74]. The Supreme Court rejected this argument, stating [75]:

> [C]onclusions and methodology are not entirely distinct from one another. Trained experts commonly extrapolate from existing data. But nothing in either *Daubert* or the Federal Rules of Evidence requires a district court to admit opinion evidence . . . connected to existing data only by the *ipse dixit* of the expert. A court may conclude that there is simply too great an analytical gap between the data and the opinion proffered.

Absent a *nexus* between the methodology and the witness' conclusions, the expert's opinions are subject to the exclusionary rule.

In *Black v. Food Lion, Inc.* [76], a case analyzing the validity of a physician's purported differential diagnosis, the plaintiff slipped and fell in defendant's grocery store. The plaintiff's physician diagnosed the plaintiff with "fibromyalgia syndrome." The defendant challenged the admissibility of the physician's causation opinion, which linked the plaintiff's fall with her medical condition. The Fifth Circuit Court found the physician's testimony "[did] not bear the necessary indicia of intellectual rigor, and that the trial court abused its discretion in admitting the physician's testimony" [77].

Applying the *Daubert* factors, the Fifth Circuit Court found that the physician's theory of causation had not been verified by testing or peer-reviewed, and that the physician had acknowledged that fibromyalgia is idiopathic. "On its own terms, [the physician's] opinion includes conjecture, not deduction from scientifically-validated information" [78]. The Court also found that the physician's theory had failed to gain general acceptance within the medical profession. "Mere conjecture does not satisfy the standard for general acceptance. . . . Finally, [the physician's] theory of causation, which has not been verified or generally accepted, also has no known potential rate of error" [79].

The Court also found fault with the physician's methodology, which amounted to obtaining a history from plaintiff, ruling out prior or subsequent "causes" of fibromyalgia, performing or reviewing physical tests, and deducing that the accident was the only plausible remaining cause. "This analysis amounts to saying that because [the physician] thought she had eliminated other possible causes of fibromyalgia, even though she does not know the real 'cause,' it had to be the fall at Food Lion. This is not an exercise in scientific logic but in the fallacy of *post-hoc propter-hoc* reasoning, which is as unacceptable in science as in law" [80]. Diagnostic fallacy is subject to the exclusionary rule. There may simply be "too great an analytical gap" between an expert's conclusions and the purported methodology; unsubstantiated etiologic inferences fall in this crevasse.

Evaluation and Conclusion: "To a Reasonable Degree of Medical Certainty"

An expert witness' conclusion must be stated "to a reasonable degree of medical certainty" in a majority of jurisdictions. The civil burden of proof requires that the probability that the purported cause resulted in the adverse health outcome exceed 50 percent. The standard of "reasonable medical certainty" reflects an objectively well-founded conviction that the likelihood of one cause is greater than any other; it does not mean the expert is personally absolutely certain of the cause or that cause is discernible to a certainty [81].

SUMMARY

The Federal Rules of Evidence, construed by *Daubert* and its progeny, prescribe an evidentiary paradigm for the resolution of medicolegal disputes. Expert witness testimony must assist the jury in understanding the evidence and determining disputed medicolegal facts. An intelligent evaluation of complex or novel scientific testimony is dependent upon the testimony of witnesses with specialized knowledge. The expert's testimony must be sufficiently based on *reliable* facts and data; the testimony must be the product of *reliable* principles or methods; and the witness must have *reliably* applied the principles and methods to the facts of the case.

Daubert articulated a two-prong test to determine whether expert scientific testimony is admissible: The testimony must relate to "scientific knowledge" *and* "assist the trier-of-fact to understand or determine a fact in issue. Four nonexclusive factors were enumerated to gauge reliability: (1) whether the theory or technique can be or has been tested, (2) whether the theory or technique has been the subject of peer review and publication, (3) whether the actual or potential rate of error is known, and (4) whether the theory or technique is "generally accepted" in the relevant scientific community.

These four factors are "illustrative" not "exhaustive," and courts should also consider whether the proposed testimony grows "naturally and directly" from research conducted independently of litigation or, alternatively, whether there is objective evidence that the testimony is based on scientifically valid principles.

Although the crux of the *Daubert* inquiry is directed toward the scientific methods employed and not the conclusions generated, too great an "analytical gap" may exist between the purported methodology and an expert's opinion. Unreliable conclusions, even if ostensibly premised on scientific methods, may be excluded.

The NIOSH Guide [48] is a logical methodology for collating and evaluating the medical, epidemiologic, and technical information relevant to the determination of work relatedness of an occupational or environmental injury or illness through probabilistic means. The Guide describes an evidence-based method for the determination of work relatedness:

1. Consideration of evidence of disease
2. Consideration of epidemiologic data
3. Consideration of evidence of exposure
4. Consideration of validity of testimony
5. Consideration of other relevant factors
6. Evaluation and conclusion

The Guide is recommended as an algorithm for the determination of work relatedness of disputed adverse health consequences. The Guide is also useful in preparing for deposition and trial testimony. Each element of the Guide may incorporate additional valid and reliable methods for evaluation of the scientific evidence. Conscientious application of the Guide promotes conformance to legal principles governing the admissibility of scientific evidence, provided there is a nexus between an expert witness' methods and the conclusions reached.

REFERENCES

1. *Daubert v. Merrell Dow Pharmaceuticals, Inc.,* 509 U.S. 579, 113 S. Ct. 2786, 125 L. Ed. 2d 469 (1993).

2. *Daubert,* supra.

3. *Frye v. U.S.,* 54 App. D.C. 46, 293 F. 1013 (D.C. Cir. 1923).

4. Huber PW. *Galileo's Revenge: Junk Science in the Courtroom 2.* New York: Basic Books, 1991.

5. Emmerick FR, Note. *The Supreme Court Strengthens the Discretionary Powers of the District Court in Admitting Expert Testimony:* Daubert v. Merrell Dow Pharmaceuticals, Inc., 3 Widener J. of Pub. L., 105 (1994).

6. *Frye,* supra.

7. *Id.* at 1013.

8. *Id.* at 1014.

9. *Id.*

10. Cooper T. Admissibility of expert testimony—*Daubert* revisited: expert testimony—its admissibility in Pennsylvania state and federal courts. *Pennsylvania Bar Assn Quarterly* 2002:57–75 [citations omitted].

11. Geise SN, Note. *Daubert v. Merrell Dow Pharmaceuticals: The Admissibility of Expert Scientific Testimony Under the Federal Rules of Evidence Does Not Require "Frye Finding"*, Oh.N.U.L.Rev. 1061 (1994).

12. R. 104 F.R.E.

13. *Daubert v. Merrell Dow Pharmaceuticals, Inc.,* 43 F.3d 1311, 1316 (C.A. 9 1995); see also *General Electric Co. v. Joiner,* 522 U.S. 136, 118 S.Ct. 512, 139 L.Ed. 2d (1997); *In re Paoli Railroad Yard PCB Litig.,* 35 F.3d 717 (C.A. 3 (Pa.) 1994); *In re TMI Litig.,* 193 F.3d 613 (C.A. 3 (Pa.) 1999).

14. R. 26 F.R.C.P.

15. R. 702 F.R.E.

16. *Thomas v. Newton Int'l Enterp.,* 42 F.3d 1266, 1269 (C.A. 9 (Cal.) 1994); see also *Wood v. Stihl, Inc.,* 705 F.2d 1101, 1104–1105 (C.A. 9 (Wash.) 1983).

17. *Rosen v. Ciba-Geigy Corp.,* 78 F.3d 316, 318 (C.A. 7 (Ill.) 1996).

18. R. 702 F.R.E.

19. R. 703 F.R.E.

20. R. 403 F.R.E.

21. *Id.*

22. See R. 601 F.R.E.

23. *Daubert,* 951 F.2d 1128 (C.A. 9 (Cal.) 1991) vacated 113 S.Ct. 2786 (1993).

24. *Daubert,* 113 S. Ct. at 2791.

25. *Id.* at 2792.

26. *Daubert,* 951 F.2d at 1131.

27. *Certiorari (cert.)* Most commonly used to refer to the Supreme Court of the United States, which uses the writ of *certiorari* as a discretionary device to choose the cases it wishes to hear. See 28 U.S.C.A. 1254. The Supreme Court denies most writs of *certiorari.* Hence, granting of *cert.* is reserved for issues of seminal importance. *Black's Law Dictionary,* 6th Ed. St. Paul, MN: West Publishing, 1991.

28. *Daubert,* 113 S. Ct. at 2794.

29. *Id.*

30. *Id.* at 2795.

31. *Id.* at 2795 (citing R. 702 F.R.E.).

32. *Id.*

33. *Id.* at 2797.

34. *Id.*

35. *See* R. 403 F.R.E.

36. *Daubert v. Merrell Dow Pharmaceuticals, Inc.,* 43 F.3d 1311 (C.A. 9 1995) [On remand, *Remand,* literally: "to send back," the act of an appellate court when it sends a case back to an inferior court and orders the inferior court to conduct limited new hearings or an entirely new trial or to take some further action in accord with the ruling on appeal. See 28 U.S.C.A. 2106.]

37. *Id.* at 1315.

38. *Id.* at 1316.

39. *Id.* at 1316–1317.

40. *Id.*

41. *Id.* at 1317–1318 (citing Huber PW. *Galileo's Revenge* at 209).

42. *Id.* at 1318 (citing *Daubert,* 223 S. Ct. at 2797).

43. *Id.* at 1318.

44. *See Webster's New International Dictionary.* Cambridge, MA: G. & C. Merriam Co., Riverside Press, 1924; ***see also*** *Black's Law Dictionary,* 5th Ed. St. Paul MN: West Publishing, 1979.

45. R. 401 Fed. R. Evid.

46. See *Webster's New International Dictionary,* Cambridge, MA: G. & C. Merriam Co., Riverside Press, 1924.

47. Sackett D, Straus SE, Richardson WS, et al. *Evidence-Based Medicine: How to Practice and Teach EBM.* New York: Churchill Livingstone, 2000.

48. Kusnetz S, Hutchison M (eds). *A Guide to the Work-Relatedness of Disease,* Rev'd Ed. NIOSH Pub. No. 79-116 (1979).

49. *Id.* at iii.

50. *Id.* at iv.

51. *Id.* at 2.

52. *Id.* at 3.

53. *Id.* at 4.

54. *Id.* at 4.

55. American College of Occupational and Environmental Medicine, Glass L (ed). *Occupational Medicine Practice Guidelines: Evaluation and Management of Common Health Problems and Functional Recovery of Workers,* 2nd Ed. Beverly Farms, MA: OEM Press, 2004.

56. R. 26 F.R.C.P.

57. R. 35 F.R.C.P.

58. R. 30 F.R.C.P.

59. *See* Venes D (ed), *Taber's Cyclopedic Medical Dictionary,* 20th Ed, Philadelphia: F.A. Davis, 2005.

60. *See* Venes D (ed), *Taber's Cyclopedic Medical Dictionary,* 20th Ed, Philadelphia: F.A. Davis, 2005.

61. *Amorgianos v. Nat'l RR. Passenger Corp.,* 137 F. Supp. 2d 147 (2001).

62. *Id.* at 163 (citing *Kuhmo Tire Co., Ltd. v. Carmichael,* 526 U.S. 137. 142(1999)).

63. *Id.* (citing *Weisgram v. Marley Co.,* 528 U.S. 440, 455 (2000)).

64. *Id.* at 163.

65. *Id.* at 167–168.

66. *Id.* at 168.

67. See *Id.* at 168; see also Hill AB. *The environment and disease: association or causation. Proc Royal Soc Med* 1965;58:295–300; Susser MW. Judgment and causal inference. *Am J Epidemiol* 1977;105:1–15; Rothman KJ (ed), *Causal Inference,* Chestnut Hill, MA: Epidemiological Resources, 1988.

68. *Amorgianos,* 137 F. Supp. 2d at 163; *See* also *Moore v. Ashland Chemical,* 151 F.3d 269, 278-79 (5th Cir. 1990) *en banc, cert.* den'd, 119 S. Ct. 1454 (1999).

69. *See* fn. 53–59, *supra.*

70. *See* A Guide to the Work-Relatedness of Disease, Rev'd Ed., Kusnetz, S. and Hutchison, M., Eds., NIOSH Pub. No. 79-116 at 21 (1979).

71. *General Electric Co. v. Joiner,* 522 U.S. 136, 118 S. Ct. 512, 139 L. Ed. 2d 508 (1997).

72. *General Electric Co. v. Joiner,* 864 F. Supp. 1310, 1329 (N.D. Ga. 1994).

73. *General Electric Co. v. Joiner,* 78 F.3d 524, 533 (C.A. 11 (Ga.) 1996).

74. *Daubert,* 522 U.S. at 146.

75. *Id.* (citations omitted).

76. *Black v. Food Lion, Inc.,* 171 F.3d 308 (C.A. 5 (Tex.) 1999).

77. *Id.* at 312.

78. *Id.* at 313.

79. *Id.*

80. *Id.*

81. *Robinson v. Group Health Assoc. Inc.,* 691 A.2d 1147 (D.C. 1997).

CHAPTER

25

Improving Health and Safety Program Quality

with Jeffrey S. Lee

During the past few decades business writers have increasingly emphasized the importance of quality in determining the success or failure of a company [1,2]. One of the eight attributes of successful companies discussed in *In Search of Excellence* is that of staying close to the customer by providing exceptional service and quality products [1]. Indeed, at the time of the book's publication, the authors noted that the leading companies had almost an "obsession" with quality. Whether dealing with a product—or, in the case of most health and safety activities, a service—these organizations placed zealous emphasis on quality. In some instances, this emphasis made little financial sense in the short run, as when products with "minor" defects were discarded rather than being placed on sale. In the long run, however, the quality standards resulted in increased sales that made the organizations recognized leaders. Examples included Procter & Gamble, Caterpillar, Deere, Hewlett-Packard, and similar "name" businesses. In other companies, when quality was allowed to decrease, the leadership position also suffered. Today, many large and small organizations have embraced quality-improvement programs in an effort to improve "customer" satisfaction and profitability. More recently, the concept of *Six Sigma,* with the goal of reducing errors to 3.4 defects per million opportunities or "products," has reinforced the emphasis on quality. In essence, the theory is that products should fall within six standard deviations (six sigma), thus reducing outliers (errors) to 3.4 per million [3].

EARLY DEVELOPMENTS

Part of the increased interest in quality reflects what has been called the *rediscovery* of W. Edwards Deming's precepts and the resultant emphasis on the concepts of quality control and Total Quality Management. As emphasized in *The Deming Management Method,* statistics can be used to understand processes, bring them under control, and improve them [4]. Processes are charted, with documentation of variations as upper and lower control limits (see Figure 25-1). Data from these efforts help define "special" causes of variation, such as a faulty machine or defective material from a supplier, and the information can be used to improve quality. Figure 25-1 shows that the clinic profiled had established control limits for acceptable variation in clinic waiting times. When the waiting time suddenly exceeded the limits, the situation was readily identified and corrective action implemented. Waiting times promptly returned to desired limits. Without tracking the waiting times and comparing them to established limits, the increase in time might not have been detected so promptly. Significant patient dissatisfaction could have resulted before clinic managers became aware of the problem. Use of control charts is the basis for Statistical Quality Control (SQC), or Statistical Process Control (SPC).

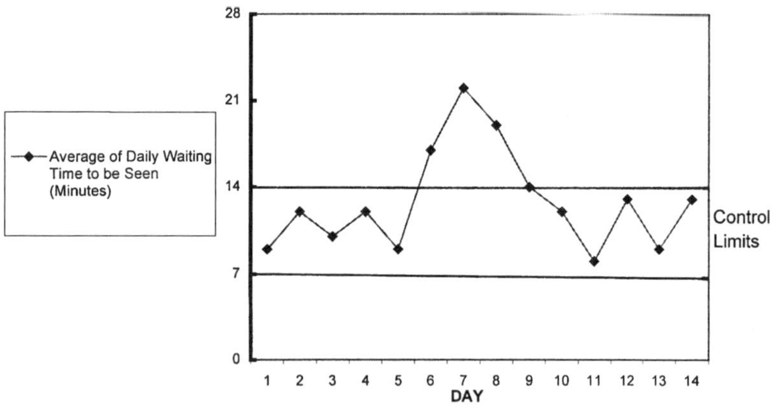

Figure 25-1. On day 6, a new medical assistant began work without adequate training in clinic procedures. Prompt recognition of the need for additional training resulted in return to desired control limits.

Shortly after World War II, Dr. Deming began working with Japanese manufacturers to improve product quality through SQC or SPC. He has been credited with revitalizing Japanese industry after the war and making it a strong competitor with other countries. Although lauded in Japan, he received little recognition in the United States until an interview on the NBC special "If Japan Can . . . Why Can't We?" in June 1980. During the interview, as described in The Deming Management Method, Dr. Deming observed that the statistical control of quality used by the Japanese would work in the United States, but "There's no determination to do it. We have no idea . . . what's the right thing to do, have no goal."

As a result of the interview, interest in "the Deming management method" rapidly expanded, and companies began implementing Total Quality Management (TQM), Continuous Quality Improvement (CQI), and similar programs.

In addition to charting variations from control limits, Dr. Deming also emphasized the importance of other charts. Flow charts depict each step in a process, for example, from the time a patient arrived at a reception desk until treatment was accomplished and the individual left the medical facility. Properly done, flow charts will document each step required, who is responsible for that step, and where the source of a problem in the final "outcome" may be. Another important chart is a Pareto chart. Examples of Pareto charting are provided in Chapter 15 (Figures 15-1 and 15-2). These charts can be particularly useful in developing priorities. As depicted in these figures, Pareto charting can define areas for focused interventions to prevent the costliest injuries. One could use a similar charting process to delineate the most frequent types of accidents, the source of the largest number of injuries producing lost workdays, or other areas of concern. The Pareto approach is especially helpful when insufficient resources limit preventive interventions, or other health and safety activities, that can be accomplished. As discussed in Chapter 15, the underlying concept is to focus limited resources in the areas that will produce the greatest returns. This approach is in contrast to attempting to "do everything," which usually results in dilution of efforts and little return on investment.

In addition to charting, Dr. Deming defined "Fourteen Points," "Seven Deadly Sins," and "Obstacles." As delineated in *The Deming Management Method* [4], the Fourteen Points are:

1. Create constancy of purpose for improvement of product and service.

2. Adopt the new philosophy in which mistakes and negativism are unacceptable.

3. Cease dependence on mass inspections as products come off line, but rather, improve the process.

4. End the practice of awarding business on the "price tag" alone.

5. Improve constantly and forever the system of production and service.

6. Institute training to help assure jobs are done correctly.

7. Institute leadership.

8. Drive out fear so that employees will ask questions or take positions to improve quality and productivity.

9. Break down barriers between staff areas and avoid competition and conflicting goals between departments.

10. Eliminate slogans, exhortations, and targets for the work force.

11. Eliminate numerical quotas, since quotas only take account of numbers, not quality or methods.

12. Remove barriers to pride of workmanship.

13. Institute a vigorous program of education and retraining in the new methods, including teamwork and statistical techniques.

14. Take action to accomplish the transformation.

The "Seven Deadly Diseases" and the "Obstacles," also described in *The Deming Management Method,* are practices and problems that interfere with improving quality of products or service. For the health and safety director, it is noteworthy that Dr. Deming recognized excess medical costs as one of the "diseases" and observed that they can be the largest single expenditure for some companies.

He also emphasized the significance of the Shewhart cycle: Plan, Do, Check, Act. After planning a quality improvement, test it in a small sample to evaluate the results, check the results, and then act to implement the improvement.

The above summary cannot, of course, describe all components of Dr. Deming's methods, and the reader is referred to his book [4] for necessary comprehensive discussion of the Deming approaches and techniques. The fact that these methods have been adopted by a number of major corporations and government agencies emphasizes their usefulness. The emphasis on improved quality of service is of particular significance to those directing all, or portions, of health and safety activities.

MORE RECENT DEVELOPMENTS

The Baldrige Criteria

In 1988, Public Law 100-107 established the Malcom Baldrige National Quality Award. Its purpose is to recognize companies that excel in quality management and achievement. Since the Baldrige Award was established, several other countries have implemented quality award programs based on the Baldrige criteria, and the European Quality Award (presented initially in 1992) is modeled on the Baldrige specifications. Originators observed that quality can be defined as meeting or exceeding customer expectations, and the best system for doing such is defined by the Baldrige criteria [2]. The Baldrige approach may be summarized by its Criteria for Performance Excellence goals:

1. Delivery of ever-improving value to customers and stakeholders, contributing to organizational sustainability

2. Improvement of overall organizational effectiveness and capabilities
3. Organizational and personal learning

These criteria build on a series of interrelated core values and concepts, as follows [5]:

1. Visionary leadership
2. Customer-driven excellence
3. Organizational and personal learning
4. Valuing employees and partners
5. Agility
6. Focus on the future
7. Managing for innovation
8. Management by fact
9. Social responsibility
10. Focus on results and creating value
11. Systems perspective

The thrust of the Baldrige effort is to identify the attributes, beliefs, and behaviors that characterize high-performing organizations.

Application of Baldrige Criteria in Health Care Organizations. The initial focus of quality-improvement programs was on larger manufacturing and service industries. Over time, it has become apparent that the quality-management concept was also applicable to health care activities. For example, the Baldrige National Quality Program has developed Criteria for Performance Excellence for Education and for Health Care programs. The Criteria for Performance Excellence goals are the same as for the basic award. The core values and concepts have changed slightly, as follows (changes noted in italics) [5]:

1. Visionary leadership
2. *Patient-focused excellence*
3. Organizational and personal learning

4. *Valuing staff and partners*
5. Agility
6. Focus on the future
7. Managing for innovation
8. Management by fact
9. *Social responsibility and community health*
10. Focus on results and creating value
11. Systems perspective

Other Organizations Concerned with Quality of Health Care

The Joint Commission on Accreditation of Health Care Organizations (JCAHO) evaluates organizational quality-management programs as part of its accreditation process. The Accreditation Association for Ambulatory Health Care (AAAHC) also emphasizes quality. AAAHC audits focus on problem identification, resolution, and evaluation by posing five questions [6]:

1. Was an important problem or concern identified?
2. Were the frequency, severity, and source of the problem measured/evaluated?
3. Were appropriate corrective actions implemented?
4. Was the problem re-evaluated?
5. If the problem persisted, were new measures developed and implemented?
6. Were the results of the investigation reported to appropriate management levels?

In 2000, the Committee on Quality of Health Care in America of the Institute of Medicine published the report *To Err is Human—Building a Safer Health System* [7]. The report found that other studies indicated between 44,000 and 98,000 patients die in hospitals each year due to medical mistakes [7]. This information gave added emphasis to improving quality in health care activities.

In the 1990s, the National Committee for Quality Assurance (NCQA) established the Health Plan Employer Data and Information Set (HEDIS). HEDIS includes performance measures to calculate a "rating" ("report card") for some important indicators of quality [8]. In addition to evaluating care provided, such as using recommended agents and amounts for control of hypertension, the report card also evaluates preventive health practices, such as smoking cessation. In spite of budgetary and other constraints, the U.S. Department of Veterans Affairs has implemented the VA Quality Enhancement Research Initiative to enhance the quality and outcomes of VA heath care [9,10]. The VA has also stressed reduction of errors by analyzing the true cause of errors, as is done in aircraft accidents, rather than blaming a single individual. Reporting of errors and "near misses" is encouraged so that problems can be identified and resolved early [11]. This approach, recommended in *To Err is Human,* complements safety professionals' efforts to ascertain the "root causes" of errors and unsafe working conditions. In essence, the safety team members systematically analyze the events, conditions, and values that could have caused the error or contributed to the hazard and then develop methods to prevent the causative factors [12].

Health and safety organizations, including the American Industrial Hygiene Association, the American Association of Occupational Health Nurses, and the American College of Occupational and Environmental Medicine have described quality management approaches to health and safety activities [13–19]. Presentations at national and international conferences have also discussed applications and benefits of quality management programs in health and safety programs. The recent increased emphases on practice guidelines, outcome measurements, and "best practices" are other manifestations of attempts by health care groups to enhance overall quality of their programs [20].

National and International Quality Standards

ISO 9000. The International Organization for Standardization (ISO), Geneva, Switzerland, develops standards that contain technical specifications and criteria to be used as rules and guidelines. ISO includes members who are the national standards bodies in over 110 countries. The United States' member is the American National Standards Institute (ANSI). In 1979, ISO formed Technical Committee 176 to develop worldwide standards for quality management and quality assurance. The committee's efforts culminated in the ISO 9000 standards, which emphasize consistently meeting specifications for quality products and services. Meeting the standards is accomplished by a number of approaches, including implementing and maintaining a systematic quality management system, and the concepts were carried over in developing subsequent ISO standards [21]. The process used in developing ISO standards is applicable to health and safety groups. For example, the American Industrial Hygiene Association (AIHA) and ANSI developed the *Occupational Safety and Health Management Systems* to assist in improving the quality of health and safety management [13].

ISO 14000. ISO 14000 established standards through the International Environmental Management System, which, as its name implies, focuses on environmental concerns. It follows the PLAN-DO-CHECK-ACT (Shewhart cycle) model used in developing the ISO 9000 standard [21].

ANSI Z10-2005. Recently, the AIHA and ANSI developed ANSI Z10-2005. Members of industry, labor, government, professional organizations, and others participated in development of the document *Occupational Health and Safety Management Systems* [13]. The standard emphasizes the need to incorporate health and safety management system principles in other quality management processes. The Shewhart cycle approach was incorporated into ANSI Z10-2005, and the standard is compatible with ISO 9000 and ISO 14000.

CRITIQUES

The many success stories of quality-management programs are tempered by adverse comments from others. For example, several senior managers contributed to the *Harvard Business Review* article "Does the Baldrige Award Really Work?" [22]. This publication was a counterpoint to a previous article, "How the Baldrige

Award Really Works" [23]. A number of contributors described successes achieved by quality-management efforts and by competing for the Baldrige Award. However, other comments were not so favorable. As one contributor noted, "The award is potentially as destructive to business as deregulation was to the savings and loan industry." A number of critics commented that some companies had expanded considerable efforts to compete for the award but had not actually incorporated quality-management practices into the organization, resulting in no lasting quality-management benefits. Buckingham and Coffman note that the Baldrige Award was once highly coveted. More recently, only a few companies bother to apply for it [24].

IMPLEMENTATION CONSIDERATIONS

The detailed guidance available in this chapter's references will assist in designing and implementing a health and safety quality-improvement program. However, managers must recognize that implementing a successful quality-management program requires considerable time, effort, and management commitment [13]. Both managers and workers must be actively involved in quality-improvement efforts, and there must be necessary support in order to implement changes. For the program to be effective, there must be management and worker commitment to "customer" satisfaction, employee involvement and empowerment, and, in particular, demonstrable results [14]. A quality-improvement effort may produce great initial enthusiasm, with posting of workplace banners, distribution of lapel pins, and full attendance at quality-training sessions. However, these activities alone will produce little change. If management commitment flags when immediate results are not produced, or if recommendations for quality improvement by workers, customers, or others are ignored or disapproved "due to funding" or other reasons, the program will gradually or abruptly disappear. The cost of such a failure may be much more than the costs of training, banners, consultants, and other components. The loss of credibility and the other impacts of the failed effort, including finger pointing to identify those responsible for the failure, may leave the health and safety organization worse off, with lowered morale and motivation, than it was before the effort. In essence, an improperly accomplished program can result in an organizational disaster [15].

EVALUATION OF QUALITY STANDARDS OF HEALTH AND SAFETY ORGANIZATIONS THROUGH NON-HEALTH-CARE INDICES

Whether established through formal training efforts or informal means, the emphasis an organization places on quality can usually be determined rather rapidly. If work is lost or delayed beyond the promised date, reports repeatedly have errors (even "minor" grammatical or technical ones), telephone messages are lost, or similar difficulties are trademarks, the quality of the unit is suspect. Similarly, notices and other postings that are several months out of date, hand-lettered signage, areas that are not clean, and other such activities or events suggest lack of attention to detail. For example, one major hospital had a large banner in the patient reception area that read, "Quality of Patient Care is our Top Priority." Unfortunately, one edge of the banner developed a tear that extended halfway down the banner, with the result that a major portion of the sign was folded upon itself. The banner remained in place for several days after the tear occurred, and it is therefore doubtful that patients or staff were convinced that quality was truly important to that organization. In these circumstances, even extensive, conscientious health and safety efforts and studies may be given little credence. The situation may not be "fair," but the health and safety director needs to establish quality standards that will help ensure the diligent efforts of the members of the organization are recognized and the products appropriately accepted.

As noted in the discussion of evaluation processes (Chapter 12) it is important to remember that "customers" and "owners" evaluate the quality of a health and safety organization. Customers can be all the individuals affected by the health and safety unit including workers, patients, senior managers, community members, government representatives, and others. Customer surveys, formal and informal discussions with those using the health and safety services, interactions with "owners" and other managers, and similar interchanges and feedback can help the health and safety director evaluate how users perceive the quality of the unit and its programs.

THE RESPONSIBILITY FOR ORGANIZATIONAL QUALITY

The documented successes of properly designed and implemented quality-management programs underscore the fact that they can bring numerous benefits to health and safety organizations. The descriptions of both achievements and failures emphasize that the success of the program depends heavily on the manager's job performance and the manager's *commitment* to the program. Appropriate commitment can help all health and safety members enjoy the satisfaction of being in a well-recognized quality organization.

REFERENCES

1. Peters TJ, Waterman RH Jr. *In Search of Excellence.* New York: Warner Books, 1982.
2. George S, Weimerskirch A. *Total Quality Management: Strategies and Techniques Proven at Today's Most Successful Companies.* New York: John Wiley & Sons, 1994.
3. Brue G. *Six Sigma for Managers.* New York: McGraw-Hill, 2002.
4. Walton M, Deming WE. *The Deming Management Method.* Penguin, 1988.
5. National Institute of Standards and Technology. Available at: www.quality.nist.gov; accessed April 19, 2007.
6. Schultz D. Accreditation of occupational health services. In: McCunney RJ, *A Practical Approach to Occupational and Environmental Medicine,* 3rd Ed. Philadelphia: Lippincott Williams & Wilkins, 2003.
7. Committee on Quality of Health Care in America. *To Err is Human—Building a Safer Health System.* Washington, DC: National Academy Press, 2000.
8. National Committee for Quality Assurance. Available at: web.ncqa.org/tabid/441/Default.aspx; accessed April 23, 2007.
9. U.S. Department of Veterans Affairs. Quality enhancement research initiative. Available at: www.hsrd.reserarch.va.gov/OUERI/program.cfm; accessed April 23, 1007.
10. Armstrong B, Levesque O, Perlin JB, et al. Reinventing the Veterans Healthcare Administration: focus on primary care. *J Healthcare Manag.* Nov/Dec 2005;50(6): 399–408.
11. Hann D, Poell-Cope G, Collins S, et al. Understanding nurses' behaviors in reporting errors and near misses. Available at: www.research-va.gov/meetings/2001/HSRD2001AMab70.htm; accessed April 23, 2007.
12. Prieskop FG. Occupational safety. In: LaDou J, *Current Occupational & Environmental Medicine,* 4th Ed. New York: McGraw-Hill, 2007.

13. American Industrial Hygiene Association and American National Standards. *Occupational Health and Safety Management Systems.* Fairfax, VA: American Industrial Hygiene Association, 2005.

14. Yarborough CM III. System for quality management. *J Occup Med* 1993;35: 1096–1102.

15. Moser R Jr. Quality management in occupational and environmental health programs—benefit or disaster? *J Occup Med* 1993;35:1103–1105.

16. Yarborough CM III. Strategic quality planning. *J Occup Med* 1994;36:414–418.

17. Rudolph L. A call for quality (editorial). *J Occup Environ Med* 1996;38:343–344 and response to letter. *J Occup Environ Med* 1997;39:501–503.

18. Widfeldt AK, Widfeldt JR. Total quality management in American industry. *AAOHN J* 1992;40:311–318.

19. Rooney E. TQM/CQI in business and health care—an overview. *AAOHN J* 1992;40:319–325.

20. American College of Occupational and Environmental Medicine, Glass LS (ed). *Occupational Medicine Practice Guidelines—Evaluation and Management of Common Health Problems and Functional Recovery of Workers,* 2nd Ed. Beverly Farms, MA: OEM Press, 2004.

21. Tibor T with Feldman I. *ISO 14000—A Guide to the New Environmental Management Standards.* Chicago: Irwin Professional Publishing, 1996.

22. Debate: Does the Baldrige award really work? *Harvard Bus Rev* 1992;70(1): 126–147.

23. Garvin DA. How the Baldrige plan really works. *Harvard Bus Rev* 1991;69(6):80–93.

24. Buckingham M, Coffman C. *First Break All the Rules.* New York: Simon & Schuster, 1999.

CHAPTER

26

Meeting Health and Safety Integrity and Ethical Challenges

with Jeffrey S. Lee

The director of a health and safety organization plays a major role in setting standards for the unit. While other individuals contribute, the person in charge typically provides good or bad examples that are copied by the other health and safety members. This process occurs regardless of whether the director wants to perform this role or not. In turn, the standards that are established by examples of behavior will help determine whether the health and safety unit is considered mediocre, acceptable, or one of the leading organizations in the field. Two areas that are particularly significant in setting standards are integrity and ethics. There is considerable overlap of these two areas in actual practice, research, and academic settings, but division for discussion will facilitate emphasis on items that more appropriately fall into one or another category. While it obviously is not possible to cover these areas in any depth in a single chapter, the following discussion should highlight some aspects of interest to the health and safety manager.

INTEGRITY

Integrity, implying trustworthiness and incorruptibility, and *honesty,* implying refusal to lie, steal, or deceive, are closely related areas of concern for the health and safety director. Through leadership and personal example, the manager of a health and safety program can help set the integrity standards for the health and safety unit.

Although some might suggest that integrity and honesty cannot be maintained in corporate or business environments, evaluation of outstanding companies demonstrates that these attributes can be cornerstones for success. The authors of *In Search of Excellence* quote Ogilvy's observation "In the best institutions, promises are kept no matter what the cost in agony and overtime" and then go on to emphasize the importance of loyalty and integrity in guiding activities of the best-run companies [1]. The "win/win" approach to solving problems and developing agreements emphasizes that the effort should result in maximum benefits for all involved. Covey, in *The 7 Habits of Highly Successful People,* discusses the computer software president who permitted a new bank president to cancel a contract [2]. Even though the software company had major financial difficulties and the contract was legally enforceable, the company president agreed to cancel the $84,000 contract and returned the deposit. The decision was made using the win/win concept, which is based on acceptance of the need for a contract or agreement to benefit everyone or there was no deal. Three months later, the bank signed a new contract with the software company for $240,000.

An important method to help establish integrity and honesty standards is to ensure that your word is your bond. Once an agreement is reached, it is the responsibility of the health and safety director to live up to the agreement. This requirement exists even for minor aspects such as meeting with a colleague at an agreed time as well as for major program activities, such as completing a health and safety audit by the time specified. If unforeseen circumstances necessitate changing an agreement, everyone involved must know the reasons for the change and agree to it. Similarly, standards are established by the way in which expense account reimbursement requests or other payment billings are prepared. In addition, the manager needs to be able to say "I don't know" or "I am sorry" when appropriate, rather than trying to bluff or provide an excuse for an inappropriate action. Taken together, the attitudes and actions of the health and safety manager play major roles in determining the levels of honesty and integrity manifested by others in the unit.

There are many opportunities to make "compromises" that will affect an organization's integrity standards. "Rounding down" exposure results that are just a few points to the right of the decimal so standards are met may lead, as it did in one organization, to the same process even though the excess readings were to the *left* of the decimal. Adding some expenses not actually incurred to a reimbursement request to "make up" for items not previously reimbursed can set the stage for serious loss of credibility. "Forgetting" to report a day's vacation, taking home a few reams of paper since "some work is always done at home," and similar examples say much about the integrity orientation of the unit. An individual may believe such practices are not noticed by colleagues, but it would be a very rare organization in which such lapses were not common knowledge. Interactions between health and safety professionals are based on trust, and if members of a unit cannot be certain of the validity of colleague's statements, that organization will probably have severe operational problems.

In essence, it is absolutely essential that health and safety managers be trusted, not only by upper management and colleagues but also by those workers the managers serve. Of course, blatant examples of lack of integrity can have long-lasting effects on the organization. Lack of trust may result in inappropriate legal claims against the organization, unnecessary and unjustified complaints to OSHA or other government agencies, and unwarranted fear of health and safety hazards. As discussed in Chapter 13 on leadership, loss of trust can have a demoralizing influence on all of the members of the health and safety team. (Recall the discussion in that chapter where a senior manager blamed a budget cut for a subordinate unit on the parent organization when the cut was, instead, under the total purview of the manager.) Once trust is lost, it is difficult, if not impossible, to regain. Conversely, high levels of trust can be correlated with highly effective health and safety programs that benefit workers/patients, their families, others supported by health and safety activities, managers, and members of the health and safety organization.

Occupational and environmental medicine may pose particular challenges for practitioners and those engaged in research face similar concerns. The current debate over global warming, with accusations that the government inappropriately edited reports of its scientists to downplay the significance of findings supporting the concept, is only one example. Unfortunately, this situation is apparently not unique, with reports of scientific advisory committees being manipulated for political purposes by government agencies and whistleblowers being sued and experiencing personal attacks [3]. Since 1992, the Office of Research Integrity (ORI) has reviewed scientific misconduct investigations in research funded by the U.S. Public Health Service [4].

ETHICS

The health and safety professional is often in the challenging position of trying to do, at the same time, what is best for a population, for an individual patient, and for the corporation, organization, or other business government or private activity employing the professional. Meeting these diverse, and sometimes conflicting, requirements, can pose important ethical challenges.

Health and safety professional organizations have attempted to provide ethical guidelines to assist their members in dealing with ethical dilemmas. For example, the American Medical Association, American College of Occupational and Environmental Medicine, American Association of Occupational Health Nurses, and the American Board of Industrial Hygiene have developed codes of ethics. The most recent versions of the codes can be accessed at the sites provided in the references [5]. While the codes can assist those facing ethical problems, a guideline is obviously just that. No guideline, code, or set of rules can provide *specific* guidance to resolve all the ethical problems the health and safety professionals will face.

Walsh has long discussed the problems of divided loyalties in the practice of occupational medicine, wherein the provider owes allegiance to both patients and the company hiring the provider [6]. Walsh, in her book, *Corporate Physicians: Between Medicine and Management,* discusses the problems of those who are both corporate officers and physicians [7]. She notes the difficulty of balancing humanistic concerns with economic requirements of businesses and discusses the need for the physician to help develop recognition that the mission of the organization should be consistent with the needs of individual workers. Such a corporate vision can help prevent the clashes and ethical problems that occur in companies without such an orientation. Recall the previous discussion of the significance of being *of* the corporation rather than merely *in* the corporation (see Chapter 1).

A major difficulty in ethical dilemmas is that the issues typically are not black and white. As noted by Sir Adrian Cadbury in his *Harvard Business Review* Ethics Prize Winner article, ethical guidelines do not always point in the same direction [8]. Even such simple guidance as "to always tell the truth" can pose a problem if doing so will hurt someone's feelings unnecessarily. In business interactions, clinical settings, relationships with government organizations, and other interactions, no simple formulae exist for making ethical decisions, and so we must base decisions on our own codes of conduct, as these have been developed through our professional education and life experiences.

Cadbury emphasized that the people making up an organization who are responsible for its conduct set the organization's standards. Regardless of what is said about ethics, an organization's ethics are manifested by its actions. He provides two rules of thumb regarding ethics—that openness and ethics go together, and actions are unethical if they cannot stand scrutiny. If an action or decision would cause embarrassment if it became public knowledge, there are associated ethical problems. Ethical concerns can pose serious problems for the health and safety manager, but, as Cadbury observes, not making hard decisions can be the least ethical choice. Postponing difficult decisions may help avoid legal challenges, but the result of not making decisions can have adverse and long-lasting effects on the organization and on society.

The health difficulties due to asbestos provide a tragic example of what can occur if ethics are compromised. The series of *New Yorker* articles by Paul Brodeur documented a number of ethical problems [9]. These included delays in making industry-supported research studies public, attempts to avoid responsibility, cover-ups, and similar actions. Regrettably, it appears that some health and safety members were involved in some of these events. The consequences of these inappropriate events underscore the adverse impacts that can result from failure to maintain humanistically oriented ethical standards. In this instance, thousands of workers have developed disease, and more will do so over the coming years.

In addition, companies have been forced into bankruptcy, and the livelihood of numerous other workers jeopardized. This situation is a true *ethical failure*. A similar circumstance existed as cigarette manufacturers attempted to suppress knowledge of the hazards of cigarette smoking.

It is not unusual for health and safety professionals to be faced with ethical concerns that cannot be readily resolved. For example, health and safety members of our organization noted a serious lead exposure while evaluating another function of a manufacturing facility. When we brought the problem to the attention of management, we were advised that we were only there to consult on methods to avoid musculoskeletal problems and should ignore the lead exposure. When further discussion made it apparent that management would not change its position, we terminated our efforts and advised that we felt we were obligated to report the problem. Subsequently, the company agreed to modify the work process to prevent the exposure, and we worked with management on a "no-cost" basis to assist.

In contrast to the situation with asbestos, tobacco, and the lead exposure, corporations can take the lead in maintaining high ethical standards to protect the health of workers. For example, various investigations, including animal studies, indicated that the primary health concerns of vinyl chloride were narcotic effects and, as reported in the 1960s, skin and finger changes similar to scleroderma and Raynaud's syndrome. In 1973, evidence of liver toxicity was reported in workers in Germany, and in 1974 angiosarcoma was reported in workers involved in the manufacture of polyvinyl chloride. Prompt investigation and reporting of the problem by the corporate management concerned led to institution of procedures to prevent future problems due to worker exposure. In similar situations, other companies have also taken aggressive action to identify and prevent risks. For example, the reports of increased colon cancer among one group of oil refinery workers resulted in the company's promptly implementing a major epidemiologic study. The study clearly showed there was no increased risk in the workers, but the willingness of the corporation to conduct the study, even if it could result in adverse impacts on the organization, is typical of our recent experiences (at the Rocky Mountain Center) with a number of corporations.

Although some situations may have relatively clear and achievable resolutions, others may not. For example, OSHA requires a report of any worker who is treated with a prescription medication for a work-related condition. Anti-inflammatory drugs, such as ibuprofen, can be obtained over the counter in reduced-strength formulations. A health provider may believe that the worker's condition requires full prescription-strength dosage (e.g., 800 mg of ibuprofen three times a day). Over-the-counter preparations are 200 mg tablets. It is tempting to advise the patient to take four of the 200 mg tablets three times a day and avoid the necessity to file an OSHA report. Since some plant managers are evaluated on the number of OSHA reports filed, such action "supports" management while still providing the patient the necessary treatment. The disquieting aspect is that the labeling on the 200 mg tablets indicates that more than 400 mg should not be taken at one time unless under the direction of a physician. In order to exceed the 400 mg dosage, the treating health provider has had to effectively "prescribe" the higher dose. Questions regarding this area are among the most frequent we receive when presenting short courses on basic aspects of occupational health and safety. Some participants believe that once the higher levels have been prescribed, the prescribed regimen should be recorded on the OSHA log regardless of the size of the tablets used. Others argue that "in the real world" such a practice may result in the company using other, less qualified health providers. This dilemma typifies many of the ethical challenges facing health and safety professionals, for which there are no clear-cut, "easy" answers [10]. Samuels, in his discussion of ethics, notes that occupational health and safety ethical issues such as confidentiality in reporting medical conditions interfering with work, discrimination as a result of genetic testing, ownership and confidentiality of worker medical records, and similar situations may not be easily resolved [11]. His observation—"Complex situations . . . almost never yield simple choices, even when the medical judgments and the ethical questions are clear"—emphasizes the difficulty many ethical questions pose.

Ethical concerns are not limited to practice settings; they can occur in any health and safety activity. For example, a graduate academic program required that all members of a student's supervisory/thesis committee be listed as co-authors on publications even though some had not participated in research design, in the research, or in report writing. Their primary activity was correcting typographical errors in the final report. This practice was not in accord with specifications of many professional journals or with those of the International Committee on Medical Journal Ethics [12].

More recently, Capozzi and Rhodes discussed the concept of "lying for the patient's good." They present the example of a patient with a knee injury who requests that the physician report the injury as work related even though the physician considers such an etiology as unlikely. They suggested that the report to the workers' compensation carrier be reported as "according to the patient . . ." as a way around the ethical dilemma [13]. A number of letters to the editor challenged this approach as unethical, although one noted "the physician should find another way to secure medical care for the patient"—advice that may not be particularly helpful [14].

The health and safety professional is dedicated to caring for the well-being of others. Kortsha comments that this concept can serve as a useful yardstick when caught between opposing financial or political interests [15]. A health and safety professional's findings and recommendations can cost a company or organization part or all of its profits. It is consequently essential that the recommendations be based on sound science and the best available knowledge.

As noted by Cullen and Rosenstock in a prior edition of their textbook, the health and safety practitioner "has an ethical obligation to be as informed as possible and, perhaps more importantly, to recognize when his or her own knowledge or experience is insufficient to resolve the problem and consultation is needed" [16]. It also must be recognized that there can be honest disputes in standard setting and other attempts to establish "safe" levels. The ban on saccharin in the United States but not in Canada, even though Canadian researchers reported increased cancer rates in rats and mice, is only one example. In developing recommendations, the health and safety professional must be able not only to balance differing study results but also to separate the essential from the merely desirable [15]. In many situations, such processing emphasizes that much of our practice is still more "art" than science.

Recent emphasis on cost-benefit considerations has added to the difficulty of making clear-cut ethical decisions. Workers' employment benefits the workers, their families, and society. These benefits can be weighed against the costs of reducing worker risk of illness or injury. For example, reducing emissions to "zero-risk" levels, if such is actually possible, could mean plant closures that would force workers into high-risk occupations or unemployment. Unemployment is a known risk factor for illness and injury, so the attempt to protect workers could result in an outcome opposite from that intended. Cost-benefit approaches attempt to evaluate such potentials, but often leave us with difficult ethical challenges. All of us value human lives, yet the fallacious position "I don't care if it costs $10 million, if it saves one life it's worth it" has caused excessive expenditures to save a few lives. Unfortunately, the fact that all resources are limited meant that the funds could not be used for other efforts that would have saved multiple more lives. Although the preceding discussion is oversimplified, it is presented in an effort to emphasize that ethical choices and answers are often not clear-cut. Goodman and Frumkin, in their discussion of ethics in international health and safety activities, provide an excellent series of "real world" examples of health and safety ethical problems [17]. The discussion of each example highlights the scope of dilemmas that the health and safety manager faces in attempting to resolve the situation.

Significant discussions about changing the Code of Ethics of the American College of Occupational and Environmental Medicine also highlight ethical challenges [18–20]. As noted by Guidotti and others, these challenges can be expected to continue, and globalization will produce additional concerns [21–23]. In an excellent discussion of OEHS ethical concerns, Brandt-Rauf and colleagues noted that 69 percent of respondents in a survey of occupational physicians reported ethical problems in their practice [24]. These problems were most frequently due to conflicting responsibilities to patients and to management, the "double-agent" issue. A similar survey conducted today could be expected to produce similar results.

The significance of honesty, integrity, and ethics was emphasized in a series of articles titled "IBD's 10 Secrets to Success" in *Investor's Business Daily* [25]. The authors analyzed leaders and successful people in numerous different occupations. The analysis delineated 10 traits that characterize successful people. Particularly pertinent to this discussion is number 10: "Be honest and dependable; take responsibility; otherwise numbers 1–9 won't matter" [25]. As Kortsha observes, health and safety professionals have the "obligation to do their best, to offer solid professional counsel unbent by partisan pressures" [15]. By helping establish integrity and ethical standards, the health and safety director can benefit the health and safety unit, the organization, and the workers. While often difficult, establishing and maintaining appropriate standards can not only help programs and those the health and safety professionals serve but also engender a sense of pride and satisfaction in the members of the health and safety unit.

REFERENCES

1. Peters TJ, Waterman RH Jr. *In Search of Excellence.* New York: Warner Books, 1982.

2. Covey SR. *The 7 Habits of Highly Effective People: Restoring the Character Ethic.* New York: Simon & Schuster (Fireside), 1989.

3. Tong S, Olsen J. The threat to scientific integrity in environmental and occupational medicine. *Occup Environ Med* 2005;62(12):843–846.

4. Reynolds SM. ORI findings of scientific misconduct in clinical trials and publicly funded research, 1992–2002. *Clin Trials* 2004;1(6):509–516.

5. Sites for Codes of Ethics:

American Association of Occupational Health Nurses
http://aaohn.org

American College of Occupational and Environmental Medicine
http://acoem/

American Industrial Hygiene Association
www.aiha.org

American Medical Association
www.ama-assn.org

American Society of Safety Engineers
http://asse.org/

Association of Occupational and Environmental Clinics
http://aoec.org/

Board of Certified Safety Professionals
http://bcsp.org/

6. Walsh DC. Divided loyalties in medicine: the ambivalence of occupational medicine practice. *Soc Sci Med* 1986;23(8):789–796.

7. Walsh DC. *Corporate Physicians: Between Medicine and Management.* New Haven, CT: Yale University Press, 1987. (Available as secondhand copies, April 2007.)

8. Cadbury AH. Ethical managers make their own rules. *Harvard Bus Rev* 1987;65(5):69–73.

9. Brodeur P. Annals of law: the asbestos industry on trial. *New Yorker* 1985.

I. A Failure to warn. June 10, pp 49–101;

II. Discovery. June 17, pp 45–111;

III. Judgement. June 24, pp 37–77;

IV. Bankruptcy. July 1, pp 36–80.

10. McCunney RJ, Brandt-Rauf P. Ethical conflicts in the private practice of occupational medicine. *J Occup Med* 1991;32:80–82.

11. Samuels SW. Principles for ethical practice of environmental and occupational medicine. In: Rom WN (ed), *Environmental and Occupational Medicine,* 2d Ed. Boston: Little Brown and Company, 1992.

12. International Committee on Medical Journal Ethics. Uniform requirements for manuscripts submitted to biomedical journals: writing and editing for biomedical publication, 2004. Available at: www.icmj3.org/#author; accessed September 8, 2005.

13. Capozzi JD, Rhodes R. Lying for the patient's good. *J Bone Joint Surg Am* 2004;86:187–188.

14. Brett AS, Perez-Skanz JR, Levin PE. (Three letters to the editor.) *J Bone Joint Surg Am* 86-A(8):1826–1827.

15. Kortsha GX. Ethics. In: Garrett JT, Cralley LJ, Cralley LV (eds), *Industrial Hygiene Management.* New York: John Wiley & Sons, 1988.

16. Cullen MR, Rosenstock L. Principles and practice of occupational and environmental medicine. In: Rosenstock L, Cullen MR (eds), *Textbook of Clinical Occupational and Environmental Medicine.* Philadelphia: W.B. Saunders, 1994.

17. Goodman K, Frumkin H. Ethical issues in international occupational health. In: Fleming LE, Herzstein JA, Bunn WB III (eds), *Issues in International Occupational and Environmental Medicine.* Beverly, MA: OEM Press, 1997.

18. Rothstein MA. A proposed revision of the ACOEM code of ethics. *J Occup Environ Med* 1997;39:616–622.

19. Teichman RF. ACOEM code of ethical conduct (editorial). *J Occup Environ Med* 1997;39:614–615.

20. Brodkin CA, Frumkin H, Kirkland HK, et al. and the AOEC Board of Directors. Choosing a professional code for ethical conduct in occupational and environmental medicine (editorial). *J Occup Environ Med* 1998;40:840–842.

21. Guidotti TL. Ethics and skeptics: what lies behind ethical codes in occupational health. *J Occup Environ Med* 2005;47(2):168–175.

22. Voyi K. Is globalisation outpacing ethics and social responsibility in occupational health? *Med Lav* 2006 Mar–Apr;97(2):376–382.

23. Brodkin CA, Frumkin H, Kirkland HK, et al. and the AOEC Board of Directors. Choosing a professional code for ethical conduct in occupational and environmental medicine (editorial). *J Occup Environ Med* 1998;40:840–842.

24. Brandt-Rauf PW, Brandt-Rauf SI, Fallon LF Jr. Management of ethical issues in the practice of medicine. In: Fallon LF Jr, Dickerson OB, Brandt-Rauf PW (eds), *State of Art Rev Occup Med* 1989;4(1).

25. Achstatter GA. IBD's 10 secrets to success. *Investor's Business Daily* 1999; January 4(15):186.

ADDITIONAL READING

Bartolome F. Nobody trusts the boss completely—now what? *Harvard Bus Rev* 1989; 67(2):135–142.

Berger JT, Rosner F. The ethics of practice guidelines. *Arch Int Med* 1996;156:2051–2056.

Council on Ethical and Judicial Affairs. *Code of Medical Ethics—Current Opinions with Annotations* (150th Anniversary Ed). Chicago: American Medical Association, 1997.

Goodpaster KE, Piper TR (eds). *Managerial Decision Making and Ethical Values.* Boston: Harvard Business School, 1989.

Samuels SW. Moral questions of occupational and environmental medicine. In: Rom WN, Markowitz S (eds), *Environmental and Occupational Medicine,* 4th Ed. Philadelphia: Lippincott, Williams & Wilkins, 2007.

Wynia MK, Cummins DS, VanGeest JB, et al. Physician manipulation of reimbursement rules for patients—between a rock and a hard place. *JAMA* 2000;283(14):1858–1865.

APPENDIX

A

Items for New Health and Safety Program Managers to Consider

Note: Many of the following items will be accomplished simultaneously, and both the extent of each effort and its sequence will vary, depending on the size and organizational position of the health and safety unit.

1. Obtain/review organizational charts for the corporation or other organization and the health and safety unit.
2. Obtain/review most recent corporate or organization annual report (if applicable).
3. Obtain/review/evaluate pertinent recent health and safety reports and evaluations, such as:
   a. Reports of any OSHA inspections.
   b. Pending OSHA, Clean Air Act, Clean Water Act, Americans with Disabilities Act, or other government citations.
   c. In-house or external evaluation or audit reports.
   d. Pending litigation.
   e. Current health and safety plans and objectives.
   f. Current health and safety budget.
   g. Annual OSHA incidence and severity reports.

4. Meet with key health and safety staff, including those on-site and at any geographically separate units.
5. Meet with human resources/personnel members to obtain information on personnel policies/contracts.
6. Meet with budget department members to ascertain budgeting procedures and requirements.
7. Meet with immediate supervisor to discuss expected activities and preliminary plans, goals, and objectives.
8. Work with:

   a. Supervisors
   b. Key on-site and off-site health and safety members
   c. Advisory Committee
   d. Worker and union representatives
   e. Others as appropriate

   to develop preliminary plans, goals, objectives, and program evaluation procedures.
9. Work with immediate supervisor to arrange meetings with senior management to discuss and obtain input on health and safety program activities.
10. Establish appropriate schedule for interaction and communication with senior management, representatives of populations served, health and safety staff members, Advisory Committee, and others as needed.

APPENDIX B

Examples of Budget Categories

The following are some of the budget categories that often have to be considered by those in charge of health and safety programs.

1. Personnel
   a. Salaries
      1. Professional
      2. Staff, technicians, or other support personnel
   b. Bonuses (if applicable, e.g., incentives based on number of patients seen, total billings, timeliness of reports)
   c. Benefits
      1. Social Security
      2. Other retirement-plan contributions
      3. Group insurance, other life insurance
      4. Health insurance
      5. Other
   d. Overtime (if projections of overtime are permitted)
   e. Consultants
2. Professional Development
   a. Continuing education, other courses conducted off-site (e.g., include registration/tuition, travel, reimbursement for meals and lodging, local transportation)

b. Library text and software replacements
c. Journal and data base subscriptions
d. Development courses conducted in-house, including honoraria, travel, other expenses of speakers
e. Other

3. Equipment (frequently defined by an organization as all items above a certain dollar amount, e.g., $500.00)
   a. Includes administrative and professional items, such as personal protective equipment, computers, laboratory equipment, and new health and safety equipment.
   b. Includes both new and replacement items (justification typically required).

4. Supplies

   Office supplies, professional supplies, software, printing or duplicating charges, film or video tapes, maintenance contracts on office and laboratory equipment, other. Note that some items may be in the equipment category, depending on costs.

5. Communications
   a. Telephone—installation, maintenance, local access, long distance
   b. Information technology equipment and support
   c. Facsimile transmission
   d. Radios
   e. Postage

6. Vehicles

   Replacement costs and regular/preventive maintenance

7. Travel

   Include transportation, lodging, meals, tips, etc. (The organization may establish a maximum per day for lodging and meals.)

8. Insurance

   Malpractice and other liability insurance, fire, theft, etc. (the organization may be self-insured).

9. Facilities, Rent, and Utilities

   Depending on the organization, charges may be made for rent, utilities, janitorial services, parking, and similar items. New building programs may be included here or as a separate item in the organizational budget.

10. Special Programs

    May include health promotion/disease prevention programs, research projects, and similar efforts. Each special program may require a detailed "sub-budget" containing many categories listed earlier.

11. Subcontracts and Reimbursements

    Costs to obtain services, such as laboratory analyses, medical procedures, or other activities not accomplished "in-house."

12. Other

    Data processing charges, medical examinations of staff, public relations/advertising, support for visitors, funding for unit retreats or "off-sites," etc.

Note: Some organizations will request projected budgets for one or more follow-on years. In such situations, it is important to use a reasonable "inflation factor" to help adjust for increased costs in the out-years.

APPENDIX

C

Examples of Audit Instruments for Occupational and Environmental Health and Safety Program Components

Medical Audit and Consultation Program—Asbestos

Plant Location: Address:	Plant Medical Services Contact:	Phone No: Fax No: E-mail:
Medical Auditor:	Date of Audit:	
Program Audited: Asbestos (OSHA Construction Standard CFR 1926.1101)		

Program Criteria: (weight 2x)	**C/NC/O/NA***	**Comments**
1. Surveillance program managed by / under direction of licensed physician		
2. Inclusion criteria defined: a. Class I, II, III asbestos work for 30 days per year b. Exposed over PEL for 30 days per year c. Employee does not meet surveillance inclusion criteria		
3. Exit examination if inclusion criteria not met (optional)		
4. Exams performed by / under direction of licensed physician		
5. PFTs performed by trained personnel		
6. Medical & work history with emphasis on pulmonary, cardiovascular and gastrointestinal systems		

* C = Compliant
NC = Not compliant
O = Optional
NA = Not applicable

	Patient Case No.									
Patient Criteria: (weight 1x)	**1**	**2**	**3**	**4**	**5**	**6**	**7**	**8**	**9**	**10**
1. Annual exam within 30 days of due date										
2. Initial exam within 10 days following 30th day of exposure										
3. "Initial Medical Questionnaire"—Part I										
4. "Periodic Medical Questionnaire"—Part II										
5. Physical examination directed to pulmonary and gastrointestinal systems										
6. Height, weight, blood pressure and resting heart rate										
7. Pulmonary function tests including FVC and FEV_1										
8. Chest x-ray at discretion of physician										
9. Physician's Written Opinion (PWO) within 30 days of completed file to physician										
Overall Compliance Rate (#C minus #NC ÷ #C) = %										

Specific Recommendations (see Executive Summary):

Signature of Medical Auditor ______________ Date ______________

Signature of Plant Medical Services Representative ______________ Date ______________

(My signature indicates that I have reviewed and discussed this Occupational Health Services audit report with the medical auditor.)

Medical Audit and Consultation Program—Bloodborne Pathogens

Plant Location: Address:	Plant Medical Services Contact:	Phone No: Fax No: E-mail:
Contact Person	Date of Audit:	
Program Audited: Bloodborne Pathogens (29 CFR 1910.1030) (Program review only)		

Program Criteria	C/NC/O/NA*	Comments
1. Exposure determinations defined: a. Job classifications—all employees have occupational exposures b. Job classifications—some employees have occupational exposures		
2. Tasks and procedures defined		
3. Occupational exposures immediately reported and evaluated by Medical Dept. or Safety/IH		
4. Emergency medical providers trained in protection techniques against bloodborne pathogens		
5. Disposable medical materials segregated, labeled, red bagged, and prepared for disposal as infectious waste		
6. Disposal of infectious waste contracted by Medical, Safety/IH to certified collection and disposal service		
7. Collection and disposal of infectious waste scheduled by Medical Dept. based on volume generated at each collection site		
8. Copy of Exposure Control Plan and OSHA final rule for bloodborne pathogens available to affected employees in Medical and Safety Departments		
9. Exposure Control Plan reviewed and updated annually		
10. Universal Precautions and Work Practices attached to Exposure Control Plan		

* C = Compliant O = Optional
NC = Not compliant NA = Not applicable

Program Criteria	C/NC/O/NA*	Comments
1. Handwashing facilities readily available to affected employees		
2. Contaminated needles and sharps discarded in appropriate containers		
3. Personal protective equipment provided at no cost to affected employees		
4. Food, eating, drinking, smoking, etc., prevented in work areas with high likelihood of occupational exposures to bloodborne pathogens		
5. Food and drink not kept in refrigerators, freezers, shelves, cabinets, etc., where blood or other potentially infectious material is present		
6. Biohazard warning label affixed to refrigerator/containers which occasionally contain potentially infectious materials		
7. Makes Hepatitis B vaccine and post-exposure evaluation available at no cost to all employees potentially affected or at risk of occupational exposure		
8. Declination statement available to employees who decline Hepatitis B vaccine		
9. Maintains permanent record for each person with possible occupational exposure to bloodborne pathogens		
10. Makes available a record of exposure incidents upon request to Ass't. Sec. of Labor OSHA and Director of NIOSH		
Overall Compliance Rate (#C minus #NC ÷ #C) = %		

Recommendations: (see Executive Summary):

____________________________ ________________

Signature of Medical Auditor Date

____________________________ ________________

Signature of Plant Medical Services Representative Date

(My signature indicates that I have reviewed and discussed this Occupational Health Services audit report with the medical auditor.)

Medical Audit and Consultation Program—Hearing Conservation

Plant Location: Address:	Plant Medical Services Contact:	Phone No: Fax No: E-mail:
Medical Auditor:	Date of Audit:	
Program Audited: Noise / Audiometric Testing (29 CFR 1910.95)		

Program Criteria	C/NC/O/NA*	Comments
1. Audiometric testing available at no cost to all employees whose exposures equal or exceed 8-hour TWA of 85 dB as defined by Industrial Hygiene		
2. Baseline audiogram on all new employees assigned to a Hearing Conservation Area (HCA) or within 6 months of transferring into a HCA		
3. Audiometric tests performed by licensed or certified audiologist, otolaryngologist or technician certified by CAOHC		
4. Audiometric tests are pure tone, air conduction, hearing threshold examinations with minimum test frequencies at 500, 1000, 2000, 3000, 4000, and 6000 Hz		
5. Audiometric tests conducted with audiometer that meets specifications of, and maintained and used in accordance with, ANSI spec. S3.6-1969		
6. Audiometric examinations administered in room meeting CFR 1910.95 requirements		
7. Audiometric calibration: a. functional operation checked before each day's use b. checked acoustically every year in accordance with CFR 1910.96 Appendix E c. exhaustive calibration performed at least every two years in accordance with ANSI spec. S3.6-1969		

* C = Compliant   O = Optional
NC = Not compliant   NA = Not applicable

	Patient Case No.									
Patient Criteria: (weight 1x)	**1**	**2**	**3**	**4**	**5**	**6**	**7**	**8**	**9**	**10**
1. Baseline audiogram performed within 6 months of first exposure at or above action level as defined and notified by Industrial Hygiene										
a. baseline audiogram preceded by at least 14 hours without workplace (and non-workplace) noise										
2. Annual audiogram obtained and compared to baseline audiogram to determine if valid and if standard threshold shift (STS) has occurred										
a. If STS has occurred, employee re-tested in 30 days with re-test becoming annual audiogram										
b. If STS has occurred, employee notified in writing within 21 days of determination										
Overall Compliance Rate (#C minus #NC ÷ #C) = %										

Recommendations (see Executive Summary):

______________________________ ________________

Signature of Medical Auditor Date

______________________________ ________________

Signature of Plant Medical Services Representative Date

(My signature indicates that I have reviewed and discussed this Occupational Health Services audit report with the medical auditor.)

Medical Audit and Consultation Program—Respiratory Protection

Plant Location: Address:	Plant Medical Services Contact:	Phone No: Fax No: E-mail:
Medical Auditor:	Date of Audit:	
Program Audited: Respiratory Protection (29 CFR 1910.134)		

Program Criteria	C/NC/O/NA*	Comments
1. Surveillance program managed by/under direction of licensed physician or other health care professional (PLHCP)		
2. Administers OSHA medical questionnaire and conducts follow-up medical examinations for employees who may wear respiratory protection equipment as part of their essential job functions prior to assignment		
3. Confidential biennial medical questionnaire administered by a physician or other licensed health care professional (PLHCP) during the employee's normal working hours, or at a time and place convenient to the employee. The medical questionnaire shall be administered in a manner that ensures that the employee understands its content		
4. Immediately refers any employee who experiences unusual difficulty breathing while being fitted for or while using a respirator for further medical evaluation. (If fit-testing performed by medical unit)		
5. Record retention system to keep spirometry records for at least 20 years		

* C = Compliant
NC = Not compliant
O = Optional
NA = Not applicable

	Patient Case No.									
Patient Criteria: (weight 1x)	**1**	**2**	**3**	**4**	**5**	**6**	**7**	**8**	**9**	**10**
1. Biennial (every two years) medical evaluation performed using OSHA medical questionnaire or an initial medical examination that obtains the same information as the medical questionnaire										
2. Follow-up medical exam provided for employee who gives positive response to any of questions 1 through 8 in Section 2, Part A of the questionnaire, or whose medical exam demonstrates the need for a follow-up medical exam. The follow-up medical exam shall include any medical tests, consultations, or diagnostic procedures that the PLHCP deems necessary to make a final determination.										
3. Provides notification to employee, supervisor, and training department as to the employee's respirator fitness-for-duty status and any necessary restrictions regarding respirator use										
4. Provides opportunity for employee to discuss the questionnaire and examination results with the PLHCP at the employee's request.										
Overall Compliance Rate (#C minus #NC ÷ #C) = %										

Recommendations: (see Executive Summary):

____________________________ _______________

Signature of Medical Auditor Date

____________________________ _______________

Signature of Plant Medical Services Representative Date

(My signature indicates that I have reviewed and discussed this Occupational Health Services audit report with the medical auditor.)

Medical Audit and Consultation Program—Lead

Plant Location: Address:	Plant Medical Services Contact:	Phone No: Fax No: E-mail:
Medical Auditor:	Date of Audit:	
Program Audited: Lead (29 CFR 1910.1025)		

Program Criteria	C/NC/O/NA*	Comments
1. Surveillance program managed by / under direction of licensed physician		
2. Initial exam within 60 days from time of assignment to lead exposure area		
3. Detailed work and medical history including: a. Past lead exposure or any toxic substance b. History of drugs, alcohol, tobacco, and medications routinely taken c. History of GI, hematologic, renal, CV, reproductive, and neurological problems		
4. Physician examination with emphasis on teeth, gums, hematologic, GI, CV, renal, and neurological systems		
5. Laboratory tests including: a. Blood pressure b. Hemoglobin & hematocrit c. Red cell indices d. Peripheral smear cell morphology and differential e. Blood lead level		

* C = Compliant O = Optional
NC = Not compliant NA = Not applicable

Program Criteria	C/NC/O/NA*	Comments
f. Blood ZPP g. BUN h. Serum Creatinine i. UA—routine and microscopic		
6. Pregnancy/male fertility testing if requested by exposed employee		
7. Other tests as deemed necessary by physician		
8. Blood lead and ZPP performed by CDC licensed lab		
9. Periodic examination if blood lead at or above 40 μg/dl during preceding 12 months		
10. Symptomatic evaluation ASAP after complaint of symptoms associated with lead intoxication, problems wearing respirator, or concerns about procreating a healthy child		

	Patient Case No.									
Patient Criteria: (weight 1x)	**1**	**2**	**3**	**4**	**5**	**6**	**7**	**8**	**9**	**10**
1. Physician's written opinion within 30 days of any lead examination										
2. Blood lead and ZPP performed at least every 6 months for employees in lead surveillance program										
3. Appropriate medical removal and intervention for any employee with blood lead at or above 30 μg/dl										
Overall Compliance Rate (#C minus #NC ÷ #C) = %										

Recommendations: (see Executive Summary):

Signature of Medical Auditor ____________ Date ____________

Signature of Plant Medical Services Representative ____________ Date ____________

(My signature indicates that I have reviewed and discussed this Occupational Health Services audit report with the medical auditor.)

APPENDIX

D

Summary of Parliamentary Procedures [1,2]

To Do This	You Say This	May You Interrupt the Speaker?	Must You be Seconded?	Is the Motion Debatable?	What Vote is Required?
Adjourn the meeting*	"I move that we adjourn"	No	Yes	No	Majority
Amend a motion	"I move that this motion be amended by"	No	Yes	Yes	Majority
Ask for votes by actual count to verify a voice vote.	"I call for a division of the house"	No	No	No	No vote
Complain about noise, room temp, etc.*	"I rise to a question of privilege affecting the assembly"	Yes	No	No	No vote—Chair rules
Complain about personal affront or similar item	"I rise to a question of personal privilege"	Yes	No	No	No vote—Chair rules
Consider something out of its scheduled order*	"I move we suspend the rules and consider"	Yes	Yes	No	2/3 vote
End Debate	"I move (or call for) the previous question"	No	Yes	No	2/3 vote

To Do This	You Say This	May You Interrupt the Speaker?	Must You be Seconded?	Is the Motion Debatable?	What Vote is Required?
Have something studied further	"I move that we refer matter to a committee"	No	Yes	Yes	Majority
Introduce business (a primary motion)	"I move that ___"	No	Yes	Yes	Majority
Object to procedure or to a breach of rules*	"Point of Order"	Yes	No	No	Chair decides
Postpone consideration of something	"I move that we postpone this matter until"	No	Yes	Yes	Majority
Recess the meeting	"I move that we recess until ___"	No	Yes	No	Majority
Reconsider something already disposed of*,**	"I move we reconsider our action relative to"	Yes	Yes	Yes	Majority
Request information*	"Point of Information"	Yes	No	No	No vote
Suspend further discussion of something*	"I move that we table this item or motion"	No	Yes	No	Majority
Take up a matter previously tabled*	"I move to take from the table"	No	Yes	No	Majority
Vote on a ruling by the chair*	"I appeal the chair's decision"	Yes	Yes	Yes	Majority

*Action not amendable, ** must be moved by one who voted for the prior winning action.

1. De Vries MA. *The New Robert's Rules of Order,* 2d ed. New York, Signet, 1998.

2. Robert HM III, Evans WI, Honemann DH, et al. *Robert's Rules of Order Newly Revised,* 9th ed. Cambridge, MA: Perseus, 2000.

Index

Absenteeism, 231
Academic community, consulting in, 274
Access to Employee Exposure and Medical Records Standard, 355
  provisions of, 358–359
Accidents, reacting to, 42–43
Accountability, in team building, 126
Accounting, importance of, 144–145
Accreditation, agencies dealing with, 409
Accreditation Association for Ambulatory Health Care (AAAHC), 409
Adapting-dealing personality, 110, 111
Advocating a health and safety program, 219–221
After Action Review (AAR), 53
Agenda, for meeting, 318
"Agree-er," role in meetings, 321
Americans with Disabilities Act (ADA), 3, 16
  defenses provided by, 361
  enforcement of, 359
  and HPDP programs, 258
  implications of, 361–362
  management according to, 162, 163–165
  management issues caused by, 100
  provisions of, 360
Amorgianos doctrine, 391–394
ANSI Z10-2005, 411
"Apples and oranges," comparing, 131
Application, for employment, 87–88
Asbestos
  case study of, 421–422
  sample audit instruments for, 436–437
AT&T, HPDP programs at, 252
Audience
  of communication, 301–302
  of media release, 339–340
Audit, program
  described, 168–169
  example of, 168
  sample instruments for, 435–445
  steps in, 174
  utility of, 169
Audit instruments (sample)
  for asbestos, 436–437
  for bloodborne pathogens, 438–439
  for hearing protection, 440–441
  for lead, 444–445
  for respiratory protection, 442–443
Autocratic leadership, 179–180

Balance sheet, projected, 153
Baldrige criteria, 407–408
  application to health care organizations, 408–409
Begin with the learner, 95
Billing, types of, 144
Biologic terrorism, 41–42, 44
Bird flu, 43
Blood pressure screening programs, 249
Bloodborne pathogens, sample audit instruments for, 438–439
Blueprint for Health, 210–211
Bona fide occupational qualifications, 364
Bradford Hill criteria, 393–394
Brainstorming, 195
Break-even analysis, 152
Budget cycle, 132–133
Budget fallout, 139–140

Budget monitoring and evaluation, 137–139
Budgeting, 129
  approval process in, 136–137
  categories for, 431–433
  costing for, 133–134
  dealing with constraints in, 135
  fallout, 139–140
  framework for, 134
  monitoring and evaluation, 137–139
  priorities in, 134–136
  process of, 133–136
Bureaucratic leadership, 180
Business plan
  company summary, 147, 148
  components of, 147
  confidentiality statement, 146
  executive summary, 146, 147
  financial plan, 147, 152–153
  management summary, 147, 152
  market analysis, 147, 149–150
  services delivery plan, 147, 148
  strategy and implementation summary, 147, 150–152
Business plans, 132
Business ratios, 153
Buying patterns, analyzing, 149

Calendar, keeping, 286–287
  difficulties with, 287–288
Call for the question, 322
Cancer screening programs, 249
Case management, 214
Cash flow
  problems with, 154–155
  projected, 152
Causation, evidence of, 386–389
CBRNE agents (terrorism), 41
CDC Category A agents, 41–42
Cellular phones, 48, 310
Chain of authority, 43, 46
Charisma, 179
Chemical, biologic, radiologic, nuclear, and explosive agents (CBRNE), 41
Cholesterol programs, 249
Circumstantial evidence, defined, 386
Civil procedure, rules of, 389
Clarity, of communication, 303
Clean Air Act, 369–370
Command, chain of, 43, 46
Command, Control, Communication, and Coordination (4 Cs model), 40
Commitment, in team building, 126
Communication
  admission of not knowing, 312–313
  audience of, 301–302
  clarity and conciseness in, 303–304
  coordination in, 310–311
  effective reading and listening, 308–310
  formal, 313–314
  good grammar in, 304–305
  in HPDP programs, 261
  importance of, 14, 299
  informal, 314–315
  by leader, 182
  listening with understanding, 308–309
  maintaining, 47–48
  with news media, 337–350
  one-way and two-way, 305
  organization of, 302–303
  procedures for, 47
  purpose of, 300–301
  pyramid of, 337–338
  recognizing failure of, 313
  requirements for, 299–300
  research for, 305–306
  resources for, 48–50
  revision and editing in, 306–307
  in risk management, 218–219
  spell-check hazard in, 307–308
  in team building, 118
  with supervisors and management, 311–312
  See also Meetings; Presentations
Community relations, 188–189
Company summary, 147, 148
Competition, 149
  analysis of, 150–151
Compliance, with legal and regulatory requirements, 16
Compliments, distinguished from positive feedback, 94
Conceptual skills
  importance of, 13
  types of, 13–14
Conciseness, of communication, 303–304
Confidentiality
  statement, 146
  of worker's compensation claims, 367
Conflict, in team building, 126
Conserving–holding personality, 109, 111
Consulting
  activities in, 271–275
  business concerns in, 276–277
  business skills needed for, 2756
  characteristics of, 268–269
  legal services for, 276
  malpractice issues in, 373–375
  marketing of, 276–277

scope of occupational health services in, 269–270
selection of consultant, 277–278
settings for, 267–268
Contingency planning, 52
Control, in emergency, 44–45
Controlling-taking personality, 109, 110–111
Conversations, managing, 294–295
Coordination of communication, 310–311
Coordination of response
among facilities, 51
importance of, 50
of logistics, 51
Corel Presentations, 327–328
Corporate health and safety risk, 205
Cost comparisons, 131
Costs, determining, 130, 133, 144–145
Costs, health and safety, 205
direct, 205–209
indirect, 205–209
Credentials, checking, 89
Credit, who gets, 189–190
Criminal liability
under Clean Air Act, 369–370
example of, 370
under OSH Act, 369
Crises, avoiding, 31
Criticism, of employees, 95
Current status, determining, 24

Daubert standard, 378–379
establishment of, 383–385
implications of, 398
Decision making, 15
delegation of, 195
forcing a decision, 323–324
group, 319–320
importance of, 191–193
leadership and, 178, 185
participatory approach to, 193–195
in team building, 123
Delegation
of decision making, 195
effective, 121–122, 296
Deming Management Method
Fourteen Points of, 406
history of, 405
rediscovery of, 404
Seven Deadly Diseases in, 407
Democratic leadership, 180–181
Direct evidence, defined, 386
Disability, defined, 360
Disability management programs, 230
Disaster Mortuary Assistance Teams (DMORTs), 49
Disasters
accreditation for, 39–40
appropriate response to, 39
command and control in, 43–46
communications in, 47–50
contingency planning for, 52
coordination of response in, 50–52
defining threats, 40–43
planning for, 39–40
practicing for, 52–53
Discipline, in time management, 288–289
Discrimination, 16, 359–350
avoiding, 99–100
effects of, 362
gender, 363–365
Disease, evidence of, 390–301
DO-CHECK-ACT, 411
Dominator, role in meetings, 321

Editing, of communication, 306–307
Elaborator, role in meetings, 321
Emergencies
accreditation for, 39–40
appropriate response to, 39
command and control in, 43–46
communications in, 47–50
contingency planning for, 52
coordination of response in, 50–52
defining threats, 40–43
planning for, 39–40
practicing for, 52–53
Emergency Operations Centers (EOCs), 49
Emerging infections, 43
Employee handbooks, 92
Employment
application for, 87–88
interview for, 89–90
references for, 90
testing for, 90–91
Energizer, role in meetings, 320
Errors and omissions (E&O) insurance, 276, 375
Ethics
actions reflective of, 421–422
codes, 420
consequences of failure in, 422–423
dilemmas in, 423–424
divided loyalties and, 420–421, 424–425
failure, 421–422
in health programs, 16–17
issues in, 421–423

Ethics *(continued)*
journalistic, 340
lying for patient's good, 424
professional guidelines for, 420, 426
setting example for, 182
European Quality Award, 407
Evaluation
agreeing on criteria for, 161
areas used for, 162, 166
of budget, 137–139
by customers of program, 161–162
dangers of, 157–158
by director and members of program, 161
by government, 162
of HPDP programs, 259–260
importance of, 157
interrelationships between evaluators and evaluated, 160, 167
by judicial members, 162
of outcomes, 166–167
by owners of program, 161
performers of, 159–162
of personnel, 93–96
in planning, 32
of problem solving process, 201
of program, 162, 166
of progress and processes, 166
of risk management plan, 217
techniques of, 167–175
Evidence
of causation, 386–389
circumstantial, 386
defined, 386
direct, 386
of disease, 390–301
epidemiological, 391–394
of individual exposure, 394–395
of nonworkplace exposure, 395
relevant, 386
Evidence-based medicine, 385–386
defined, 386–387
types of evidence, 386–391
Exclusionary rule, 378
Executive summary, 146, 147
Exercise/fitness programs, 249
"Expert," role in meetings, 320
Expert testimony
admissibility of, 378, 380–383
challenges to, 377–378
Daubert standard, 378–379
disclosure requirement for, 389
epidemiological, 391–394
evaluation of methodology of, 396–397
evidence-based medicine and, 385–398
Federal Rules of Evidence on, 380–383
Frye standard, 379–380
general acceptance test for, 379–380
junk science and, 379, 381
level of certainty required in, 397–398
in personal injury litigation, 377
preparation for testimony, 389–390
prerequisites and qualifications for, 381, 384, 385, 389, 395–396
validity of, 395–397
Exposure
evidence of, 394–395
to industrial chemicals, 421–423

Facilitation, of team, 122
Fairness
importance of, 99
in team building, 119–120
Family Medical Leave Act, 16
coverage under, 365
employee protection under, 365–366
Favoritism, avoiding, 99–100, 119–120
Fayol, Henri (one person report to one supervisor), 75
Federal Bureau of Investigation (FBI), 50
Feedback
negative, 95–96
positive, 94
provided by leader, 185–186
specificity of, 186
to team, 120
Fetal protection policies, 363–364
Financial overview, in business plan, 148
Financial plan, 147, 152–153
Financial planning, 129
business plan, 132
determining costs and revenues, 130–131
Force a decision, 323–324
Forced analysis, 195
Forecasting, 24–25
inputs for, 25
Formal communication, 313–314
4 Cs model, 40
4 Cs +1 model, 40
Freedom of speech, 340
Frye standard, 379–380
Functional organization, 75
Funding, for new programs, 153–154

Gantt charting, 169
Gender discrimination, 363–364
types of, 364–365
General acceptance test, 379–380

Goals, 23
definition of, 57
prioritization of, 57–58
tactical vs. strategic, 23
Government
evaluation by, 162
relations with, 188–189
Grammar, importance of, 304–305
Guide to the Work-Relatedness of Disease, 387–388

H5N1 virus, 43
HAM radios, 48
Handbooks, employee, 92
Harmonizer, role in meetings, 321
Hawthorne effect, 125–126
Hazard Communication Standard, 355
provisions of, 356–357
Health
measurement of effect on productivity, 227–231, 233, 234–235
and productivity, 226–227
value-based program design for, 236–239
Health Alert Network, 49
Health Check program, 251
Health Insurance Portability and Accountability Act (HIPAA), 210
Health Management Inspection (HMI), 158
Health Plan Employer Data and Information Set (HEDIS), 410
Health policy consulting, 272–273
Health promotion and disease prevention (HPDP) programs
benefits of, 247–248, 258
concerns regarding, 252–254
communication in, 261
components of, 247, 248–250
costs of, 258–259
criteria for implementation of, 260–261
evaluation of, 259–260
evolution of, 245–246
examples of, 251–252
expectations for, 255–256
health risk assessment instruments in, 260
incentives in, 262
incorporation into comprehensive programs, 262–263
issues in implementation of, 255–262
management involvement in, 256–258
modular planning of, 255
planning requirements for, 254–255
reasons for, 245–246
regulatory issues for, 258
research on, 253–254
resource allocation for, 258
staffing of, 250
Health Risk Assessment (HRA), 210
Health and safety costs, 203–204
data sharing on, 240–241
defining, 205–209
direct, 206
estimating, 209–211
and importance of health and safety program, 219–221
indirect, 207–208
research on, 235–236
risk management for. *See* Risk management
Health and safety professionals
consulting by. *See* Consulting
expanded role of, 228–229
graduate education for, 5
interactions with news media by, 337–350
legal testimony by, 377–402
malpractice issues, 370–373
management education for, 5–6
management skills of, 1–2, 5–7
multiple constituencies of, 352
speeches and presentations by, 317–335
strategic view of, 3–4
tasks of, 2
Hearing protection, sample audit instruments for, 440–441
Honesty, 417
cynical view of, 418
establishing, 418–419
Horizontal organization, 75
Hourly rates, for consultancy, 154–155
Human skills, 11, 12
Hurricane Katrina, 40, 41, 44, 48, 51
Hygiene factors, in motivation, 124

Image control, 343
Implementation, 57–58
through MBO, 59–63
Incentives, 262
Independent medical examination (IME), 389
Industrial hygiene programs, 230
Infectious disease, 43
and terrorism, 41–42
Influenza immunization programs, 249
Informal communication, 314–315
Initiator, role in meetings, 320
Integrity
cynical view of, 418
dangers of compromising, 184, 419
defined, 417

Integrity *(continued)*
establishing, 418–419
in health programs, 16–17
issues in, 420
setting example for, 183–184, 417
International Environmental Management System, 411
Interviews, employment, 89–90
Interviews, news, 339
action steps for, 346
to affect public opinion, 346
appearance and demeanor in, 342–343
follow-up after, 345
location for, 343–344
maintaining control over, 344–345
preparation for, 342
self-evaluation after, 345
training in techniques for, 346–348
Introduction, to speech, 331–332
ISO 9000, 410–411
ISO 14000, 411

Jargon, avoiding, 304
Job descriptions, 85–86
Johnson & Johnson, HPDP programs at, 251–252
Joint Commission on Accreditation of Health Care Organizations (JCAHO), 409
Joint Information Centers (JICs), 49
Joint Operations Centers (JOCs), 49
Journalists
ethics of, 340
as generalists, 338–339
interviews with, 339
personalization of contact with, 342
philosophy of, 340
reliance on source, 338
techniques of, 345
time and space constraints of, 341
Junk science, 379, 381

Keys to success statement, 148
KITA (Kick in the ___) motivation, 123

Labor relations, 100
management in, 100–101
Laissez-faire leadership, 180
Lead exposure
case study of, 422
sample audit instruments for, 444–445
Leadership
communication in, 182
decision making by, 185
dependent on setting, 179
distinguished from management, 177–178
by example, 182–183
feedback provided by, 185–186
ideal characteristics for, 187–188
integrity and ethics in, 183–184, 417–427
of meetings, 319–325
political aspects of, 188–189
provided by management, 178
requirements for, 181–187
skills needed in, 178–179, 181
standards setting by, 182–183
styles of, 179–181
vision provided by, 181–182
Legal requirements, 15–16
See also Americans with Disabilities Act (ADA); Family Medical Leave Act; Malpractice; Occupational Safety and Health Act (OSH Act); Workers' compensation, Chapter 23
Legal testimony by OEHS professionals. *See* Expert testimony
Life Orientation model, 109–110
interactions based on, 113–114
personnel evaluation based on, 112–113, 122–123
and work environments, 114–115
Line, defined, 71
Listener, role in meetings, 321
Listening, effective, 308–310
Live for Life program, 251–252
Log, keeping, 283, 286, 289

Major accidents, 42–43
Malcolm Baldrige National Quality Award, 407
critiques of, 411–412
Malpractice
by consulting and administrative services, 373–375
in course of employment, 370–371
by independent contractors, 371–372
preventing issues of, 372–373
risk avoidance, 374
Management
business planning by, 146–153
communication in, 1, 299–315
decision making by, 15, 191–202
entrepreneurship in, 143–156
ethics and integrity in, 16–17
financial functions in, 129–141
function of, 1, 11–14
implementation by, 57–63
initial considerations for, 429–430

leadership and, 177–190
legal and regulatory issues in, 15–16, 351–375
literature on, 6–7
motivation in, 117–127
negotiation in, 34–35
of personality issues, 107–116
of personnel issues, 85–104
planning by, 19–36
prioritization by, 21, 33–34
problem solving in, 14–15, 191–202
time, 281–298
by walking around (MBWA), 186–187, 294
Management expertise
communication skills, 14
conceptual skills, 13–14
development of, 5–7
importance of, 1–2
people skills, 12–13
technical skills, 11–12
Management by Objectives (MBO), 24
advantages of, 60–61, 63
and budgeting, 135
misuse of, 61–63
process of, 59–60
Management summary, 147, 152
Market analysis, 147, 149–150
Marketing strategy, 151
Materials Safety Data Sheets (MSDSs), 356
Matrix organization, 76
MBWA (Management by Walking Around)
benefits of, 186–187, 294
to learn information, 186
Media. *See* News media; Reporters; Visual aids
Medical certainty, reasonable degree, 397–398
Medical records, 358–359
Meetings
agenda for, 318
focus of, 323
followup after, 324
forcing a decision in, 323–324
introductions, 331–332
leading and facilitating, 319–325
managing, 295–296
proper procedure for, 321–322
purpose of, 317–318
roles of participants in, 320–321
scheduling for, 325
settings of, 324–325
start time of, 318–319
stop time of, 324
time limits of, 304
as time wasters, 317
Milestoning, 169, 172, 175
Misinformation (often worse than no information), 145, 199
Mission statement, 22–23, 148
Monitoring
of budget, 137–139
in risk management, 220
"Monkeys," 297
Motivation, 123, 178
ineffective means of, 123
psychological issues in, 124–125
relation to team building, 126–127
Multitasking, adverse effects of, 309, 310
Myers-Briggs evaluations, 108–109

National Committee for Quality Assurance (NCQA), 410
National Disaster Medical System, 50
National Incident Management System (NIMS), 46, 49
Needs assessment, 24, 149
Negative feedback, to employees, 95–96
Negotiation
literature on, 35–36
objectivity in, 35
preparing for, 34–35
setting for, 34
for "win-win" outcome, 34
New initiatives
business plan for, 146–153
challenges of, 143–145
funding of, 145–146
New programs, funding for, 153–155
News media
audience of, 339
as big business, 340–341
and communication pyramid, 337–338
impression made to, 342–343
interacting with, 349–350
press releases for, 348–349
time and space constraints of, 341
tips for communication with, 338–345
Noah principle, 30
Normative analysis, 209
Nutrition programs, 249

OBE'd (overtaken by events), 192
"Of" vs. "in" an organization, 3
Objectives
cost determination for, 59–60
definition of, 58
evaluation of, 60

Objectives *(continued)*
measurement of, 58–59
responsibility allocation for, 59
stating, 148
time allocation for, 59
Obstacles, removing, 122
Occupational and environmental health and safety (OEHS)
audit instruments for, 435–445
benefits of, 4
challenges of, 143–145
funding of new programs, 153–155
improving program quality of, 403–414
legal and regulatory issues, 15–16, 351–375
organizational position of, 78–79
organizing activities in, 65–83
planning new initiatives in, 145–146
practitioners of. *See* Health and safety professionals
scientific challenges in, 4
scope of, 2–3
skills in, 12
standard setting in, 16–17
Occupational Safety and Health Act (OSH Act)
coverage of, 352–353
criminal provisions of, 369
enforcement of, 354–355
history of, 352
penalties under, 355
standard setting by, 353–354
Occupational Safety and Health Administration (OSHA), 351
Occupational Safety and Health Management Systems, 411
Opportunities, identifying, 30–31
Organization, of communication, 302–303
Organization of health and safety activities, 79–82
line and staff relationships in, 71–74
organizational chart for, 66–71, 80–82
organizational concerns in, 74–78
philosophy of, 65–66
Organizational goals, 57–58
Orientation, personnel, 92
Outcomes, evaluation of, 166–167

Pandemics, 43
Paperwork, dealing with, 290–292
Pareto principle, 33
applied to quality control, 405–406
applied to risk assessment, 211–212
Parliamentary procedure, 322
summary of, 447–448
Participatory approach to decision making, 193–195
Patient assistance, 214
Patton, George, 179
People skills
importance of, 12
types of, 12–13
"Perception is reality," 120
Performance incentives, 239–241
Performance standards, 86–87
Personality issues, 107
effective management of, 107–110
Life Orientation, 109–115
in team building, 122–123
testing for, 108–109
Personnel
avoiding discrimination against, 99–100
documentation provided to, 92
evaluation of, 93, 95–96
feedback given to, 93–94, 96
labor issues, 100–101
orientation for, 92
performance standards for, 85–87
recruitment of, 87
selection of, 87–91
staffing levels, 102–103
underperformance by, 95–99
PERT (Program Evaluation and Review Techniques), 24
applications of, 171–173
components of, 170–171
computer programs in, 170
costs and benefits of, 173–174
example of, 172, 175
in team building, 120
PERT charts
example of, 173
symbols in, 171
Phone trees, 47
Physical or mental impairment, defined, 360
Plan, Do, Check, Act, 30
Planning
advisory committees in, 28–29
appropriateness and relevance of, 29–30
to avoid crises, 31
concerns in, 20
departments, 20–21
determining status and potentials, 24
dynamism of, 31
for emergencies, 39–40
evaluation in, 31–32

forecasting in, 24–25
format of, 22
to identify opportunities, 30
importance of, 19
inputs into, 26–27
overcoming obstacles to, 32
regulatory issues in, 27
resistance to, 19–20
resources devoted to, 19
steps in, 21–32
timeline for, 22
updates to, 21
Plant oversight, 271–272
Politics, leadership and, 188–189
Positive feedback, distinguished from compliments, 94
Potentials, determining, 24
PowerPoint, 327–328
Pregnancy, legal issues in, 363
Presentations
environmental factors in, 333–334
format for, 327
handouts at, 331
importance of, 317
introductions, 331–332
practicing, 332–333
purpose of, 325–326
research assistance for, 332, 334–335
settings of, 317
sound tapes in, 330
time allotment for, 326
time slot for, 326–327
videotapes in, 330
visual aids in, 327–330
Presenteeism, 231–232
Press releases, 348–349
Prevention
importance of, 225–226
and productivity, 227–231
Prioritization, 21, 58
arranging priorities, 284–285
and budgeting, 134–136
importance of, 33–34
Pareto principle in, 33
scheduling priorities, 285–286
Pro forma profit and loss statement, 152
Proactive approach to risk management, 214–215, 216
Problem personnel
failure to provide feedback to, 96
feedback to, 95
identifying problem, 96–97
remedying problems, 98–99
working with, 97
Problem solving, 14–15
advocating best solution, 200
consultancy and, 273, 274–275
developing solutions, 200
evaluation of, 201
gathering background for, 199
gathering data for, 199
implementing solution, 200
obtaining upper management support in, 198
problem definition, 196–198
schema for, 196
techniques for, 195–201
testing solutions, 200
Procrastination, 288–289
damage caused by, 191–192
Procter and Gamble, HPDP programs at, 251
Product stewardship, 272
Productivity
consulting for, 273–274
factors affecting, 231–232
health and, 226–227
maximizing potential, 233
measurement of health effect on, 227–231, 234–235
medical conditions affecting, 237–239
minimizing loss, 233
performance incentives for, 239–240
Profit and loss statement, pro forma, 152
Program currency, 15–16
Project management, 169
Projected balance sheet, 153
Projected cash flow, 152
Proposal, for new activities, 147
Purpose
of communication, 301
of meeting, 317
of presentation, 325–326
Pyramid phone trees, 47

Quality, of program
accreditation for, 409–410
Baldrige criteria for, 407–409
Deming management method and, 404–407
evaluation of, 413–414
implementation considerations for, 412–413
importance of, 403
lack of consensus regarding, 411–412
leadership to ensure, 182–183
national and international standards for, 410–411
responsibility for, 414

Radio use, in emergencies, 48
Radiologic terrorism, 42

Reactive approach to risk management, 213–214, 216
Reading, effective, 309
Recording and Reporting Occupational Injuries and Illnesses Standard, 355
 provisions of, 357–358
Recruitment, of staff, 87
 guide to, 103–104
References, for employment, 90
Regulatory issues, 15–16
 in planning, 27
 in risk management, 220
 See also Americans with Disabilities Act (ADA); Family Medical Leave Act; Occupational Safety and Health Act (OSH Act); Workers' compensation
Relevant evidence, defined, 386
Replacement, of key personnel, 43–44
Reporters
 ethics of, 340
 as generalists, 338–339
 interviews with, 339
 personalization of contact with, 342
 philosophy of, 340
 reliance on source, 338
 techniques of, 345
 time and space constraints of, 341
Research, for communication, 305–306
Research community, consulting in, 274
Respiratory protection, sample audit instruments for, 442–443
Results, in team building, 126
Retention, of staff, 87
 guide to, 103–104
Return on investment (ROI), determining, 131–132
Revenue forecasting, 151
Revenues, determining, 130, 133
Revision, of communication, 306–307
Risk management
 advocacy for, 219–221
 communication in, 218–219
 defining present costs, 205–209
 developing program of, 204–205, 213–221
 establishing objectives for, 213
 estimating potential costs, 209–211
 evaluation of plan, 217
 facets of, 216
 financial plan for, 215, 217
 identifying high risk areas, 211–212
 implementation and followup of plan, 218
 risk assessment instruments, 260
 senior management involvement in, 218
Robert's Rules of Order, 322
Rules of civil procedure, 389

Sales forecasting, 151
Salt Lake City mall attack, 44, 47, 50
SARS (severe acute respiratory syndrome), 43
Services delivery plan, 147, 148
Shewart method, 407, 411
Single-point billing, 144
Six Sigma, 16, 183
Smoking cessation programs, 249
Sneak, role in meetings, 321
Span of control, 76–77
Spell-check, 307–308
Staff, defined, 71
Staffing levels
 forecasting for, 101–102
 maintaining, 102–103
Standard Code of Parliamentary Procedure, 322
Standards setting, 16
 as part of leadership, 182–183
Statistical Process Control (SPC), 404
Statistical Quality Control (SQC), 404, 405
Strategic goals, 22
Strategic National Stockpile, 50
Strategy and implementation summary, 147, 150–152
Strengths, Weaknesses, Opportunities, and Threats (SWOT), 24
Stress management programs, 249
Supporting–giving personality, 109, 110
Surveys, 169
Systems analysis, 195

Tactical goals, 22
Target market segment strategy, 149
Team building
 communication in, 118
 decision making in, 123
 delegation in, 121–122
 essentials for, 126–127
 of existing team, 117–118
 facilitation in, 122
 fairness in, 119–120
 feedback in, 120
 importance of, 117
 participatory approach to, 118–119
 personality issues in, 122–123
Technical skills
 applications of, 12
 importance of, 11–12

Telephone, proper use of, 293–294
Television
  image control in, 343
  interview location for, 343–344
Terrorism, 41–42
  response to, 51–52
Testing, for employment, 90–91
Threats
  natural disasters, 41
  terrorism, 41–42
Time management
  assessing use of time, 282–283
  avoiding monkeys (transfer of responsibility), 297
  components of, 281
  controlling time wasters, 289–297
  delegating, 296
  goal orientation of, 282
  importance of, 281
  prioritization, 283–289
  saying no, 297
  self-study for, 297
  working files, 292
Time wasters, 289
  extraneous activities, 289
  managing conversations, 294–295
  managing meetings, 295–296, 317
  paperwork, 290–292
  repeat crises, 289–290
  telephone, 293–294
To Err is Human—Building a Safer Health Care System, 409, 410
Tobacco, 422
TOPOFF I, 44, 48
Tort liability, 368
Total quality control, 183, 405
Total Quality Management, 16, 183
Triangle approach, 159–161
Trust, in team building, 126

Underperforming personnel
  failure to provide feedback to, 96
  feedback to, 95
  identifying problem, 96–97
  remedying problems, 98–99
  working with, 97

VA Quality Enhancement Research Initiative, 410
Value-based program design, 236–239
Verbosity, 304
Vertical organization, 74
Vinyl chloride exposure, 422–423
Vision, importance of, 181–182
Vision, Mission, Goals planning, 22–24
Visual aids
  appropriate density of use of, 330
  caveats regarding use of, 330–331
  font size and typeface for, 328
  proper use of, 327–328
  slides in, 328–329

Withdrawer, role in meetings, 321
Workers' compensation, 16
  confidentiality and, 367
  health and safety programs and, 230
  function of, 366–367
  vs. tort liability, 368
Workplace safety standards, 230
World Trade Center attacks, 51
Written employment application, 87–88